FLUORESCENT BIOMOLECULES

Methodologies and Applications

FLUORESCENT BIOMOLECULES

Methodologies and Applications

Edited by

David M. Jameson

Department of Pharmacology
University of Texas Southwestern Medical Center at Dallas
Dallas, Texas

and

Gregory D. Reinhart

Department of Chemistry and Biochemistry
University of Oklahoma
Norman, Oklahoma

PLENUM PRESS • NEW YORK AND LONDON

Library of Congress Cataloging in Publication Data

Fluorescent biomolecules: methodologies and applications / edited by David M.
Jameson and Gregory D. Reinhart.
    p.   cm.
    ''Proceedings of the international symposium in honor of Gregorio Weber's seven-
tieth birthday, held September 9–12, 1986, in Bocca di Magra, Italy''—T.p. verso.
    Includes bibliographies and indexes.
    ISBN-13: 978-1-4684-5621-9   e-ISBN-13: 978-1-4684-5619-6
    DOI: 10.1007/978-1-4684-5619-6

    1. Fluorescence spectroscopy—Congresses. 2. Biomolecules—Analysis—Congres-
ses. 3. Bioluminescence—Congresses. 4. Fluorescence microscopy—Congresses. I.
Jameson, David M. II. Reinhart, Gregory D. III. Weber, Gregorio.
[DNLM: 1. Biochemistry—congresses. 2. Spectrometry, Fluorescence—methods—
congresses. QD 96.F56 F646 1986]
QP519.9.F56F58 1989
574.19'285—dc19
DNLM/DLC   88-36898
for Library of Congress   CIP

Proceedings of the International Symposium in Honor of
Gregorio Weber's Seventieth Birthday,
held September 9–12, 1986, in Bocca di Magra, Italy

Softcover reprint of the hardcover 1st edition 1989

© 1989 Plenum Press, New York
A Division of Plenum Publishing Corporation
233 Spring Street, New York, N.Y. 10013

GREGORIO WEBER

(Photo by Juliet Weber)

PREFACE

This volume is based on an international symposium held during September 9-12, 1986 in Bocca di Magra, Italy. The intent of the organizers was to bring together expert practitioners of fluorescence spectroscopy, particularly as applied to biological systems, to assess recent developments in the field and discuss future directions. At the same time the meeting was intended to honor the singular and outstanding scientific career of Gregorio Weber on the occasion of his seventieth birthday.

Gregorio Weber is truly *the* pioneer in the application of fluorescence methods to biochemistry and biophysics. A complete list of his scientific contributions to fluorescence and to protein biochemistry is beyond the scope of this preface. Suffice it to say that since his initial landmark articles on fluorescence, published in the late 1940's and early 1950's, Gregorio Weber has continued to make seminal contributions to both the theory and practice of fluorescence and, contrary to many who might be tempted to rest on their laurels, he shows no signs of slackening his pace.

In addition to his more obvious tangible contributions to the scientific field, Gregorio Weber has made equally valuable contributions of another type. Specifically, he has had the most profound impact, both professionally and personally, on generations of young scientists. To paraphrase and extend an observation made by Dr. Brand in his lecture, Gregorio Weber has consistently shown by example how scientists ought to interact with each other, namely with courtesy, respect, selflessness, good humor and generosity. All of us privileged to have known and worked with "The Professor" will always be grateful for the experience.

The organizing committee, which consisted of N. Acerbi, A. Finazzi-Agro, E. Gratton, D. M. Jameson, G. Motolese, R. A. Ricci and N. Rosato, wishes to thank all those involved and acknowledge the financial support generously provided by I.S.S., Inc., SLM Instruments Inc., Abbott Laboratories, Perkin-Elmer, LTD, Fluorescence Unlimited and Molecular Probes. The Italian C.N.R. also provided generous financial assistance.

Special thanks are due to Alessandro Finazzi-Agro and Nicola Rosato who coordinated local arrangements with Monastero Santa Croce, the meeting site, and to Franco Bassani who contributed greatly to the initial meeting organization and also generously provided excellent wine from his own vineyard for each participant. We also thank R. A. Ricci, President of the Italian Physical Society for the opening remarks. The chairmen of the various sessions, including S. Lehrer, R. A. Ricci, G. Hervé, M. Eftink, A. Szabo and R. Gennis, are gratefully acknowledged.

The meeting could not have succeeded without the special efforts of Amedeo Lia and Guido Motolese who made generous contributions, not only financially, but also of their time and expertise to many of the logistical and social arrangments, including the dinner banquet. We are also indebted to the many individuals who assisted so expertly in the numerous secretarial, organizational and transportational tasks including Frederica Scaraglio, Glauco Bionducci, Marco Bassani, Claudio Furno, Francesco Giovannani, Alessandro Carrozzi and Mrs. Nicotra.

Ms. Vickey Thomas is most gratefully acknowledged for her assistance in many stages of this meeting ranging from early preparations to manuscript proofreading. We also thank Ms. Mary Born from Plenum Press for her numerous helpful suggestions. Finally, we thank, most sincerely, Ms. Jo Hicks for her virtually heroic efforts in retyping all the manuscripts and tolerating the idiosyncracies of the book editors.

David M. Jameson
Gregory D. Reinhart
Enrico Gratton

Dallas, Texas
September 29, 1988

CONTENTS

LUMINESCENCE: GENERAL CONCEPTS AND APPLICATIONS TO THE STUDY OF SOLIDS

Franco Bassani

Scuola Normale Superiore
I-56100 PISA
Italy

GENERAL CONCEPTS AND EARLY HISTORY

Luminescence is the general term used to indicate emission of light by matter. Luminescence phenomena have always been known to man: lightning, the aurora borealis, light emission by bacteria in the sea or by decaying organic matter are common natural phenomena. However, it was a specific artificial effect observed by accident by a shoemaker named Vincenzo Casciarolo in 1603 that started the scientific inquiry into the problem of luminescence.

Casciarolo made a living as a cobbler but was a man with ideas, and his dream was to produce gold by baking the appropriate stones. He used to go on Monte Paderno, a mountain near Bologna to collect the heaviest stones he could find, and he baked them at night using coal. One day in 1603 he observed that some of the stones he had baked (probably barium sulfate) emitted a purple-blue light at night. He also observed that this quality was lost after some time, but it returned again after the stone had been exposed to daylight.

This light-emitting stone was called "lapis solaris" or more commonly "petra luminifera bononiensis", and it immediately attracted the interest of the scientists of the time. Galileo was fascinated by it; he displayed the phenomenon at the Lincei Academy and sent samples of that stone to reputed scientists all over Europe, some samples also to Lagalla, who described the "lapis solaris" in his book on natural philosophy in 1612, "De phenomenis in orbe lunae".

1

A full monograph was published in 1640 by Liceti with the Greek title "Lithephosphorous", which means light-bearing stone, and because of this the term phosphorescence was later used.

The early explanations were all systematically disproved by Galileo, who though blind and ill, in his letter to the Duke Leopoldo of Tuscany, dictated a beautiful description of the phenomenon and of the facts to be explained; his views can be summarized as follows: "It must be explained how it happens that the light is conceived into the stone, and is given back after some time, as in childbirth".

From this description we may observe that he was closer to an understanding of luminescence than later great physicists like Grimaldi, Boyle and Newton, who studied luminescence in special liquid solutions and interpreted the phenomenon as reflection and transmission of light. The concept of absorption and subsequent emission eluded them completely.

It was not until 1852 that Stokes interpreted the phenomenon along the lines of the Galileo description. Making use of previous experiments by Bruester and Haüy he formulated the law that light was absorbed in the violet and ultraviolet region of the spectrum and emitted in the blue or red region, at longer wavelength (the Stokes shift). Already Johann Wolfgang Goethe, in his book on colors, had reported experiments which proved that the emitted light from the luminescent material was stronger when it was exposed to the invisible region just before the violet rays coming from a prism. This was used as a proof for the existence of ultraviolet radiation.

The Stokes law was disputed for several decades, but was finally accepted after experiments on many substances (For a history of luminescence see Newton-Harvey, 1957). The true understanding of the phenomenon was nevertheless a long time to come; it required quantum mechanics and the concept of spontaneous emission.

EINSTEIN'S THEORY OF SPONTANEOUS EMISSION

Semiclassical Theory

The basic concept for the understanding of luminescence was introduced by Einstein in 1917 and can be stated as follows: "Any excited state of the system, characterized by an electronic energy E_i, higher than the ground state energy E_o, decays spontaneously (i.e. in the absence of any radiation) to the lower energy state E_o by emitting a photon of angular frequency ω such that

$$\hbar\omega = E_i - E_o \tag{1}$$

where h is Planck's constant and $\hbar = h/2\pi$."

The probability rate of such a spontaneous decay process is denoted by A_{io}, and the number of spontaneous transitions per unit time is expressed in terms of the number of excited states $n_i(t)$ by the rate equation

$$\frac{dn_i}{dt} = -A_{io}n_i(t) \tag{2}$$

where $n_i(t)$ is the number of electrons in the state E_i, at time t.

In the absence of any external radiation and of other processes which change the population of the state E_i, we obtain a definition of the lifetime τ_i of the state E_i by writing the solution of equation (2)

$$n_i(t) = n_i(o)e^{-A_{io}t} = n(o)e^{-t/\tau i}. \tag{3}$$

The number of excited states decays exponentially with time, with a lifetime given by $\tau_{io}=1/A_{io}$ (natural lifetime). If the state E_i can be depopulated by other processes, such as spontaneous decays to other lower states E_f, collision processes with probability rate C_i, Auger processes with probability rate A_i, equation (3) is still valid, but the lifetime is correspondingly modified

$$\tau_i^{-1} = A_{io} + \sum_f A_{if} + C_i + A_i. \tag{4}$$

Luminescence with a finite lifetime τ_i is observed only when spontaneous emission is significant.

Besides spontaneous emission, it was well-known at the time of Einstein that absorption took place in the presence of external radiation of appropriate frequency. The probability rate of this process is proportional to the energy density of the stimulating radiation and can be written as

$$P_{oi} = B_{oi} u(\omega) \tag{5}$$

where $u(\omega)$ is the energy per unit volume per unit frequency in the interval ω to $\omega + d\omega$. Quantum mechanics requires that emission be also induced by the presence of radiation with probability rate $P_{io} = P_{oi}$ of equation (5).

Einstein was able to obtain both spontaneous and stimulated emission from the requirement that at thermal equilibrium $u(\omega)$ be given by Planck's expression (3)

$$u(\omega) = W(\omega)\bar{n}(T) = \left(\frac{\hbar\omega^3}{\pi^2 c^3}\right)\left(\frac{1}{e^{\hbar\omega/kT}-1}\right) \tag{6}$$

and that the statistical distribution between states be in the Boltzmann ratio

$$\frac{n_i}{n_o} = e^{-(E_i-E_o)/kT} = e^{-(\hbar\omega/kT)} \tag{7}$$

Substituting these two conditions (6) and (7) into the rate equation at equilibrium

$$\frac{dn_i}{dt} = B_{oi}u(\omega)n_o - B_{io}u(\omega)n_i - A_{io}n_i = 0 \tag{8}$$

Einstein obtains $B_{io} = B_{oi}$, and in addition the important relation between stimulated emission and spontaneous emission probability

$$A_{io} = B_{io}\left(\frac{\hbar\omega^3}{\pi^2 c^3}\right) \tag{9}$$

This is an expression which has been obtained by using thermal equilibrium, but must be valid in general. As a consequence the total probability of emitting radiation from an excited state E_i to the ground state E_o, as obtained from the last two terms of (8), is

$$P_{io} = A_{io}(n(\omega)+1) \tag{10}$$

where $n(\omega)$ is the occupation number of state ω (number of photons of frequency ω present); in conditions of thermal equilibrium $n(\omega)$ is given by the Bose-Einstein expression

4

$$\bar{n}(\omega) = \frac{1}{e^{(\hbar\omega/kT)}-1} \qquad (11)$$

We then obtain from equation (10) that the emission of light consists of two terms, one gives the stimulated emission and is proportional to the photon number, $n(\omega)$, the other gives the spontaneous emission and is independent of any other radiation present.

The Einstein theory explains luminescence in terms of the only parameter A_{io} and explains the relative roles of spontaneous emission and stimulated emission (for a detailed discussion of Planck's and Einstein's contributions see D. der Haar, 1967), $n(\omega)$ smaller or greater than 1 respectively. At given temperatures and frequencies, spontaneous emission prevails if $\hbar\omega \gg kT$, thermally stimulated emission prevails if $\hbar\omega \ll kT$. Even at high frequency, however, we may have prevalence of stimulated emission only when an external intense beam of light is present as in the laser.

Full Quantum Theory

The full explanation of the origin of spontaneous emission and the microscopic prescription to compute the probability rate A_{io} can be obtained from quantum mechanics (for a full account of the quantum theory of absorption and emission of light, see Louden, 1983). One must also include the quantization of the electromagnetic field first developed by Dirac and Heisenberg in the twenties (D. der Haar, 1967, chapter 4). This can be achieved by simply writing the vector potential $\vec{A}$ of the electromagnetic fields in terms of the creation operators $a_k^+(\omega)$ and of the destruction operators $a_k(\omega)$

$$\vec{A} = \sum_k (2\pi\hbar c^2/V\omega_k)^{1/2}(a_k e^{i\vec{k}\cdot\vec{r}} + a_k^+ e^{i\vec{k}\cdot\vec{r}})\hat{\varepsilon}_k \qquad (12)$$

where $\hat{\varepsilon}_k$ is the polarization vector and the functions a_k and a_k^+ can be treated as classical variables with their time-dependence in classical electrodynamics, or as quantum operators in quantum electrodynamics. In the second case the Bose commutation properties and state orthonormalities imply that a_k and a_k^+ operate on the photon states $|n_{k1}...n_k...\rangle$ as follows:

$$a_k|...n_k...\rangle = (n_k)^{1/2}|...n_k-1..\rangle \qquad (13)$$

$$a_k^+|...n_k..\rangle = (n_k+1)^{1/2}|...n_k+1..\rangle$$

from which the name of destruction and creation operators is apparent, because n_k is the number of photons present in state of wave vector $\vec{k}$ and frequency ω_k.

We can also write an expression similar to (12) for the electric field

$$\vec{E} = -\frac{1}{c}\frac{\partial\vec{A}}{\partial t} = i\Sigma_k(2\pi\hbar\omega_k/V)^{1/2}(a_k e^{i\vec{k}\cdot\vec{r}} - a_k^+ e^{-i\vec{k}\cdot\vec{r}}) \tag{14}$$

with the same properties (13) for the photon creation and destruction operators.

The photon Hamiltonian is immediately obtained from expression (14) by recalling the Bose commutation properties $[a_k, a_k^+]=\delta_{kk'}$ and the properties (13). We obtain

$$H = V\sum_k E_k^2/4\pi = \sum_k(a_k^+ a_k + \tfrac{1}{2})\hbar\omega_k \tag{15}$$

which gives for every mode the eigenvalues $(n_k + \tfrac{1}{2})\hbar\omega_k$, appropriate to the level $|..n_k..\rangle$ of the harmonic oscillator, or equivalently to a number n_k of photons of mode k and frequency ω_k.

Once the photon Hamiltonian is considered, the transition between states is determined by the electron-photon interaction term, which we recall to be given to lowest order by

$$H' = \frac{e}{mc}\vec{A}\cdot\vec{P}\approx e\vec{E}\cdot\vec{r} \tag{16}$$

where the second equality implies the use of the electric dipole approximation, which for electrons in atoms, molecules and solids is fully adequate, since we consider only optical transitions and not spin flip effects (paramagnetic resonance).

The interaction between the electrons, whose wave functions are denoted by $\vartheta(\vec{r},\vec{R})$ ($\vec{r}$ denoting the electron and $\vec{R}$ the nuclear positions respectively), with eigenvalues E_e, and the photons, whose wave function per mode is given by $|...n_k...\rangle$ and whose energy is $(n_k + \tfrac{1}{2})\hbar\omega_k$, produces a transition probability rate for decay from an excited state to the ground state. This is given by the Fermi golden rule

$$P_{io} = \frac{2\pi}{\hbar} |<0|H'|i>|^2 \delta(E_o^T - E_i^T) \tag{17}$$

where i denotes the initial excited state and o the final ground state, and E^T is the total energy, including that due to the photons.

We obtain immediately the emission probability considering the wave functions

$$|i> = |\vartheta_i>|..n_k..> \tag{18}$$

and

$$|0> = |\vartheta>|...n_k + 1...> \tag{19}$$

with the related eigenvalues $E_i + n_k \hbar\omega_k$ and $E_o + (n_k+1)\hbar\omega_k$, and substituting expression (12), or expression (14) in H'. From the contribution a_k^+ in (12), or in (14), we obtain the same result as from Einstein's theory

$$P_{io} = A_{io}(n_k + 1) \tag{20}$$

but we now have the prescription to compute A_{io}

$$A_{io} = \left(\frac{e}{m}\right)^2 \frac{4\pi^2}{\omega_k} |<\vartheta_o|\hat{\varepsilon}\cdot\vec{P}|\vartheta_i>|^2 \delta(E_o - E_i + \hbar\omega_k)$$

$$\tag{21}$$

$$= 4\pi^2 e^2 \omega_k |<\vartheta_o|\hat{\varepsilon}\cdot\vec{r}|\vartheta_i>|^2 \delta(E_o - E_i + \hbar\omega_k)$$

The dipole matrix element in (21) $<\vartheta_o|\hat{\varepsilon}\cdot\vec{r}|\vartheta_i>$, or the similar expression $<\vartheta_o|\hat{\varepsilon}\cdot\vec{P}|\vartheta_i>$, give the oscillator strength of the transitions.

Both spontaneous and stimulated emission appear, and luminescence is fully explained. Also the lifetime τ_{io}, due to the spontaneous emission, can now be computed as the inverse of expression (21). Of course other processes can modify the lifetime as explained before (see equation (4)), and the study of lifetimes, with their departure from the expected (21), sheds light on Auger processes and collision effects.

EFFECT OF LATTICE VIBRATIONS AND STOKES SHIFT

While lifetime and luminescence phenomena are fully explained by

Quantum Theory and the Einstein Theory of spontaneous emission, to explain the Stokes shift required a further idea, i.e. the introduction of lattice vibrations into the transition process. This was done for diatomic molecules by Frank and Condon in the twenties (Frank, 1926) and was extended to solids by F. Seitz in the thirties (Seitz, 1936; Seitz, 1939).

The essential point is to consider the fact that the equilibrium distance between the nuclei (minimum in the electronic energy) is generally different for the ground state and for the excited states, and that during the optical transitions the nuclei do not move from their equilibrium position, because their motion requires longer time than electron motion. Consequently the energy difference in absorption is larger than in emission, as shown in Figure 1.

A more precise picture results by considering explicitly the vibrational eigenstates which result from the electron energy curves, which act as potential curves for the nuclear motions. The vibrational states of the nuclei are then harmonic oscillator functions $\chi_n(\vec{R})$, with different centers of oscillations for the ground state and for excited states, and energies $\hbar\omega_v(n+\frac{1}{2})$ (Califano, discussion this meeting).

As indicated by (21), the transition probabilities depend on the dipole matrix elements, which in turn are proportional to the overlap between harmonic oscillator wave function centered on different points.

$$<\vartheta_o|\hat{\varepsilon}\cdot\vec{r}|\vartheta_i> = <\chi_{om}(R)|\chi_{in}(R)><\psi_o|\hat{\varepsilon}\cdot\vec{r}|\psi_i> \qquad (22)$$

The wave functions ϑ of equation (21) have been separated into the product of the electronic function ψ and the $\chi(\vec{R})$ which depends on the nuclear coordinates.

This means that in absorption we start from the lowest vibrational ground state (m = 0) and tend to go to states which have their maximum at the same value of $\vec{R}$ (n>>1). In emission we start from state n=0 because the vibrational interaction thermalizes the electrons very fast (10^{-12} s), and consequently we go preferentially to states with m>>0.

The situation is illustrated schematically in Figure 1 for molecules and in Figure 2 for solids.

The emission broadening is explained by the convolution of a large

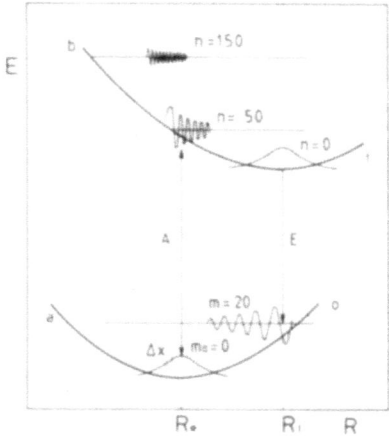

Figure 1. Schematic representation of the energies of a diatomic molecule as functions of internuclear distance.

The ground state 0 and the excited state i have their minima at different equilibrium positions. The Frank-Condon principle is illustrated considering the line A which gives the absorption energy and E which gives the emission energy.

number of vibrational lines which are not generally separated. In this way the Stokes shift is explained, and it is understood why a larger value of the Stokes shift is usually accompanied by a larger broadening of the emission band.

It is also explained why in some cases, as in Figure 3, radiationless

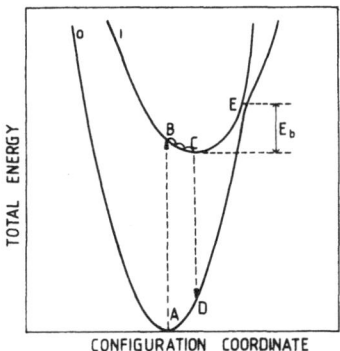

Figure 2. Schematic representation of ground and excited states of a localized center in a solid as a function of configuration coordinates (for instance, the distance of crystal atoms from the impurity). The Frank-Condon principle explains the Stokes shift. Only a very weak luminescence takes place with a transition from B to A, because phonon interaction in the excited state brings the electron from B to C in a time of the order of 10^{-12} s.

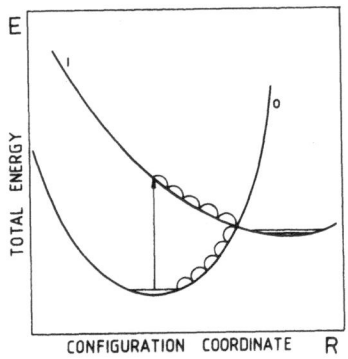

Figure 3. Configuration Coordinate Diagram for the case when radiationless decay to the ground state occurs.

decays occur, and in other cases, as in Figure 4, the decay is partly luminescent and partly radiationless, the temperature playing an important role in the lifetime of the excited luminescence state.

The explanation of the Stokes shift had been until now rather qualitative in solids and "a fortiori" in large biomolecules. However, in solids quantitative calculations can be made (for a description of optical properties of localized states in solids see Bassani and Pastori-Parravicini, 1975 and Dexter, 1958), and I wish to show a new analysis by Czaja, Testa, Quattropani and Schwendiman (Testa et al., 1987), which proves quantitatively the above described theory for the particular case of luminescence due to iodine impurities in AgBr. In this case the crystal is ionic and the vibrational states which are most strongly coupled to the

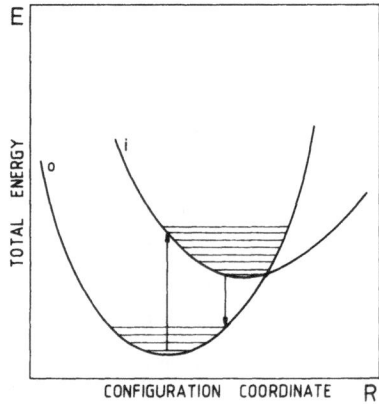

Figure 4. Configuration Coordinate Diagram for the case when luminescence occurs only at low temperature. At high temperature excited vibrational states can be occupied, and they decay radiationlessly to the ground state (temperature quenching of the luminescence).

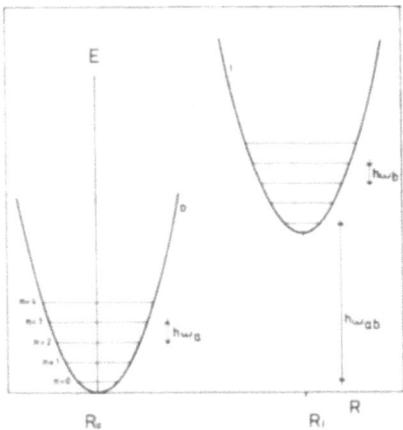

Figure 5. Schematic representation of electronic and vibrational states of excitons trapped to iodine in AgBr:I. The energies of the optical phonons $\hbar\omega_a$ of the ground state are the same as $\hbar\omega_b$ of the exciton state because they are both crystal phonons.

electronic states are the longitudinal optical phonons (L. O. phonons), whose energies are well-known. The theoretical picture is given in Figure 5, and the transition probability for spontaneous emission can be computed. The agreement between theory and the experimental luminescence band is striking, as shown in Figure 6. The structure of the different phonon transitions appears very clearly; up to eleven phonon replicas are resolved in the broad emission band.

It is unfortunate that the lifetimes associated with the different structures have not been measured. Lifetime measurements and better resolutions are very important in future solid state studies.

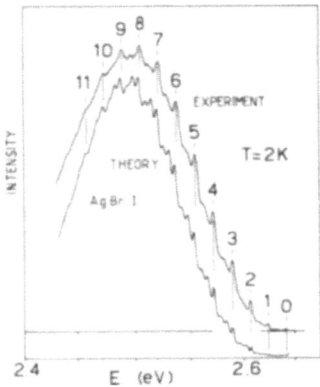

Figure 6. Emission intensity due to exciton recombination in AgBr:I. Theory and experiment are in good agreement, and 11 phonon replicas are singled out. (Testa et al., 1987)

STUDY OF LOCALIZED STATES IN SOLIDS

So far I have spoken of the theory of luminescence and indicated how
its spectral analysis gives information on localized states in solids
(Bassani and Pastori-Parravicini, 1958). It is apparent however that the
lifetimes of excited states involve all the details of excited state
processes and provide other information on the geometrical arrangements
about the impurity centers.

I will not go into the problem of how lifetimes are measured. This
also has a long history: special machines for time resolving luminescence
were already made in the 19th century; famous was the Wiedemann machine
(Wiedemann, 1888). Other people will treat this problem at this meeting.
I only mention that, mostly because of the contribution by G. Weber in the
past decades, one can now measure lifetime values much smaller than a nanosecond with accuracy of 10 picoseconds (Gratton, et al. 1984) by using
periodically modulated excitation and measuring the phase shift by the
cross correlation method.

Lifetime measurements of localized states in solids (color centers,
impurities, defects, surface states, etc.) will certainly be of great
interest for understanding the microscopic nature of such centers.

I cannot go into a review or a detailed analysis. A first experiment
with phase sensitive detection using synchrotron radiation was performed by
Grassano et al. (1986) and will be presented at this meeting by Piacentini.

I will only show the first lifetime measurement carried out on a
classical luminescent center, Tl in KI, by De Piccoli (De Piccoli and
Bassani, unpublished) with the I.S.S. phase-fluorometer; this has produced
a new and unexpected result which is shown in Figure 7, where phase shift
and demodulation as a function of the frequency of the excitation light are
given. The fluorescent decay can be understood only with a model which
uses two different lifetimes, a long time of 183 ns and a much shorter time
of 3.3 ns. This requires a new model for the Tl centers which is shown in
Figure 8. The room temperature luminescence takes place from the two
minima at X; the lower is dipole forbidden in first order and is
responsible for the long lifetime, the higher is allowed and is responsible
for the short lifetime. Only at high temperature the higher state is
partly populated, and its fractional component is much smaller than that
with the longer lifetime.

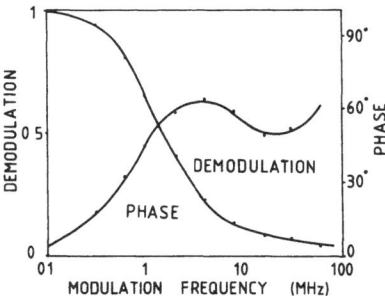

Figure 7. Measurement of phase shift and demodulation ratio for luminescence of KI:Tl at room temperature, as functions of the modulation
frequency of the excitation light. The experimental points are well fit by
a double exponential decay with τ_1 = 183 ns and τ_2 = 3.3 ns, and fraction
f_1 = 0.940 and f_2 = 0.060. A single lifetime would give a completely
different behavior.

Still another case of extreme interest is that of self-trapped
excitons, i.e. excitations of the perfect crystal which become localized
because of the lattice relaxation about them. Such localization may occur
by preserving the lattice symmetry or by symmetry breaking (two atoms
approach each other in the [110] direction in solid rare gases). Different
lifetimes for the two localized excitons have been found by Zimmerer and
Woodruff in argon and xenon (Roick et al., 1984).

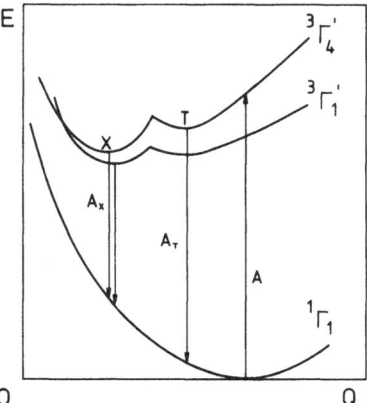

Figure 8. Proposed model for the excited states of thallium in KI relevant
in the luminescence process. The spin orbit allowed excited state is the
$^4\Gamma_4'$, while the dipole forbidden state is the $^3\Gamma_1'$. Both states have two
minima corresponding to different relaxations of the configuration
coordinates which do not preserve the full cubic symmetry. The barrier
from the T to the X minimum can be overcome only at high temperature.

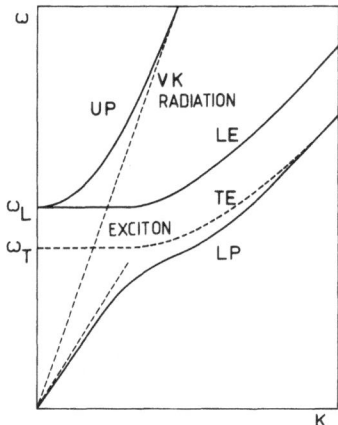

Figure 9. Dispersion of the optical modes in a crystal (polaritons) near an excitation frequency of the medium ω_T. The dotted lines give the radiation dispersion law $\omega = vk$ and the dispersion law of a transverse exciton (T.E.), whose coupling gives the polariton modes. The longitudinal exciton (L.E.) is also shown, though it is not coupled to the optical modes and is not expected to give luminescence.

STUDY OF INTRINSIC EXCITED STATES (POLARITONS)

Excited states in solids are obtained by optical excitation and consist of an electron and a hole bound to each other by the Coulomb attraction and moving through the crystal with translational energy

$$E_{ex} = \frac{\hbar^2 k^2}{2(m_e + m_h)} \tag{23}$$

where m_e and m_h are the effective masses of electrons and holes respectively.

Luminescence from excitons has been observed but their lifetimes have not been measured except in very special cases like cuprous oxide (Cu_2O) where the transition is first order forbidden and τ is very long (Trauernicht and Wolfe, 1986).

I wish to show how lifetime measurements can give an experimental verification of the new state which originates from the exciton when the interaction with the electromagnetic field is considered. This new state is called "polariton" and is displayed in Figure 9. At frequency ω, such that $\hbar\omega$ is close to the excitation energy, the exact eigenvalues of the Hamiltonian which contains the exciton, the radiation field and their mutual interaction is the "polariton state", with the peculiar dispersion relation displayed in Figure 9. At frequencies ω larger than the exciton longitudinal frequency two different polariton states exist, the upper polariton (UP) and the lower polariton (LP) (for a general review of polaritons in solids, see Bassani and Andreani, 1986).

The polariton picture implies two facts. First of all, the group velocity in the medium in the frequency region close to the exciton energy is much smaller than what is expected for radiation, where it coincides with the phase velocity. Secondly, in the absence of scattering by phonons or impurities, no luminescence should be produced, because $\vec{k}$ remains a good quantum number for the polariton, and the momentum $\hbar\vec{k}$ is conserved. Luminescence is however produced as diffused light due to scattering of the polaritons with phonons, impurities, or surfaces, which change the polariton momentum. This vindicates the ideas of Newton, which we recalled in the historical introduction.

Measurements of phase shifts with periodically modulated excitation light, both in transmission and in diffusion should give information on group velocity and on the scattering mechanism. The lifetime of the polariton in this respect takes the new meaning of "time during which the polariton remains free" and is similar to the collision time of free particles in transport theory.

ACKNOWLEDGEMENTS

I wish to acknowledge the collaboration of Fabio De Piccoli, who made the measurements on KI:Tl, and of A. Quattropani, who communicated results on AgBr:I before publication. Special thanks are due to Dr. Guido Motolese of the I.S.S. Company, who allowed extensive use of the Greg 200 phase fluorometer.

REFERENCES

Bassani, F., and Andreani, L., Quantum Theory of Polaritons to appear in Excited States Spectroscopy, Proceedings of the E. Fermi School (North Holland, Amsterdam, 1986).

Bassani, F., and Pastori-Parravicini, G., 1975, Electronic States and Optical Transitions in Solids, Pergamon Press, Oxford.

Dexter, D. L., 1958, in:Solid State Physics, Vol. 6, F. Seitz and D. Turnbull, eds., Academic Press, NY.

Einstein, A., 1917, Zur Quantentheorie der Strahlung, <u>Phys. Z.</u>, 18:121.

Frank, J., 1926, Elementary Processes of Photochemical Reactions, <u>Trans. Faraday Soc.</u>, 21:536.

Grassano, U. M., Piacentini, M., and Zema, N., 1986, <u>Il Nuovo Cimento D</u> 7:379.

Gratton, E., Jameson, D. M., and Hall, R., 1984, Multifrequency Phase and Modulation Fluorometry, <u>Ann. Rev. Biophys. Bioeng.</u>, 13:105.

der Haar, D., 1967, The Old Quantum Theory, Pergamon, Oxford.

Loudon, R., 1983, The Quantum Theory of Light, Clarendon Press, Oxford.

Newton-Harvey, E., 1957, History of Luminescence, From the Earliest Times Until 1900, American Philosophical Society, Philadelphia (reprint 1980).

Planck, M., 1900, Verh. d. Deutschen Physikal. Gesellschaft, <u>Verh. dt. Phys. Ges.</u>, 2:202 and 237.

Roick, E., Gaethke, R., Gurtler, P., Woodruff, T. O., Zimmerer, G., 1984, Observation of Surface-Sensitive Luminescence in Solid Argon; Relation to Self-Trapping and Relaxation of Exitons, <u>J. Phys. C.</u>, 17:945.

Seitz, F, 1936, Elektrische Leitfähigheit von Mechanisch Beanspruchten Seignettesalz-Einkristallen, <u>Z. Physik</u>, 101:234.

Seitz, F., 1939, An Interpretation of Crystal Luminescence, <u>Trans. Faraday Soc.</u>, 35:74.

Testa, A., Czaja, V., Quattropani, A., and Schwendimann, P., 1987, Light Emission from Impurity States in Solids: Conventional and Coherent Phonon Descriptions, <u>J. of Phys.</u>, <u>C</u>, (Solid State) 20:1253.

Trauernicht, D. P., and Wolfe, J. P., 1986, Drift and Diffusion of Paraexitons in Cu_2O: Deformation-Potential Scattering in the Low-Temperature Regime, <u>Phys. Rev. B.</u>, 33:8506.

Wiedemann, E., 1888, Uber Fluorescenz und Phosphorescenz, I. Abhandlung, <u>Ann. Physik und Chemie</u>, 34:446.

PROTEIN DYNAMICS: FLUORESCENCE LIFETIME DISTRIBUTIONS

Enrico Gratton[*], J. Ricardo Alcala[*] and Franklyn G. Prendergast[†]

[*]Laboratory for Fluorescence Dynamics, Department of Physics
Urbana, IL   61801   USA
[†]Department of Biochemistry, Mayo Foundation, Rochester, MN
55905   USA

INTRODUCTION

It is now well established that proteins in their native conformation can exist in a large number of subconformations slightly different one from the other (Lakowicz & Weber, 1973; Austin et al., 1975; Careri et al., 1975, 1979; Karplus & McCammon, 1983). Subconformations originate from small structural fluctuations around the main conformation. The protein structure is very flexible and it allows rotations around the phi and psi angles of the polypeptide chain and around the C-alpha carbon on the side chain. The stabilization of one particular native structure depends on a large number of interactions which are affected by solvent, ions of the medium and chemico-physical parameters. Some parts of the protein structure are more stable than others due to a more favorable interaction between the amino acid residues and they form structural domains. Frequently, these domains are associated with secondary structural elements such as alpha helical segments or beta sheets. The connections or loops between domains are generally more flexible. Within a domain, a side chain exposed to the solvent can have large rotational freedom and fast motions can result with rates comparable to the rates of motion of the residue in the solvent. Alternatively, a side chain at the interface between two distinct domains can move only if the domains separate enough to allow the side chain to rotate. The resultant motion of this side chain is then modulated by the relatively slow motion of the two domains. A hierarchy of motions is established in which there is an interrelation between slow and fast motions (Frauenfelder & Gratton, 1985; Ansari et al., 1985). The structural organization of the protein, made of domains and side chains with different motional characteristics, is at the origin of the complexity of the protein structure and dynamics (Careri et al., 1975, 1979). Subconformations are an expression of this complexity and the study of subconformations and their dynamics give us the opportunity to investigate the characteristics of protein motions, in particular their hierarchical organization. The experimental evidence for the existence of subconformations has been obtained mainly from low temperature studies of the CO rebinding after photolysis in myoglobin (Austin et al., 1975). From these studies the interconversion rates between subconformations have not been measured directly but rates can be deduced from the energy barrier between conformations. Strictly speaking, the distribution of energy barriers refers to the CO rebinding process and is generated by the difference between protein molecules. It is not clear, from the energy distribution plot, how two subconformations can be connected and how they interconvert one into the other. However, the existence of subconformations is well established and the characterization of their energy distribution have been obtained. X-ray temperature factors have provided a direct correlation between protein coordinates and subconformations (Frauenfelder et al., 1981), but the direct observation of interconversion between subconformations has rarely been obtained. Also a number of experiments have established the existence of different gross conformations in proteins, but generally these conformations differ by a large energy and the equilibrium conformational fluctuations are extremely slow. In this latter case the interconversion between conformations is triggered by some external factor such as binding of a substrate or by the alteration of some external chemico-physical condition. However, there is interest in studying equilibrium fluctuations between subconformations at physiological temperatures and to characterize the modalities of the interconversion between subconformations. In principle, to characterize subconformations, we should be able to determine the protein mean atomic coordinates. Alternatively, we can use an appropriate probe which will change its properties depending on the protein conformation. To follow the interconversion between conformations, the change of the probe properties must be followed as a function of time. Equilibrium fluctuations are quite difficult to observe since they occur at random and there is no phase relationship among the protein molecules observed. Relaxation methods have been successfully applied to overcome this problem (Ansari et al., 1985). The main assumption behind the use of relaxation methods to study equilibrium fluctuations is that the external perturbation to the system should be relatively small. The fluctuation-dissipation theorem then relates the rate of equilibration of the system to the rate of spontaneous fluctuations. However, to apply relaxation techniques, we must drive the system under observation out of equilibrium using an appropriate external perturbation. This perturbation is not always simple to devise and in some cases can also be difficult to apply without perturbing excessively the system, such as is the case of detection of the change in equilibrium fluctuations in the presence or absence of an enzyme substrate. To approach this problem, we have initiated a careful study of the fluorescence emission from the naturally fluorescent amino acids in proteins: the tryptophan and the tyrosine residue. The basic assumption is that we should be able to associate a single fluorescence decay rate to a well-defined protein configuration. The value of the decay rate depends on a number of factors and we have only a partial knowledge of how the protein conformation affects the value of the decay rate of the tryptophan or the tyrosine residue (Liebman & Prendergast, 1985; Haydock & Prendergast, 1986; Alcala et al., 1987c). To overcome this problem, the idea is to map the conformational space into the decay-rate space using an unknown function. The tryptophan residue shows a large variability of the decay rate among proteins (Longworth, 1971; Lumry & Hershberger, 1978; Beechem & Brand, 1985). The lifetime varies from few picoseconds for some strongly quenched residues to about 10 nanoseconds for the longest observed fluorescence lifetime. This extreme sensitivity of the lifetime value makes the tryptophan residue particularly sensitive to the protein environment and the decay rate axis, where we want to map the protein subconformations, extends over several decades. We will assume for the following that a different value of the tryptophan decay rate corresponds to a different protein subconformation although several subconformations can share the same value of the decay rate. The problem of measuring the rate of interconversion between subconformations can then be approached using fluorescence. However, there are some limitations, due to the intrinsic lifetime of the excited tryptophan residue.

To illustrate how we plan to determine the interconversion rate between conformations, consider a system which can have only two states (Alcala et al., 1987b). If the tryptophan residue has two different values of the fluorescence lifetime in the two protein conformations, namely 1 ns for the A conformation and 10 ns for the B conformation (figure 1) then in the absence of conformation interconversion a double exponential decay should be observed with characteristic lifetimes of 1 and 10 ns. The relative value of the preexponential factors should be proportional to the fraction of molecules in each conformation. The condition of very slow interconversion between subconformations can be obtained by lowering the temperature of the sample. As the interconversion rate is turned on, by

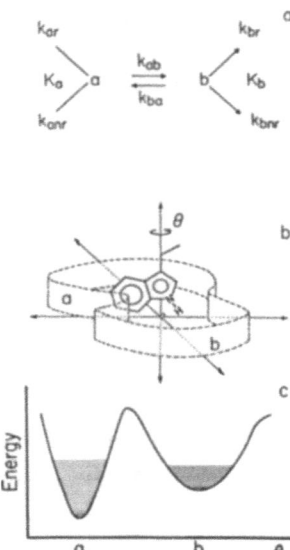

Figure 1. Schematic representation of a tryptophan residue in a protein: two-well potential energy surface.

increasing the temperature, the values of the two characteristic decay rates of the system are modified (figure 2) as well as the values of the preexponential factors. The numerical example shows that, under normal experimental conditions, interconversion rates from $10^{11}s^{-1}$ to $10^{6}s^{-1}$ can be observed using an accurate determination of the fluorescence decay from proteins. The time window available is determined by the value of the lifetime of the two subconformations. In this example we have considered only two protein subconformations. Following our view of protein structure and dynamics, we should expect a multitude of subconformations and then the problem of determining the interconversion rate between subconformations becomes more complicated, although some of the basic features illustrated by the two-state model still remain. For example, at low temperature, we should observe a decay which is the sum of a large number of exponentials, each corresponding to a different protein subconformation. In practice we should observe a continuous distribution of lifetime values. At higher temperature, as the interconversion rate increases, the distribution should narrow. In the example of figure 3 five different subconformations have been used with lifetime values equally distributed in the interval from 1 ns to 5 ns and equally populated. The interconversion rate between subconformations was the same for all subconformations and interconversion was allowed only between two adjacent subconformations which were ordered from

20

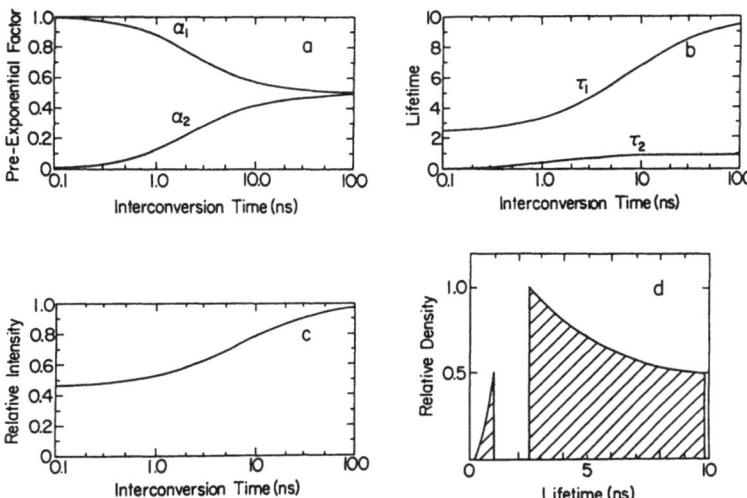

Figure 2. Evolution of a two-state fluorescent system as a function of the interconversion time between the two states. The population equilibrium constant was 2. The lifetimes of the isolated states were 1 ns and 10 ns, respectively. Evolution of a) preexponential factors. b) lifetime values. c) average intensity (apparent quantum yield). d) lifetime distribution from a uniform distribution of rates for the two-state system represented in a-c.

1 ns to 5 ns. A simulation of this "sequential" system was then performed and the decay was fitted using a continuous distribution of lifetimes with gaussian shape. At low temperature the distribution shown in figure 3a should be observed. When the interconversion between subconformations is increased to a rate of $10^9 s^{-1}$ the distribution shown in figure 3b should be observed. When the interconversion rate becomes very large compared with the decay rate of the individual subconformation, the lifetime distribution collapses to a single decay rate (figure 3c). Two major features can be observed from this simulation. A relatively wide distribution at low temperature, due to a multitude of subconformations, will narrow when the interconversion rate is increased and the center of the distribution will shift to lower lifetime values (figure 4).

In this contribution we report a qualitative analysis of the fluorescence decay from a selected number of single tryptophan residue proteins. Our aim is to analyze the decay using continuous distribution of lifetime values and to find the behavior of the lifetime distribution as a function of temperature. The results will be compared with the qualitative prediction of the model presented above. The proteins were selected on the basis of their purity, the knowledge of the x-ray structure and the location of the tryptophan residue in the protein.

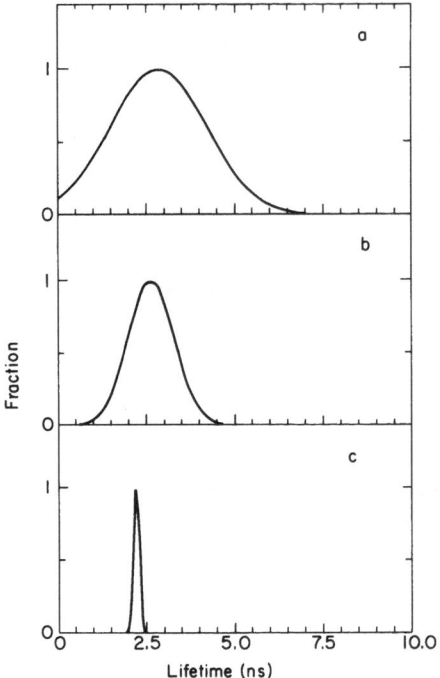

Figure 3. Sequential model for a system with a number of potential wells in a protein. a) Apparent lifetime distribution with no interconversion between the wells. b) Apparent lifetime distribution when the interconversion rate is of the same order of magnitude of the fluorescence decay rate. c) Apparent lifetime distribution when the interconversion rate is much faster than the fluorescence decay rate.

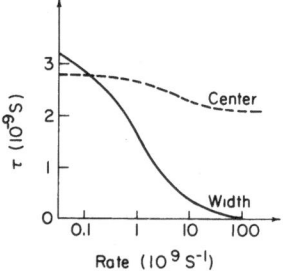

Figure 4. Evolution of the center and width of the apparent distribution as a function of the interconversion rate between the potential wells.

DISTRIBUTION OF LIFETIMES IN THE FREQUENCY DOMAIN

The derivation of the equations used for the fits reported in this work are briefly presented. In previous papers we have discussed the minimum resolvable width for a lifetime distribution of a given shape and the possibility to recover the distribution shape using probability density functions (Alcala et al., 1987a).

The fluorescence intensity may be described by the following equation:

$$I_F(t) = \sum_i f_i \tau_i^{-1} e^{-t/\tau_i} \qquad i = 1 \ldots n \qquad (1)$$

where n is the total number of independent components of the decay. In (1), f_i is the fraction of light contributing to the total fluorescence by the i-th component and the pre-exponential factor is proportional to the product $f_i \tau_i^{-1}$. In the limit when the number of components is large, equation 1 can be expressed in integral form:

$$I_F(t) = \int_0^\infty f(\tau) \tau^{-1} e^{-t/\tau} d\tau \qquad (2)$$

The distribution of lifetime values is contained in the function $f(\tau)$, which can be parameterized for fitting purposes. The above equation in the time domain is transformed to the frequency domain using the sine and cosine transforms (Weber, 1975).

$$S(\omega) = 1/N \int_0^\infty I(t) \sin(\omega t) \, dt \qquad (3)$$

$$G(\omega) = 1/N \int_0^\infty I(t) \cos(\omega t) \, dt \qquad (4)$$

$$N = \int_0^\infty I(t) \, dt \qquad (5)$$

The values of phase and modulation to be compared with the measured values, corresponding to the lifetime distribution determined by $f(\tau)$, are obtained from the following relationships:

$$P = \tan^{-1}(S/G) \qquad (6)$$

$$M = (S^2 + G^2)^{1/2} \qquad (7)$$

METHODOLOGY

The fluorescence decay was measured by use of multifrequency phase fluorometry (Gratton & Limkeman, 1983) on an instrument which utilizes the harmonic content of a mode-locked, synchronously pumped and frequency doubled dye laser as an excitation source. The operational principles of this instrument and some applications to protein studies have been reported (Alcala et al., 1985). This instrument can provide a resolution of about 0.1% in phase and modulation values and tunable ultraviolet excitation in the 285-310 nm range, with a line width of 0.06 nm, approximately. In the reference cell a solution of p-terphenyl in cyclohexane was used to correct for possible "color errors" (Lakowicz et al., 1980). A lifetime of 0.99 ns was assigned to this reference. An excitation polarizer at 55° with respect to the horizontal was used to eliminate possible polarization effects. The emission was observed after two WG-350 filters to eliminate scattered radiation from the laser. The modulation frequencies employed were in the 1 MHz to 400 MHz range. Generally 8 to 16 different modulation frequencies were used for each data set. Data were collected until the standard deviation from each measurement of phase and modulation were at the most 0.2° and 0.004, respectively. The wavelength used in this work to excite the protein was 300 nm, which corresponds to the red edge of the absorption of tryptophan. At this wavelength tyrosine residues are not excited, which minimizes heterogeneity of the emission due to excitation of more than one species and also minimizes energy transfer from tyrosine to tryptophan residues (Valeur & Weber, 1979). Additionally, 300 nm excitation selects a population of tryptophan residues not likely to be affected by dipolar relaxation processes occurring during the excited state lifetime (Lakowicz & Cherek, 1980; Macgregor and Weber, 1985). This excitation wavelength does not eliminate the risk of tyrosinate excitation and emission, but any possible effects of the latter are mitigated by the low quantum yield of tyrosinate emission (Creed 1984a,b) and also its apparently rare occurrence in proteins. Further, from their spectral characteristics, there is no reason to suspect the contribution of tyrosinate fluorescence in any of the proteins studied.

The sample temperature was controlled using an external bath circulator. The temperature of the sample was measured prior and after each measurement in the sample cuvette using a digital thermometer. Solutions were not deoxygenated.

For the data fits a non-linear least-squares algorithm for the

24

multiexponential analysis (Lakowicz et al., 1984) and a simplex algorithm
(Caceri & Cacheris, 1984) for the distributional analysis were used. In
both cases, the function to be minimized was the reduced χ^2 defined by

$$\chi^2 = \Sigma_j\{[(P_j^c - P_j^m)/\sigma_j^P]^2 + [(M_j^c - M_j^m)/\sigma_j^M]^2\}/(2n-f-1) \qquad (8)$$

where P_j^c and M_j^c are the calculated values of phase and modulation,
respectively, according to relations (6) and (7); P_j^m and M_j^m are the
measured values of phase and modulation. The standard deviations of phase
and modulation are σ_j^P and σ_j^M, respectively; f represents the number of
degrees of freedom and n the number of modulation frequencies. An acceptable fit should give a value of χ^2 close to unity. However, to expedite
calculations, the value of σ_j^P and σ_j^M were assumed to be the same for each
measurement j, so that the common value of the standard deviation can
factorize out from equation (Lakowicz et al., 1984). A value of 0.2° and
0.004 was used for the standard deviation of phase and modulation determination, respectively. These values were very close to the experimental
standard deviations since data were acquired at each modulation frequency
until the desired standard deviation was reached. The values of the
parameters obtained in the fit are independent of the common value assumed
for the standard deviations (Lakowicz et al., 1984). Using this procedure,
the values of the χ^2 can be different from one even for a good fit.
However, in this work we compare the values of the χ^2 obtained using
different models but using the same values for the standard deviation.

Each enzyme preparation was subject initially to SDS-polyacrylamide
gel electrophoretic analysis and had high specific activity. RNAse T1 was
prepared by dialysis vs 0.025 M MOPS, 0.125 M KCl, pH 7.0 (1 liter x 2
changes). An extinction coefficient of 19 ($\varepsilon_{1\%,1\ cm}$) was used to determine
the concentration of RNAse T1. Scorpion neurotoxin variant 3 (NTV3) was
the kind gift of Dr. D. D. Watt (University of Nebraska) and was >98% pure
by SDS-gel and HPLC analysis. Phospholipase A2 (PLA2) was purified from
porcine pancreas and was the gift of Dr. H. S. Hendrickson (St. Olaf
College, Northfield, MN). Azurin, from Pseudomona aeruginosa was a gift of
Dr. A. Finazzi-Agro, University of Rome, Italy.

RESULTS

Bimodal uniform, gaussian and lorentzian probability density functions
were used to fit the fluorescence decay data from NTV3, PLA2, and RNAse T1

Table 1. Parameter Fits of Probability Density Functions at 25°C

Function	F1	C1	W1	C2	W2	χ^2
PLA2 (unimodal)						
Lorentzian	1.00	2.40	2.79			10.65
Gaussian	1.00	2.50	5.98			0.77
Uniform	1.00	3.21	5.92			1.45
PLA2 (bimodal)						
Lorentzian	0.47	1.37	1.21	4.33	0.38	0.79
Gaussian	0.81	3.01	4.93	5.91	2.94	0.75
Uniform	0.93	3.32	5.20	6.20	3.35	0.70
RNAse-T1 (unimodal)						
Lorentzian	1.00	3.31	0.78			6.64
Gaussian	1.00	3.52	2.69			5.31
Uniform	1.00	3.55	4.29			10.84
RNAse-T1 (bimodal)						
Lorentzian	0.93	3.60	0.05	1.07	0.02	2.79
Gaussian	0.83	3.67	0.55	1.34	0.90	2.94
Uniform	0.92	3.62	0.65	1.07	0.33	2.71
NTV-3 (unimodal)						
Lorentzian	1.00	0.44	0.28			16.47
Gaussian	1.00	0.49	0.70			37.27
Uniform	1.00	0.52	0.85			38.78
NTV-3 (bimodal)						
Lorentzian	0.92	0.43	0.05	4.69	2.46	2.59
Gaussian	0.89	0.43	0.25	5.01	7.32	3.18
Uniform	0.91	0.43	0.24	4.84	7.62	2.86

C - center (in nanoseconds). W - full width at half maximum (in nanoseconds). F - fraction

(Table 1). The statistical significance of the distributional fit compared to the fit obtained using a sum of exponentials has been discussed elsewhere (Alcala et al., 1987c). Figure 5 shows the results obtained for the fits using a bimodal lorentzian distribution. The lifetime distributions obtained are clearly asymmetric. The center of gravity of all bimodal probability density functions monotonically shifted to lower lifetime values and the overall distribution became narrower as the temperature was increased (data not shown). This trend is more evident in azurin (data not shown) since lower temperatures were investigated. The center and the width of the distribution in this case are relatively constant for temperatures lower than 250K. Also at temperatures higher than 285K the changes are less pronounced. The observed sigmoidal behavior indicates that for azurin the two regimes, the slow and the fast interconverting, are both evident.

26

DISCUSSION

Our interpretation of the tryptophan fluorescence decay assumes that lifetime heterogeneity arises from a multitude of different conformations. One must distinguish between the actual value of the average fluorescence lifetime and the distribution of lifetimes about that average value. Neither from the present data nor from the distribution analysis can we draw any conclusion regarding the factors determining the actual value of the lifetime. These factors are dictated by the influence of the environs on the tryptophan excited state, while the distribution is determined by the subconformations of the main conformation. The distributional approach we present is dominated by the assumption that it is possible to map the multitude of subconformations into the decay rate space. A narrow distribution can have two different origins. In one case, if only restricted motions about the main conformation are possible, the lifetime distribution will be narrow because the conformational space sampled is very limited. On the other hand, the environments can be very homogeneous and consequently the multitude of subconformations will map to the same value of the decay rate. Only a correlation between rotational motions of the tryptophan residues (both amplitude and rate) and the width of the lifetime distribution can resolve this alternative interpretation. The other assumption of the distributional approach is that the interconversion between conformational substates is the major determinant of the changes observed as the temperature is varied. At higher temperatures, the motion of the indole ring will allow the fluorophore to sample a range of subconformations. The shape of the distribution, therefore, will depend on the particular mapping function of the conformational space into the rate space, and should be directly obtained from the low temperature experiment.

In general, a broad distribution should be expected for tryptophan residues free to move through large angles. Such situation is more likely for residues on the surface of proteins. From the analysis of the x-ray structure, the tryptophan in PLA2 lies on the surface of the protein fully exposed to solvent. The same conclusion has been drawn by Ludescher et al. (1986), who also suggested the existence of a distribution of states to explain the apparent multiexponentiality of the fluorescence decay in PLA2. We also note that the steady-state anisotropy values are quite low for tryptophan fluorescence of PLA2.

A tryptophan which is "fully" exposed to solvent does not necessarily have a high degree of mobility. For example, some tryptophan residues show strong non-bonded interactions with the protein matrix such that they appear to lie on the "surface" of the protein. Examples of this situation are found in W15 of the liver alcohol dehydrogenase and W48 of staphylococcus nuclease. The tryptophan of NTV3 suffers a number of restrictions on its motion that can cause a relatively narrow distribution of lifetimes,

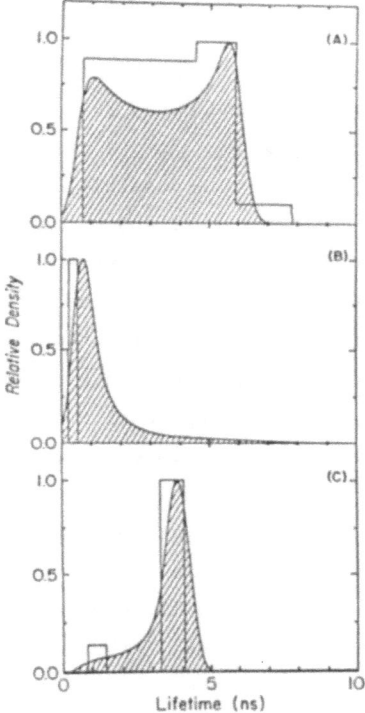

Figure 5. Lifetime distribution for three single tryptophan proteins: PLA2, RNAse T1 and NTV3.

the short average fluorescence lifetime being determined by the peculiar electronic character of the fluorophore's environs. Our analysis indicates that the mobility of the tryptophan in RNAse T1 is also highly restricted based on the relatively narrow fluorescence lifetime distribution. Indeed, James et al. (1985a,b) have suggested that, at room temperature, the tryptophan fluorescence intensity decay in RNAse T1 is mono-exponential.

TEMPERATURE EFFECTS

The fluorescence decay is determined by a distribution of lifetime values characteristic of the set of environments of the tryptophan residue in the protein. However, as the temperature is raised, the set of amplitudes and decay times which characterizes the overall fluorescence then will no longer resemble the set of static environments to which the excited residues are exposed in the frozen protein. The two general trends of the characteristic parameters of the lifetime distribution observed as the temperature was changed can be qualitatively explained as follows. As the temperature increases, the rate at which the protein fluctuates among different conformations increases. The localized environment in the protein matrix experienced by the tryptophan residue during its excited lifetime becomes more homogeneous which leads to a more homogeneous decay, with the result that the distribution narrows. This effect is illustrated with the example of figures 3 and 4. Broad distributions as those seen in PLA2 should therefore narrow substantially at higher temperature.

Intrinsically narrow distributions, caused either by a restricted motion of the tryptophan residue or by a homogeneous set of environments, are not likely to be markedly influenced in width by increasing the temperature. In the case of NTV3 and RNAse T1 the shape of the lifetime distribution as a function of temperature remains relatively narrow at low temperature.

CONCLUSIONS

The fluorescence lifetime distribution observed for the four proteins reported here is determined by the multitude of conformational substates and by the dynamics of the protein as experienced by the tryptophan residue. The shape of the lifetime distribution changed markedly from one protein to another suggesting either that the range of conformations is very different in different proteins or that the mapping function of the conformational space into the decay rate space is different. The results show that the average lifetime and the width of the distribution decrease their value with increasing temperature. The temperature effects observed indicate that, at temperatures higher than 25°C, the tryptophan residues fluctuate among different substates at rates comparable or higher than the excited state fluorescence lifetimes. The suggestion that there are conformational substates in proteins is obviously not new (Lakowicz and

Weber, 1973; Austin et al., 1975), and our approach based on fluorescence lifetime distributions may provide a good method for the study of conformational substates and of the energetics of such substates providing further validation of the conformational substates model (Gratton et al., 1986). The fluorescence approach offers the possibility to quantify the inter-conversion rates and therefore, can access the time scale of the events predicted by the molecular dynamics calculations (Karplus & McCammon, 1983). Further development of the distribution model will require correlation of its predictions with the structure of the proteins under study and with the results of molecular dynamics calculations, in particular with the dynamics of the indole ring in proteins (Ichiye & Karplus, 1983). The sensitivity of the apparent structural fluctuations to temperature and to ligand binding and sites on the protein remote from the fluorophore's location, and the role of preferential interactions of the fluorophore with the protein matrix in determining the form of the distribution, all need to be examined. Also a more complete understanding of the factors which determine the actual (average) value of the fluorescence lifetime of indole moieties is essential (Chen, 1976; Liebman & Prendergast, 1985).

REFERENCES

Alcala, J. R., Gratton, E., and Jameson, D. M., 1985, Multifrequency Phase Fluorometer Using the Harmonic Content of a Mode-Locked Laser, Anal. Instrum., 14:225.

Alcala, J. R., Gratton, E., and Prendergast, F. J., 1987a, Resolvability of Fluorescence Lifetime Distributions, Biophys. J., 51:587.

Alcala, J. R., Gratton, E., and Prendergast, F. J., 1987b, Fluorescence Lifetime Distributions in Proteins, Biophys. J., 51:597-604.

Alcala, J. R., Gratton, E., and Prendergast, F. J., 1987c, Interpretation of Fluorescence Decay in Proteins Using Continuous Lifetime Distributions, Biophys. J., 51:925.

Ansari, A., Berendzen, J., Bowne, S. F., Frauenfelder, H., Iben, I.E.T., Sauke, T. B., Shyamsunder, E., and Young, R. D., 1985, Protein States and Proteinquakes, Proc. Natl. Acad. Sci. USA, 82:5000.

Austin, R. H., Beeson, K. W., Eisenstein, L., Frauenfelder, H., and Gunsalus, I. C., 1975, Dynamics of Ligand Binding to Myoglobin, Biochemistry, 14:5355.

Beechem, J. M., and Brand, L., 1985, Time Resolved Fluorescence Decay in Proteins, Ann. Rev. Biochem., 54:43.

Careri, G., Fasella, P., and Gratton, E., 1975, Statistical Time Events in

Enzymes: A Physical Assessment, <u>CRC Crit. Rev. Biochem.</u>, 3:141.

Careri, G., Fasella, P., and Gratton, E., 1979), Enzyme Dynamics: The Statistical Physics Approach, <u>Ann. Rev. Biophys. Bioeng.</u>, 8:69.

Caceri, M. S., and Cacheris, W. P., 1984, Fitting Curves to Data, The Simplex Algorithm is the Answer, <u>Byte</u>, May, 340.

Chen, R. F., 1976, The Effect of Metal Cations on Intrinsic Protein Fluorescence, <u>in</u>: "Biochemical Fluorescence", Vol. 2, R. F. Chen and H. Hedelhoch, eds., Marcel Dekker, NY.

Creed, D., 1984a, The Photophysics and Photochemistry of the Near-UV Absorbing Amino Acids - I. Tryptophan and its Simple Derivatives, <u>Photochem. Photobiol.</u>, 39:537.

Creed, D., 1984b, The Photophysics and Photochemistry of the Near-UV Absorbing Amino Acids - II. Tyrosine and its Simple Derivatives, <u>Photochem. Photobiol.</u>, 39:563.

Fraunfelder, H., Petsko, G. A., and Tsernoglou, D., 1979, Temperature Dependent X-ray Temperature Diffraction as a Probe of Protein Structural Dynamics, <u>Nature</u> (London), 280:558.

Fraunfelder, H., and Gratton, E., 1985, Protein Dynamics and Hydration, <u>in</u>: "Biomembranes, Protons and Water: Structure and Translocation", Methods in Enzymology, 127:471.

Gratton, E., and Limkeman, M., 1983, A Continuously Variable Frequency Cross-correlation Phase Fluorometer with Picosecond Resolution, <u>Biophys. J.</u>, 44:315.

Gratton, E., Alcala, J. R., and Prendergast, F. G., 1986, Fluorescence Lifetime Distributions of Single Tryptophan Proteins: A Protein Dynamics Approach, <u>in</u>: "Progress and Challenges in Natural and Synthetic Polymer Research", C. Kawabata and A. Bishop, eds., Ohmshu Press, Tokyo, Japan.

Haydock, C., and Prendergast, F. G., 1986, A Model of Protein Fluorescence Incorporating Chromophore Interactions and Dynamics, <u>Biophys. J.</u>, 49:62a.

Ichiye, T., and Karplus, M., 1983, Fluorescence Depolarization of Tryptophan Residues in Proteins: A Molecular Dynamics Study, Biochemistry, 22:2884.

James, D. R., and Ware, W. R., 1985, A Fallacy in the Interpretation of Fluorescence Decay Parameters, <u>Chem. Phys. Lett.</u>, 120:455.

James, D. R., Liu, Y. S., De Mayo, P., and Ware, E. R., 1985, Distribution of Fluorescence Lifetimes: Consequences for the Photophysics of Molecules Adsorbed on Surfaces, <u>Chem. Phys. Lett.</u>, 120:460.

Karplus, M., and McCammon, J. A., 1983, Dynamics of Proteins, Elements and Function, <u>Ann. Rev. Biochem.</u>, 53:263.

Lakowicz, J. R., and Weber, G., 1973, Quenching of Protein Fluorescence by
Oxygen. Detection of Structural Fluctuations in Proteins on the
Nanosecond Time Scale, *Biochemistry*, 12:4171.

Lakowicz, J. R., and Cherek, H., 1980, Dipolar Relaxation in Proteins on
Nanosecond Timescale Observed by Wavelength-Resolved Phase Fluorometry
of Tryptophan Fluorescence, *J. Biol. Chem.*, 831.

Lakowicz, J. R., Laczko, G., Cherek, H., Gratton, E., and Limkeman, M.,
1984, Analysis of the Fluorescence Decay Kinetics from Variable-
Frequency Phase Shift and Modulation Data, *Biophys. J.*, 46:463.

Liebman, M., and Prendergast, F. G., 1985, Correlation of Protein Structure
and Luminescence: Use of Molecular Electrostatic Potentials, *Biochemistry*, 24:3384.

Longworth, J. W., 1971, Luminescence of Polypeptides and Proteins, *in*:
"Excited States of Proteins and Nucleic Acids", R. F. Steiner and I.
Weinryb, eds., Plenum Press, NY.

Lumry, R., and Hershberger, M., 1978, Status of Indole Photochemistry with
Special Reference to Biological Applications, *Photochem. Photobiol.*,
27:819.

Macgregor, R. B., and Weber, G., 1981, Fluorophores in Polar Media.
Spectral Effects of the Langevin Distribution of Electrostatic Interactions, *Ann. N. Y. Acad. Sci.*, 366:190.

Valeur, B., and Weber, G., 1977, Resolution of the Fluorescence Excitation
Spectrum of Indole into the $1L_a$ and $1L_b$ Excitation Bands, *Photochem.
Photobiol.*, 25:441.

Weber, G., 1981, Resolution of the Fluorescence Lifetime in a Heterogeneous
System by Phase and Modulation Measurements, *J. Phys. Chem.*, 85:949.

ENHANCEMENT OF TIME-RESOLVED FLUORESCENCE SPECTROSCOPY BY OVERDETERMINATION

Myun K. Han, Dana G. Walbridge, Jay R. Knutson[Γ], Paolo Neyroz, and Ludwig Brand

The McCollum Pratt Institute and The Department of Biology, The Johns Hopkins University, Baltimore, Maryland 21218 USA
[Γ]The Laboratory of Technical Development, National Heart, Lung, and Blood Institute, National Institutes of Health, Bethesda, Maryland 20892 USA

INTRODUCTION

Gregorio Weber pointed out over 25 years ago (Weber, 1961) that there are many ways in which the properties of the excited state can be utilized to study points of ignorance of the structure and function of protein molecules. He and his co-workers have led the way in finding new ways to use intrinsic and extrinsic fluorescent probes, with both steady-state and nanosecond time-resolved techniques, to investigate the static and dynamic structures of biological macromolecules. We are thankful to Gregorio, not only for pointing us in the right scientific direction, but also for showing us, by example, how scientists ought to interact with each other. This continues to make biochemical fluorescence an enjoyable field to work in.

A recent aim of the work from our laboratory has been to devise ways of combining information available from different types of fluorescence experiments to obtain more knowledge about macromolecules than is available from individual measurements alone. A variety of signals are available, and these can be linked together providing an "association and overdetermination" approach which makes fluorescence spectroscopy a more powerful technique.

Fluorescence decay curves obtained as a function of emission wavelength can be used to generate time-resolved emission spectra (TRES) (Easter et al., 1976) and decay-associated spectra (DAS) (Knutson et al., 1982). For systems which exhibit multiexponential decay originating in

33

ground-state heterogeneity, DAS represent the emission spectra associated with each individual species (Knutson et al., 1982).

For excited-state systems such as excimers, exciplexes and excited-state proton transfer, individual decay times are _not_ associated with specific emitting species but are characteristic of the system as a whole. To discuss these situations it is convenient to introduce a new term: species associated spectra (SAS) (Beechem et al., 1985b; Davenport et al., 1986a; Beechem et al., 1985c). The data matrix made up of fluorescence intensity versus time vs emission wavelength can be expressed in terms of TRES, DAS, or SAS; each of these representations has been of value in the discussion of particular systems.

As is the case with many biophysical techniques, procedures for data analysis are as important in nanosecond fluorometry as the instrumentation itself. Complex decay is the rule rather than the exception in most bio-chemical systems of interest. Real-time methods of analysis such as non-linear least squares (Grinvald and Steinberg, 1974) and the method of moments (Small and Isenberg, 1977) have proven valuable for the analysis of fluorescence decay data. Eisenfeld and Ford (1979) showed that additional information was available from the incorporation of multiple fluorescence decay data in a single analysis step. Knutson et al. (1983) developed a method for simultaneous (global) analysis of decay curves. The global analysis procedures result in increased model testing sensitivity and the ability to link parameters, with more accurate parameter recovery. While global analysis procedures have been applied in other areas of research (Turner et al., 1981; Johnson, 1983), it is only recently that the advantages of this approach in data analysis has become evident in nanosecond time-resolved fluorescence spectroscopy.

Global analysis has been applied to studies of excited-state reactions (Beechem et al., 1985c; Ameloot et al., 1986; Davenport et al., 1986a). Global analysis methods are also of value for analysis of emission aniso-tropy decay data. The fluorescence decay curves obtained through verti-cally and horizontally oriented polarizers can be analyzed simultaneously, without the need to generate "sum" and "difference" decay curves (Beechem and Brand, 1986a). The programs are able to combine, in one analysis, emission anisotropy decay data obtained as a function of temperature, viscosity, and excitation wavelength (Beechem et al., 1986b). The anisotropic shapes of protein molecules (e.g., axial ratio) have also been revealed by the combination of experiments done with several dyes (Beechem et al., 1986c).

Fluorescence anisotropy decay experiments may also be used to unravel rotational heterogeneity in biological systems. The approach is quite similar to that worked out for DAS as discussed above. In this case, anisotropy decay associated spectra (ADAS) are obtained and spectra associated with unique rotating species are resolved (Knutson et al., 1986; Davenport et al., 1986b).

In this contribution we wish to stress the use of a second time axis (reaction time) together with fluorescence decay time as a global overdetermination tool. The procedures used for this new application of global analysis are quite similar to those used for DAS, SAS, and ADAS. Fluorescence decay curves obtained at different times during the course of a chemical reaction are presumed to contain the same (but unknown) fluorescence decay-time components. Their properties (amplitudes) may change with reaction time. The values of the amplitudes are unrestricted and may be positive, zero, or negative. In the case of proteins (e.g. Trp residues), site specific information may be obtained; a feature often difficult to achieve in a system of reacting biological macromolecules.

INSTRUMENTATION

The desirable modifications to a typical pulse fluorometer have recently been described in detail (Walbridge et al., 1987a). Important capabilities for collecting this type of data include: 1) rapid data transfer between a multichannel pulse height analyzer and a computer (or an analyzer with sufficient memory to accumulate 50 or more decay curves), 2) a real time clock to mark the decay curves, and 3) a light source of constant intensity during the course of an experiment (or the ability to monitor source intensity during the course of an experiment).

A typical instrument is illustrated in Figure 1. The light source is a synchronously pumped, mode locked, cavity-dumped dye laser capable of producing 20 psec (fwhm) pulses at a frequency of 4 MHz. The combination of the high repetition rate, high intensity light source with a computer controlled, high speed data collection device results in an instrument with a wide range of experimental possibilities. The potential for rapid collection times (seconds or less) is quite feasible using this newer technology.

The program used to govern these kinetic experiments is sketched in

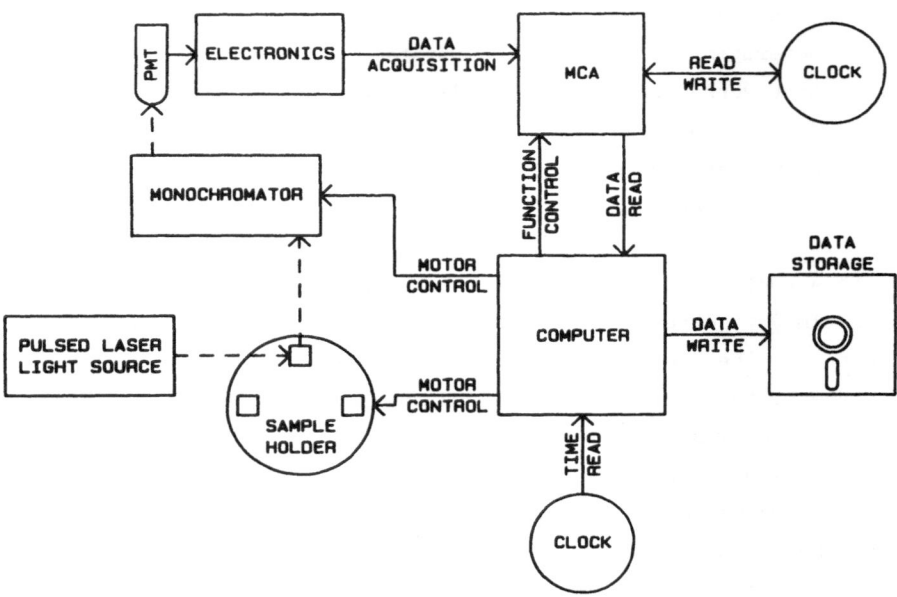

Figure 1. A schematic diagram exhibiting components of the fluorescence decay fluorometers used in most of these studies. This diagram outlines the basic elements necessary for inspecting kinetic processes using single photon counting fluorescence time-resolved techniques. The original measurements were obtained with a nitrogen flash lamp as the light source. Both of the newer instruments use a Spectra-Physics 3000 Nd:YAG synchronously pumped, mode-locked, cavity-dumped dye laser. One arrangement uses a Tracor Northern 1710 multichannel analyzer with 4Kb of memory subdivided to eight or sixteen segments to accumulate data. This MCA communicates with a Hewlett Packard A900 minicomputer through an IEEE-488 interface. The sample chamber is a PRA T-format chamber with monochromators (Photochemical Research Associates). Timing for this instrument is achieved by recording start and stop accumulation voltages directly from the MCA with a strip-chart recorder. The second arrangement uses an ORTEC ADCAM Model 918 multichannel buffer as interfaced to an IBM XT microcomputer as the data acquisition device. A Jarrell-Ash model 82-410 monochromator selects the fluorescence emission wavelength. Timing is accomplished by using the internal clocks of both the IBM XT and the ADCAM MCB. This figure is taken from Walbridge et al. (1987a) with permission from Academic Press, Inc.

Figure 2. The preceding version (which ran on a Hewlett Packard 2100 minicomputer) was described earlier (Walbridge et al., 1987a) and followed the steps of the diagram in literal detail. The current version functions in much the same manner but has been rewritten to execute on an IBM XT microcomputer.

Initially the program requests various running parameters such as file name, number of data curves, and the collection times per curve. A new

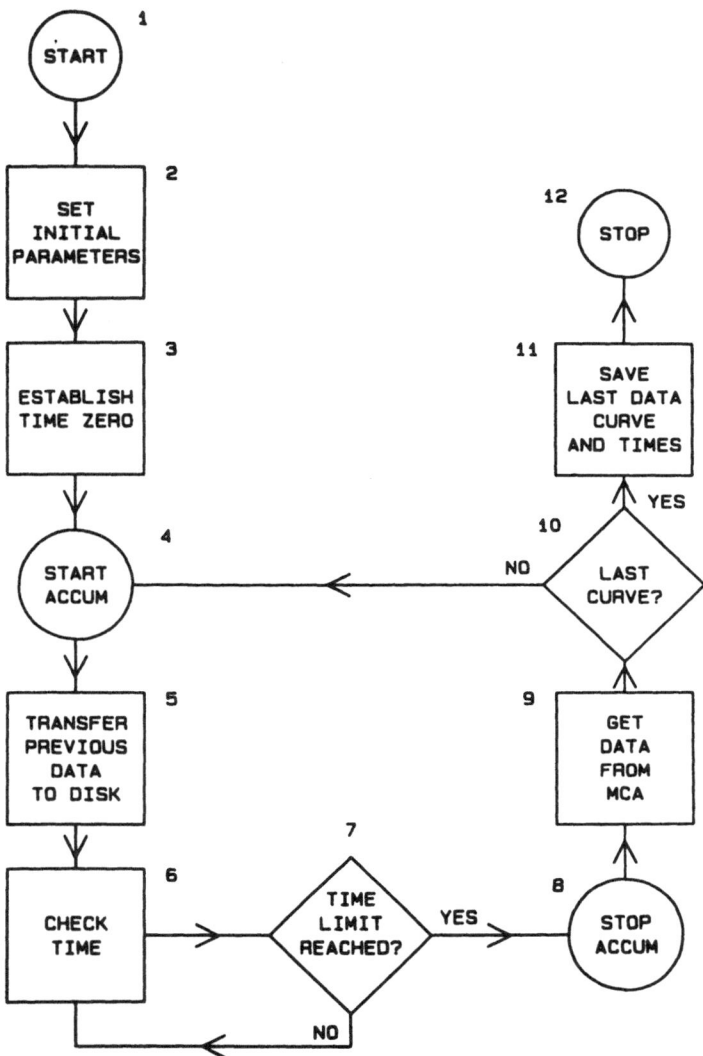

Figure 2. A flow diagram outlining one sequence of steps which can be used to collect and store an assortment of successive decay curves. A progression such as this is required for exploiting reaction processes using time-resolved fluorescence techniques. This figure is taken from Walbridge et al. (1987a) with permission from Academic Press, Inc.

addition is the ability to enter most of the set-up parameters from a file as well as the keyboard. Also, the time of collection for each decay curve can be specified; the time may be constant for all decay curves or may vary individually from one curve to another. A signal must be given to mark the origin of the reaction (step 3). A second signal commences data acquisition (step 4). At present, both signals are keyboard responses. Upon receiving these signals the program cycles through steps 4 through 10 as shown in Figure 2. Step 5 has been eliminated since the memory of a fully populated XT is sufficiently large to hold the data for 30 or more decay curves. Thus, unlike the original version, all data files are created upon completion of the experiment (step 11). The relocation of data from the multichannel buffer to the microcomputer is extremely fast. A one Kbyte segment of MCB memory can be copied to the program in under 55 milliseconds (step 9).

There is a word of caution regarding analysis of the type of data to be discussed below. Data obtained as a function of reaction time are often made up of many nanosecond decay curves, each containing only a few thousand photon counts (S/N $\approx$ 100). Fortunately, it is the total counts under the entire surface (rather than each curve) that will predominate, as long as Gaussian errors can be presumed. Unfortunately, Poisson noise (the fundamental noise process in a counted form of data) is only of Gaussian form for counts $\geq$ 50. We have observed that "thinning" the data out into many small curves can thus lead to large "lifetime defect" errors. Fortunately, as suggested by Selinger and Harris (1983), reanalysis with altered least-squares weighting restores the correct values (Walbridge et al., 1987a; Knutson, 1987). The experiments described below utilized a mode-locked laser. This provided sufficient data density such that low counts did not constitute a problem. When we extend kinetic decay experiments to "stopped flow" time scales (e.g., a few milliseconds), this analysis option will certainly be required again.

We will now present a number of examples of the application of the "associative" approach to some specific biochemical systems. As will be seen, global "kinetic decay" analysis can be further enhanced by linkage with pH, monomer/dimer fraction, or ligand titration. In some cases, the events that occur at individual sites are thereby revealed.

DENATURATION OF HORSE LADH

The intrinsic fluorescence of HLADH has received considerable attention (for review, see Beechem and Brand, 1986a). The extraction of two major decay components (and their decay association spectra, DAS) under various conditions has provided site specific "ligand perturbation" information (Brand et al., 1985). In particular, the acid denatured enzyme loses zinc (Blomquist, 1967; Heitz and Brand, 1971; McKay, 1962; Oppenheimer et al., 1967) and "unfolds", yet it retains at least biexponential

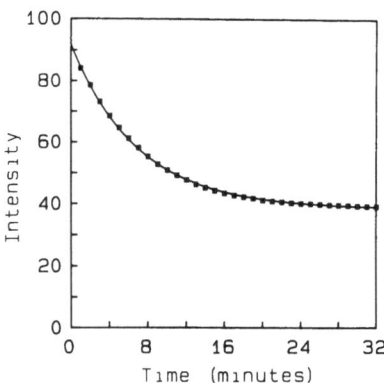

Figure 3. The acid denaturation of horse liver alcohol dehydrogenase as observed by steady-state fluorescence. The sample was prepared to 1.4 x 10^{-5} M in 0.1 M succinate buffer with a pH of 4.1 and temperature of 10°C. The steady-state fluorescence intensity was monitored by exciting at 295 nm and observing emission at 350 nm. The squares represent points digitized from the strip chart recording and are taken at each minute. The solid line is the nonlinear least squares fit to the data for a first-order rate process. The rate constant from this analysis is 0.10 min^{-1}. These data are taken from Walbridge et al. (1987a) with permission from Academic Press, Inc.

character and two distinct DAS, indicating the incomplete nature of unfolding (Knutson et al, 1982). Walbridge et al. (1982, 1987a) have studied the details of this denaturation process by collecting many brief fluorescence decay curves during the unfolding reaction. Figure 3 shows the loss in <u>total</u> tryptophanyl fluorescence intensity that occurs upon denaturation at acid pH at 10°C. Note that some initial events (e.g., the first minute) are not accounted for by the solid line fit to a first order reaction rate.

Decay curves were obtained throughout this reaction. Their individual analyses yielded (marginal) biexponential results, marked by considerable scatter in the recovered lifetimes. In the individual analyses, the only discernible trend seen was a loss of one amplitude. The entire data set was subsequently examined with global analysis methods (Knutson et al., 1983; Beechem et al., 1985). Several competing models were considered, and the simplest model consistent with the entire set is shown in Figure 4. The other models with adequate quality of fit (judged by χ^2) all required unlinking of more parameters (i.e., use of more free parameters). Further, the unlinked parameters all retained the same basic trends seen in Figure 4. The fact that the final biexponential parameters closely approached previous DAS results lends confidence to the model. The incomplete nature

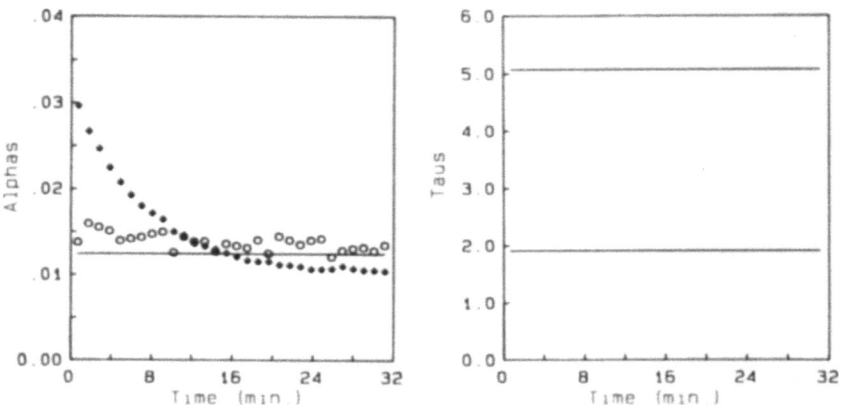

Figure 4. Fluorescence lifetime decay parameters as resolved from two component simultaneous analysis of the decay data obtained during the course of the denaturation. Several models linking one or more sets of the four parameters across the thirty decay curves were explored. The simplest model which satisfactorily describes the data links each lifetime value and the pre-exponential value for the short lifetime component. The values of these linked parameters are shown with the solid straight lines. The one free parameter (the pre-exponential term for the long lifetime parameter) decreases with time and is shown with the diamond shaped symbols. More complex models did not significantly improve the analysis and tended to point to this simpler picture. One example is shown in the figure. When the two lifetime parameters are linked with their respective amplitudes unrestrained, the amplitude for the long lifetime component continues to decrease at the same rate whereas the other is invariant with time (the open circles). Nonlinear least squares analysis of the α_2 data for a first order rate process gives a rate constant of 0.099 min^{-1} which compares favorably with the steady-state intensity decrease. These data are taken from Walbridge et al. (1987a) with permission from Academic Press, Inc.

of the "static" loss (drop in α_2 to 40%, not 0) suggests intriguing possibilities for a terminal equilibrium between "fully unfolded" ($\alpha_2 = 0$) and "partially folded" forms of the protein.

More recently, we examined the earlier events in denaturation, using a mode-locked laser source. We also ran the reaction at 7.5°C so as to slow the process. The results of the laser-based investigation (Walbridge et al., 1987b) are shown in Figure 5. Strikingly, a pair of decay times matching the "native" decays and another pair corresponding to the denatured form emerge from a global analysis. The complete domination of the "early decay curves" by native components that are lost as the denatured terms (<u>initially absent</u>) appear is of interest. The relative constancy of the "native" amplitude ratios during their decline is to be contrasted with the

40

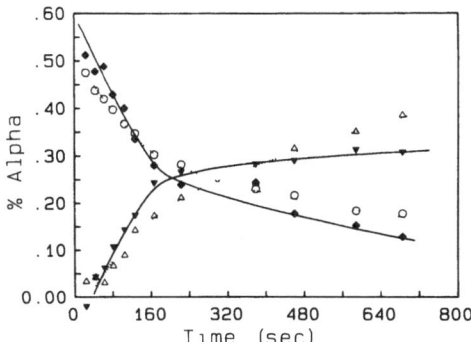

Figure 5. Simultaneous four component analysis for HLADH acid unfolding data obtained using the pulsed laser light source during the first 11 minutes of the reaction. A series of a dozen fluorescence decay curves were rapidly obtained using the instrument described in Figure 1. Since these decay curves contained significantly more counts than those obtained using the flash lamp (several thousand counts in the peak compared to several hundred), it was possible to globally analyze the data using four components and linking each of the lifetime values. The resultant lifetimes are 1.9 (Δ), 3.3 ($\blacklozenge$), 5.1 ($\blacktriangledown$), and 6.6 (o) nanoseconds. The percent alpha values for each curve are shown by the symbols. It is evident that during the first minute the native lifetimes (3.3 and 6.6 nanoseconds) predominate while the denatured lifetimes (1.9 and 5.1 nanoseconds) are virtually nonexistent. The situation changes rapidly during the next several minutes as the native lifetimes disappear and the denatured lifetimes prevail. The conditions are the same as those described in Figure 3 with the exception that this experiment was carried out at 7.5°C.

changes in α_1(2 ns)/α_2(5 ns) seen after a few minutes. These later changes presage and agree with the relative α_2(5 ns) reduction previously seen.

Thus, both phases in the denaturation process can be examined with the use of a high repetition-rate light source. Presently, kinetic <u>anisotropy</u> collections are being carried out to determine at what stage the two identical subunits disengage.

FLUORESCENCE STUDIES OF ENZYME I OF THE PTS

The phosphoenolpyruvate:glucose phosphotransferase system (PTS) of bacteria catalyzes the uptake and concomitant phosphorylation of its sugar substrates (Kundig et al., 1964; Postma and Rosemann, 1976; Meadow et al., 1984). The overall phosphorylation reaction of the PTS involves a series of phosphoryl transfer steps, which require several protein components. There are two non-sugar-specific proteins called <u>Enzyme I</u> and <u>HPr</u> and usually two sugar-specific proteins comprising (with lipid) an Enzyme II complex (III/II-B or II-A/II-B). The first protein component of the sequence, Enzyme I, accepts one phosphoryl group per monomer from phospho-enolpyruvate, the sole phosphoryl group donor, and donates the phosphoryl group to HPr (Weigel et al., 1982; Waygood, 1986).

Enzyme I exhibits a temperature- as well as concentration-dependent monomer/dimer equilibrium of identical subunits. The subunit association was studied using the ultracentrifuge (Kukuruzinska et al., 1982), as well as analytical gel chromatography (Kukuruzinska, 1984), and by nanosecond time-resolved emission anisotropy (Neyroz et al., 1987).

The DNA sequence of PTS I, the gene that encodes Enzyme I of E. coli is now known (Saffen et al., 1987). Enzyme I contains four cysteine (Cys 272, Cys 324, Cys 502, and Cys 575) and two tryptophan (Trp 357 and Trp 498) residues per monomer. All four sulfhydryl groups reacted with 5,5'-dithiobis(2-nitrobenzoic acid), DTNB. The modification of -SH groups reversibly inactivates the enzyme (Kundig and Roseman, 1971). Steady-state and time-resolved fluorescence emission anisotropy studies of didansyl-cystine-labeled Enzyme I revealed that the modified enzyme also remains a monomer (Han et al., 1986). The kinetics of the DTNB reaction with Enzyme I exhibit biphasic character. Fractional amplitudes associated with the rate constants are 25±5% for the fast and 75±5% for the slow rate. These results suggest that there is more than one class of -SH group (Han et al., 1987b).

Pyrene Maleimide Reaction with Enzyme I

Pyrene maleimide is a SH-selective reagent. It is virtually non-fluorescent in aqueous solution, but becomes fluorescent when it conjugates with proteins. There is considerable spectral overlap between tryptophan emission and pyrene absorption. These characteristics of pyrene fluorescence make it possible to monitor the reaction of pyrene with Enzyme I by measuring either the appearance or the quenching of tryptophan fluorescence upon conjugation (Figure 6).

During conjugation of pyrene maleimide to SH groups of Enzyme I, the appearance of pyrene fluorescence (shown in Figure 7) is biphasic, as is the quenching of tryptophan fluorescence. The slow rate is modified by either Mg^{2+} or Mg^{2+} plus PEP, suggesting conformational changes induced by PTS ligands (Han et al., 1987b). These changes are consistent with corre-

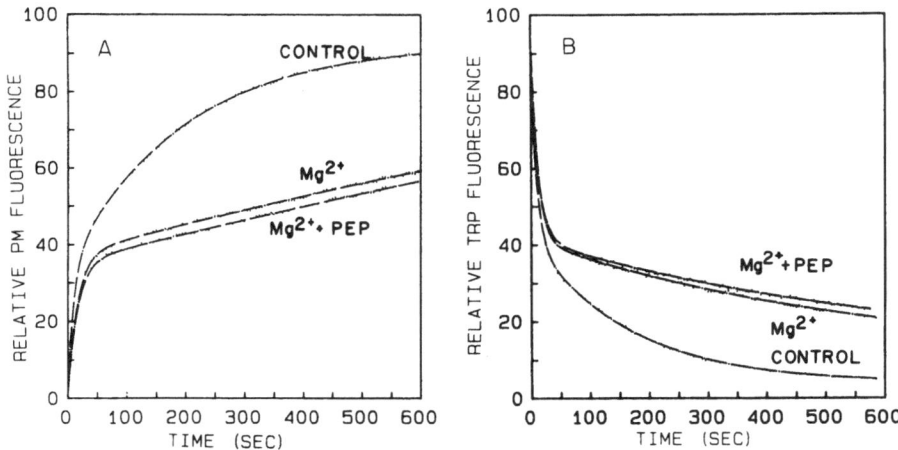

Figure 6. Kinetics of the reaction of pyrene maleimide with Enzyme I. A) Changes of pyrene fluorescence were monitored at 395 nm with an excitation of 337 nm in 100 mM potassium phosphate, pH 7.5 and 1 mM EDTA. Enzyme concentration was 0.25 mg/ml and pyrene concentration was 20 μM. Presence of 5 mM Mg^{2+} and Mg^{2+} plus PEP (2 mM) influenced the reactivity of -SH groups. B) Changes of Trp fluorescence upon reaction of pyrene to Enzyme I were monitored at 340 nm with an excitation of 295 nm. Conditions were the same as in A. Measurements were obtained with an SLM 8000 photon counting spectrofluorometer.

sponding results obtained from DTNB reactions with Enzyme I (e.g., the predominant effect of PTS ligands was on the slow rate).

Studies of the reaction of both DTNB and pyrene maleimide with the sulfhydryl groups of Enzyme I indicate more than one class of -SH groups (one of the four groups reacts $\simeq$ 10 times faster than the other three). Nanosecond time-resolved fluorescence decay studies were carried out to determine the influence of binding of pyrene maleimide with the reactive -SH group of Enzyme I.

Two types of experiments were carried out: In one set Enzyme I was reacted with 0.5 to 1.2 pyrene maleimide per monomer. The reactions were allowed to go to completion, and the decay of the tryptophan fluorescence was measured. In the second set of experiments approximately 1 mole pyrene maleimide per mole of Enzyme I monomer was reacted and a series of decay curves were obtained as a function of reaction time. New types of information about the tryptophan fluorescence of Enzyme I can be obtained from these studies.

These results obtained from the first type of experiment are summarized

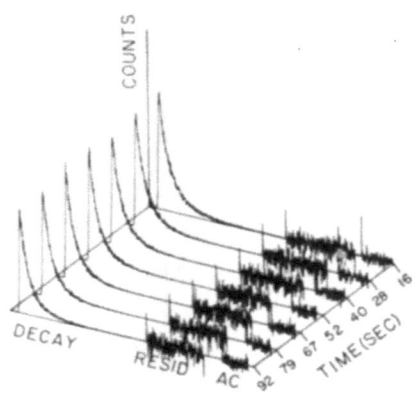

Figure 7. Decay curves, residuals, and autocorrelations for the global analysis of seven consecutive decay curves. Seven decay curves were collected at the rate of 10 sec per curve during the reaction of pyrene to Enzyme I. The experiments were performed in 100 mM HEPES, pH 7.5 and 1 mM EDTA with an enzyme concentration of 0.33 mg/ml. The measurements were begun a 6 µM PM were added. The peak counts were less than 1,000 in each curve. The seven decay curves were analyzed simultaneously, with the aid of a global procedure. Two native components, 3.7 (τ_3) and 7.3 ns (τ_4), were fixed, since the PM-labeled Enzyme I exhibits two major native components. Two other components were linked (not fixed) and the recovered lifetimes were 0.7 (τ_1) and 1.9 ns (τ_2). The global reduced χ^2 was 1.15.

Table 1. Time-Resolved Fluorescence Decay Parameters of Native and Pyrene-Labeled Enzyme I

ENZYME I	τ_1	τ_2	τ_3	τ_4
NATIVE	.6	---	3.7	7.3
PM-LABELED	.66	1.85	3.8	7.4

#SH MODIFIED	% α_1	% α_2	% α_3	% α_4	% α_{2+4}
NATIVE	16	--	41	43	43
.5 SH	22	19	35	23	42
.5 SH	22	18	36	23	41
.5 SH	16	21	38	25	46
.5 SH	16	25	36	23	48
.8 SH	21	25	41	12	37
.8 SH	18	27	42	13	40
1.2 SH	17	41	40	2	43

in Table 1. In a series of experiments performed at room temperature (dimeric conditions for Enzyme I), 0.5 to 1.2 moles of -SH group/monomer are reacted (The extent of the reaction was also determined by subsequent DTNB reaction). The entire data set was simultaneously analyzed by a global procedure with the decay constants (τ_1 to τ_4) "linked" (Knutson et al., 1983; Beechem et al., 1985b). The recovered lifetimes from this analysis are 0.7, 1.9, 3.8, and 7.1 ns.

The fluorescence decay of native Enzyme I can be resolved into a triple exponential with two major components of 3.7 ns (41%, preexponential) and 7.3 ns (43%), and a minor component of 0.6 ns (16%) (Han et al., 1987a). Thus, the decay components of 0.7, 3.8, and 7.1 ns of pyrene-labeled Enzyme I may be considered to be "native" decay components while 1.9 ns is most likely a quenched lifetime due to resonance energy transfer from Trp to pyrene.

Tracing the origins of the 1.9 ns component may provide valuable information. It may, for example, help us to distinguish the origin of the two native major decay components (especially whether or not they arise from the same tryptophan residue). In this regard, the "map" of reaction extent contained in Table 1 is of great interest. It appears that the amplitude of 7.1 ns is lost as it is exchanged to create the amplitude of 1.9 ns. On the other hand, the amplitude of 3.8 ns does not seem to be diminished. This suggests that the 1.9 ns term is a quenched decay of the 7.1 ns Trp. Moreover, it suggests that the native 3.7 ns and 7.3 ns decay terms reflect distinct tryptophan residues.

The second type of experiment (time-resolved reaction kinetics of the interaction of pyrene with Enzyme I) was performed to further clarify the changes of amplitude associated with each decay component during the course of the reaction.

For kinetic studies, rapid collection of decay curves during the reaction was essential and we achieved this with a single photon counting pulse fluorometer with a synchronously-pumped, mode-locked, cavity-dumped dye laser as the excitation source as described above. A stoichiometric amount of pyrene maleimide was added to Enzyme I (to label specifically the fast reacting -SH group) and seven consecutive decay curves were collected (10 sec per curve) during the reaction (Han et al., 1987d). Figure 8 shows these decay curves as well as autocorrelations and residuals for global analysis. The best analysis was obtained by fixing two long "native

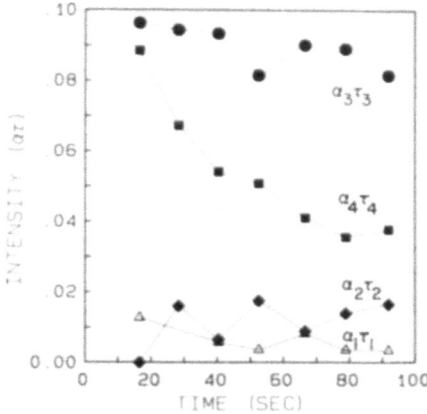

Figure 8. The intensity ($\alpha_i\tau_i$) of each individual component from Figure 7 is plotted.

lifetimes" with linkage of four decay components. Any contamination by scattered excitation light has been corrected by fixing a 0.01 ns decay term, and allowing its amplitude to vary. Figure 9 shows plots of individual intensity ($\alpha_i\tau_i$) obtained from this system wide analysis of the reaction. The intensity of the 7.3 ns component ($\alpha_3\tau_3$) was not changed noticeably.

The kinetic data appears to support the suggestion that only a 7.3 ns component is quenched upon modification of the "fast reacting" SH group. DTNB reaction with these partially pyrene-modified Enzyme I revealed that the labeling occurred only to the fast reacting -SH group (data not shown). Thus, the two major native decay components appear to originate from different tryptophan residues. This is consistent with decay-associated spectra (DAS) obtained from Enzyme I, which indicate different spectral distributions associated with the two major decay components (Neyroz et al., 1987).

These results suggest that the fast reacting -SH group may reside in very close proximity with one of two tryptophan residues. DNA sequence reveals that Cys 502 and Trp 498 are in close proximity. At this moment, it is very difficult to assign a specific residue without an X-ray crystal structure; nevertheless, it is tempting to suggest that the fast reacting sulfhydryl group is Cys 502 and the 7.3 ns component is linked to Trp 498.

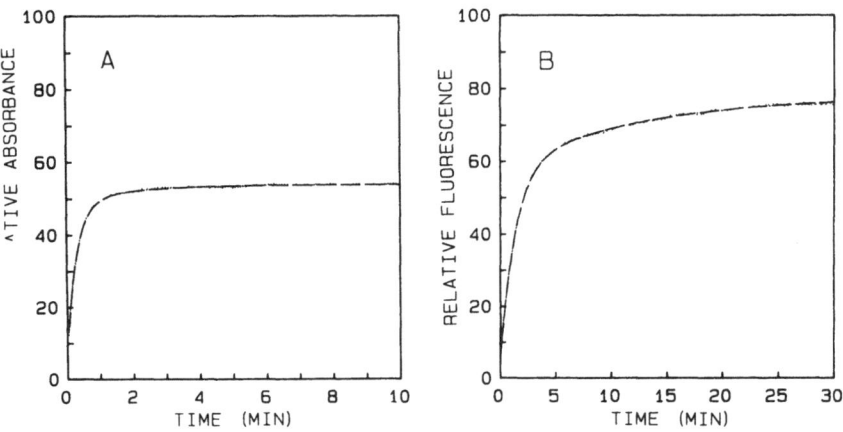

Figure 9. The release of TNB from the DTNB modified Enzyme I was measured A) by absorbance (OD_{412}) and B) by Trp fluorescence with excitation and emission wavelengths of 295 nm and 350 nm respectively at pH 6.8 (room temperature). OD_{295} of the sample was 0.1.

Kinetic Studies of the TNB Release From the DTNB Modified Enzyme I of the PTS.

The reaction of all the sulfhydryl residues of Enzyme I with SH-specific reagents inhibits both activity and dimerization (Han et al., 1986). When all four -SH groups were reacted with dansyl (by a thiol-disulfide exchange reaction with didansylcystine), fluorescence anisotropy measurements showed that this enzyme remained a monomer even at room temperature.

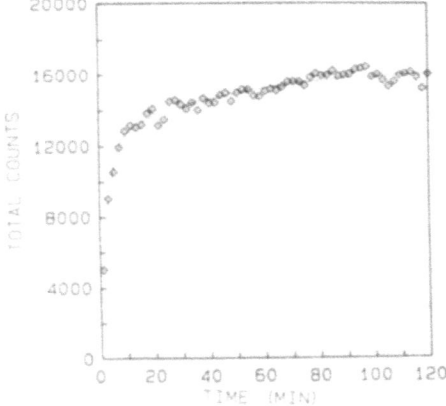

Figure 10. Total counts for each of the decay curves were obtained by integrating each decay curve, and they were plotted as a function of time. The rate constants of the reaction were obtained by a nonlinear least-squares method. The dotted line indicated the fitted data. The biphasic character is apparent, with pseudo-first-order rate constants of 0.237 min^{-1} (73%) and 0.015 min^{-1} (27%). This data is taken from Han et al. (1987a) with permission from Academic Press, Inc.

Table 2. Time-Resolved Fluorescence Decay Parameters of DTNB-Modified Enzyme I

ENZYME I	τ_1	$\%\alpha_1$	τ_2	$\%\alpha_2$	τ_3	$\%\alpha_3$	τ_4	$\%\alpha_4$	$\langle\tau\rangle$
DTNB-LABELED (2 SH)	0.4	42	2.2	29	4.3	17	7.4	13	4.7
DTNB-LABELED (3.9 SH)	0.4	61	2.2	27	4.3	9	7.4	3	3.3

When thio-2-nitrobenzoic acid (TNB) was bound to the -SH groups of Enzyme I via a thiol-disulfide exchange reaction with 5,5'-dithiobis (2-nitrobenzoic acid), DTNB, the tryptophan fluorescence was reduced compared to native Enzyme I. Resonance energy transfer is a likely quenching mechanism, since TNB absorption has good overlap with tryptophan emission (other quenching mechanisms having their origins in structural alterations related to the chemical modification cannot be ruled out, however).

The tryptophan fluorescence quenching is reversed by addition of dithiothreitol (DTT). The kinetics of the release of TNB was followed spectrophotometrically and the recovery of "native" tryptophan fluorescence was studied by both steady-state and nanosecond time-resolved measurements.

The time course of the increase in absorbance at 412 nm reflecting the release of TNB anion at pH 6.8 (room temperature) is shown in Figure 10A. Figure 10B shows the concomitant increase in tryptophan fluorescence. The release of TNB (absorption) is essentially complete within a minute. This is associated with a rapid increase in the tryptophan fluorescence intensity. This can be explained by a loss of resonance energy transfer upon release of TNB from the protein.

The fluorescence enhancement also exhibits a much slower kinetic component. This may be related to the dimerization of Enzyme I following TNB release (Han et al., 1987c). This would be in accord with the finding by Neyroz et al. (1987), that the spectrum associated with the 3.7 ns component comprises 17% of total fluorescence intensity in the native monomer form of Enzyme; on the other hand, it makes up 44% of total intensity under dimeric conditions. It is also possible that this slow fluorescence enhancement may be due to a conformational change in the monomer which must precede the dimerization step.

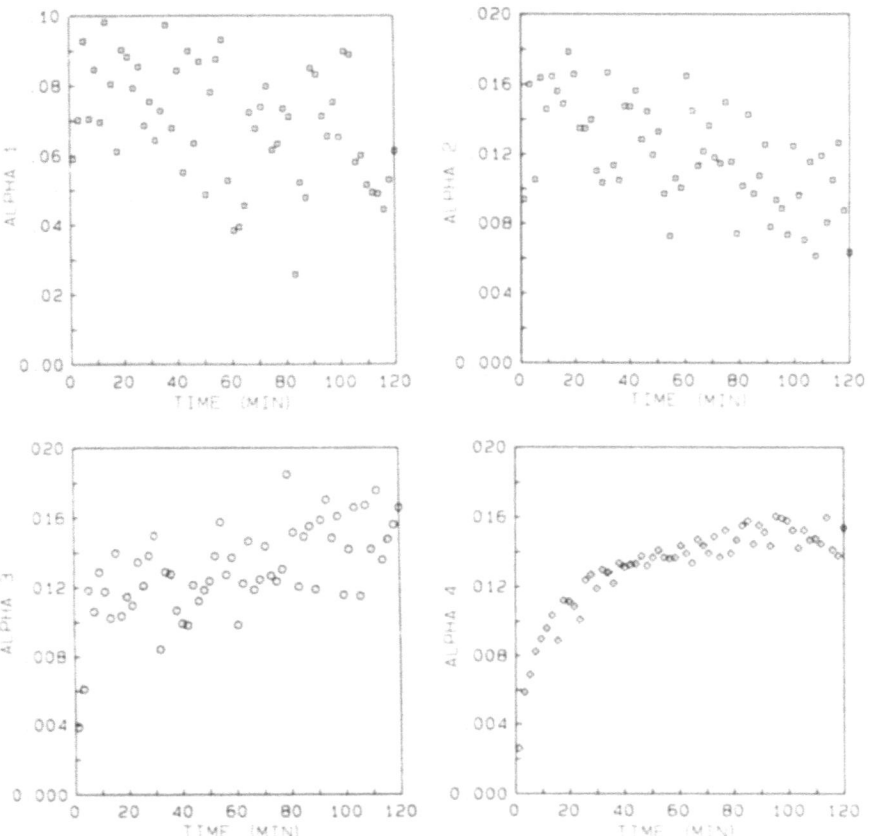

Figure 11. Four preexponential amplitudes (associated with each of four lifetime components) were plotted as a function of time. Results were obtained with a global analysis linking four lifetimes. Data were analyzed with terminal (1/fit) weighting. The recorded lifetimes were 0.07, 2.4, 3.9, 7.1 ns, in agreement with values obtained from a single precise decay curve (with high peak counts) of TNB-bound enzyme. Alpha 1,2,3, and 4 represent amplitudes associated with 0.07, 2.4, 3.9, 7.1 ns, respectively. The reduced global χ^2 was 1.03. This data is taken from Han et al. (1987a) with permission from Academic Press, Inc.

The release of TNB anion by DTT was also investigated by nanosecond time-resolved techniques (Han et al., 1987a). Sixty decay curves were collected over a two hour period (2 minutes per decay curve). The experiments were done at pH 6.8. The sixty decay curves were simultaneously analyzed with the aid of the global methodology. Since a minimum of two different lifetimes could be expected for both the initial and final states (Table 2), the global analysis was done in terms of four linked lifetime components. Decay times of 0.07, 2.4, 3.9, and 7.1 ns were obtained.

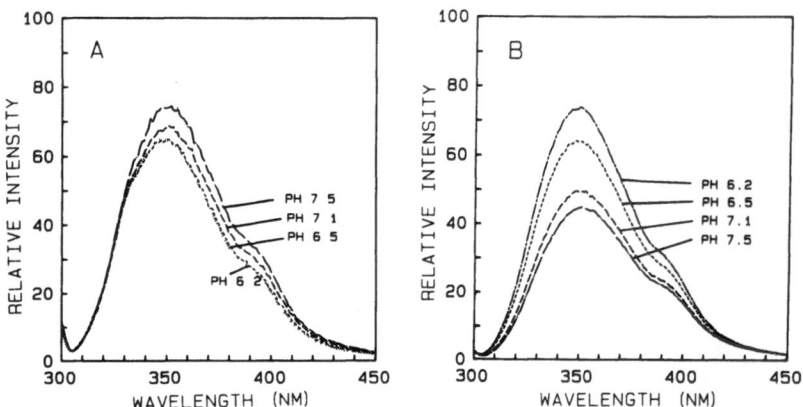

Figure 12. Technical fluorescence emission spectra of oxidized and reduced thioredoxin at various pH conditions. Emission spectra were recorded with an excitation of 295 nm at room temperature: A is oxidized and B is reduced (with B being 4.5 times the scale of A). OD_{295} of each sample was 0.1 in 100 mM potassium phosphate and 1 mM EDTA. Measurements were obtained with a SLM-8000 photon counting spectrofluorometer.

The amplitudes associated with each lifetime component were plotted as a function of reaction time and are shown in Figure 11. The amplitudes associated with lifetimes of 0.07 and 2.4 ns <u>decreased</u> during removal of the TNB. In contrast, the amplitudes associated with the two long decay components (native Enzyme I dimer) both <u>increased</u> with time. The kinetics of amplitude 3, associated with the 3.9 ns decay component and amplitude 4, associated with the 7.1 ns decay component are clearly biphasic. The higher data density that will become available as the instrumentation improves should make it possible to further unravel this complex but interesting multistep process.

FLUORESCENCE STUDIES OF THIOREDOXIN

Thioredoxin is a small, heat stable, redox active protein found in roughly 10,000 copies in all cells (Holmgren et al., 1978). The amino acid sequence (Holmgren, 1968) and the X-ray crystal structure of <u>E. coli</u> thioredoxin purified from <u>E. coli</u> is a single polypeptide chain of 108 amino acids (molecular weight = 11,700). The active site has the two redox active thiols (Cys 32 and Cys 35) and also has two tryptophans (Trp 28 and Trp 31) proximate to cysteines.

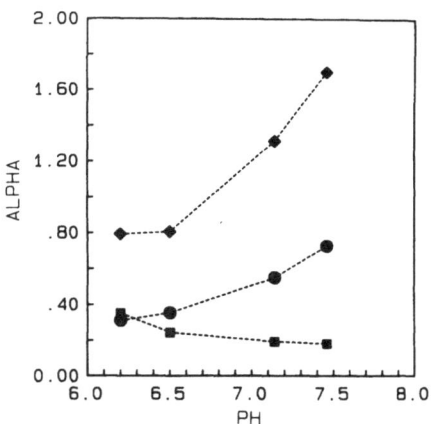

Figure 13. Plots of normalized α values of reduced thioredoxin at various pH conditions. Three decay curves were collected for each pH and all the decay curves were simultaneously analyzed linking three decay components. The recovered lifetimes were 0.4 ns, 1.5 ns, and 3.6 ns. The average amplitudes associated with each decay component were normalized to the steady-state fluorescence and plotted as a function of pH: the diamonds representing 0.4 ns, circles and squares representing 1.5 ns and 3.6 ns respectively.

Reduced thioredoxin has higher fluorescence than the oxidized form of the protein (Stryer et al., 1967). Holmgren (1972) indicated that the fluorescence enhancement upon reduction of thioredoxin is due to Trp-28.

We purified thioredoxin from an overproducer strain of E. coli. The effect of pH on the steady-state fluorescence spectra of oxidized (A) and reduced (B) thioredoxin are shown in Figure 12. The fluorescence of the oxidized thioredoxin (A) exhibits much lower fluorescence (these spectra have been amplified by 4.5-fold).

The emission of the reduced protein shows an increase in fluorescence with decreasing pH (from pH 7.5 to pH 6.2) accompanied by a minor blue shift. Holmgren (1972) reported a similar pH dependence for this form. In contrast, the oxidized (-S-S-) form of the protein exhibits comparatively little pH dependence. Increasing pH continues to correlate with a small red shift of emission, but the small intensity change is now seen to be a decrease with decreasing pH. When we examined the decay characteristics of oxidized thioredoxin, a similar insensitivity to pH was found. The analyses all yielded similar triexponential decays. The lifetime values at pH 7.1 were 0.3 ns (78%, preexponential), 0.8 ns (20%) and 4.5 ns (2%) (all at room temperature). The dominance of this weak fluorescence decay by a very short component makes it difficult to examine in detail. We soon expect to

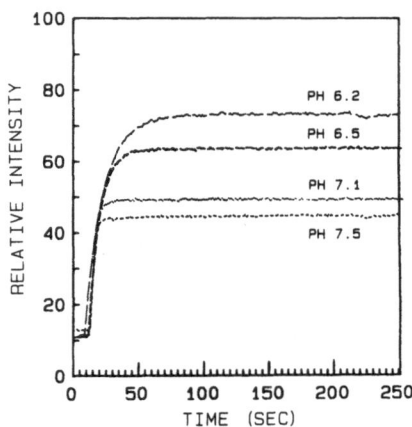

Figure 14: Steady-state kinetics for the reduction of the disulfide bond at various pH conditions. The kinetics of the reduction of the disulfide bond were recorded on the SLM-8000 photon counting spectrofluorometer at an excitation and emission wavelength of 295 nm and 350 nm respectively. The reduction was accomplished by the addition of DTT.

overcome this difficulty by replacing the photodetector with a microchannel plate that reduces time jitter an order of magnitude.

The decay characteristics of reduced $(-(SH)_2)$ thioredoxin were more interesting and informative. First, since the steady-state spectra were sensitive to pH, we examined decay behavior at four different pH values. Triplicate experiments yielded three exponentials in each case; again, little variation in the lifetimes was seen across pH. A global (linked) triexponential fit yielded the best χ^2 and the residual pattern; lifetimes of 0.4 ns, 1.5 ns and 3.6 ns were recovered. While lifetimes were pH independent (whether linked or free), <u>amplitudes</u> varied considerably with pH, as can be seen in Figure 13. The shorter lifetimes' amplitudes increase with pH, while the longest lifetime contribution decreases with pH. The latter effect clearly dominates the steady-state spectra (Figure 12). It will be interesting to identify the group(s) whose pKa has an influence on this component.

Having examined the fully oxidized and reduced forms at various pH values, we questioned whether another type of data might help us focus on the origins of the different decay behavior found in oxidized vs reduced cases. Previous studies using model peptides (Kishore et al., 1983) indicated 45% enhancement upon reduction, a finding that partly reinforced the original explanation of Holmgren (1972) that the disulfide bond was primarily responsible for the quenching of the oxidized form. Yeast thioredoxin, with

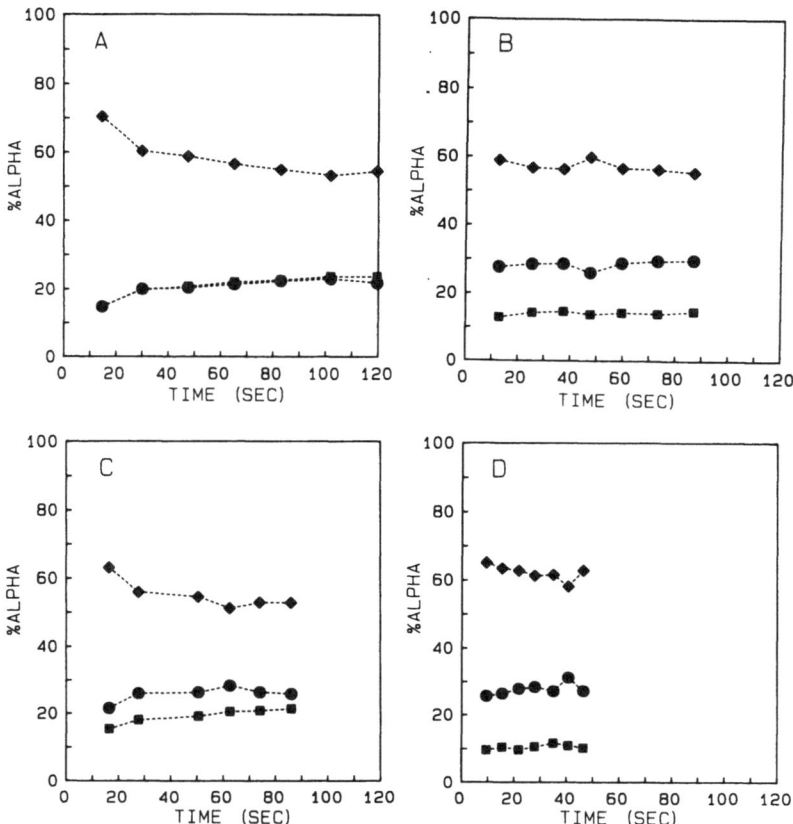

Figure 15: Time-resolved kinetics for the reduction of the disulfide bond at various pH conditions. Percent amplitudes obtained from simultaneous analysis were plotted at: A) pH 6.2, B) pH 7.1, C) pH 6.5, and D) pH 7.5. The diamonds represent the shortest lifetime component (.35 ± .1 ns) and circles and squares represent medium (1.4 ± .2 ns) and longest (3.7 ± .3 ns) lifetime components respectively.

only Trp 31 (none at position 28) exhibits a 1.2 fold increase, strengthening the general view. The fact that disulfide alone is not sufficient to account for quenching emerges from the work of Kelley and Stellwagen (1984), who find denaturation (<u>without</u> reduction) will provide a twofold increase in intensity. vandeVen et al., (1987) found lifetimes that changed in a corresponding manner. Combining the above with Holmgren's & Roberts' (1976) observation of positional changes for the aromatic residues, we conclude that other neighboring groups (neglecting -S-S-) may play important roles.

Given the success of kinetic decay measurements in other systems, we hoped they would yield new insights into this problem as well. First, the total intensity during reduction kinetics (Figure 14) confirms previous knowledge: note the large pH variation of endpoints versus the small (reversed) variation seen in initial (oxidized) values. More importantly, a significant increase in reaction rate accompanies pH increases. All reactions near completion within a minute or so; at the higher pH values only a few seconds are necessary. Even though these reactions are quite fast, we attempted to collect decay curves after manually mixing reactants in a conventional cuvette. Since we use a high repetition rate, mode-locked laser excitation source on our latest fluorometers (Han et al., 1987b; Knutson et al., 1985), we can collect moderately precise decay curves in just a few seconds. As Figure 15 shows, manual mixing "dead time" prevented us from examining virtually any of the high pH reactions. Some of the reaction progress at lower pH, however, can be recovered. The best global analyses (lifetimes linked through reaction) yielded lifetimes of .35 (± .1), 1.4 (± .2) and 3.7 (± .3) ns. The values in parentheses indicate the small variability seen across the entire pH range.

In the cases where amplitudes did change significantly _after_ the dead time, we note _increased_ contributions of the _longer_ lived components with time. This is in concert with our assertions based on previous (endpoint) data. The endpoints of these kinetic traces clearly agree with the fully reduced data shown in Figure 13. Overall, the kinetic traces of amplitude yielded consistent data, but the reaction time resolution will need improvement before this tool can help us _separate_ the reaction course into its components. Fortunately, the data acquisition time was not the limiting factor, so addition of a simple automated mixing device should soon provide what we need.

CONCLUSION

The application of fluorescence spectroscopy to biological macromolecules is enhanced if different types of experiments can be combined to test experimental data against a theoretical model. This contribution has focused on fluorescence decay measurements obtained during the time course of a chemical reaction. Reaction time is just another axis of overdetermination such as emission wavelength, pH, concentration or ligand occupancy that can be used with the aid of global analysis methods to obtain new insight.

ACKNOWLEDGEMENTS

The work on the PTS proteins is done in collaboration with Professor Saul Roseman and his group. The work on thioredoxin is done in collaboration with Professor Christian B. Anfinsen. We thank Dr. M. vandeVen for helpful discussions in regard to thioredoxin work.

This work was supported by NIH Grant GM11632 and M.K.H. was supported by NIH Training Grant 5T32 GM07231. This is publication number 1368 from the McCollum Pratt Institute.

REFERENCES

Ameloot, M., Beechem, J. M., and Brand, L., 1986, Simultaneous analysis of multiple fluorescence decay curves by Laplace transforms. Deconvolution with reference or excitation profiles, Biophys. Chem., 23:155.

Badea, M. G., and Brand, L., 1979, Time-resolved fluorescence measurements, in: "Methods in Enzymology," C.H.W. Hirs, and S. N. Timashef, eds., Academic Press, New York.

Beechem, J. M., and Brand, L., 1985a, Time-resolved fluorescence of proteins, Ann. Rev. Biochem., 54:43.

Beechem, J. M., Ameloot, M., and Brand, L., 1985b, Global and target analysis of complex decay phenomena, Anal. Instrum., 14:379.

Beechem, J. M., Ameloot, M. A., and Brand, L., 1985c, Global analysis of fluorescence decay surfaces: Excited-state reactions, Chem. Phys. Lett., 120:446.

Beechem, J. M., and Brand, L., 1986a, Global analysis of fluorescence decay: Applications to some unusual experimental and theoretical studies, Photochem. Photobiol., 44:323.

Beechem, J. M., Ameloot, M., and Brand, L., 1986b, Analysis of fluorescence intensity and anisotropy decay surfaces, in: "Excited-state Probes in Biochemistry and Biology," A. Szabo and L. Masotti, eds., in press.

Beechem, J. M., Knutson, J. R., and Brand, L., 1986c, Global analysis of multiple dye fluorescence anisotropy experiments on protein, Biochem. Soc. Trans., 14:832.

Blomquist, C. H., 1967, Structural changes associated with the inactivation of horse liver alcohol dehydrogenase, Arch. Biochem. Biophys., 122:24.

Brand, L., Knutson, J. R., Davenport, L., Beechem, J. M., Dale, R. E., Walbridge, D. G., and Kowalczyk, A. A., 1985, Time-resolved fluorescence spectroscopy: Some applications of associative behaviour to studies of proteins and membranes, in: "Spectroscopy and the Dynamics of Molecular Biological Systems", P. M. Bayley and R. E. Dale, eds., Academic Press, London.

Davenport, L., Knutson, J. R., and Brand, L., 1986a, Excited-state proton transfer of Equilenin and Dihydroequilenin: Interaction with bilayer vesicles, Biochemistry, 25:1186.

Davenport, L., Knutson, J. R., and Brand, L., 1986b, Anisotropy decay associated fluorescence and analysis of rotational heterogeneity. 2. 1,6-diphenyl-1,3,5-hexatriene in lipid bilayers, Biochemistry, 25:1811.

Easter, J. H., DeToma, R. P., and Brand, L., 1976, Nanosecond time-resolved emission spectroscopy of a fluorescence probe adsorbed to L-α-egg lecithin vesicles, Biophys. J., 16:571.

Eisenfeld, J., and Ford, C. C., 1979, A systems theory approach to the analysis of multiexponential fluorescence decay, Biophys. J., 26:73.

Grinvald, A., and Steinberg, I. Z., 1974, On the analysis of fluorescence decay kinetics by the method of least squares, Anal. Biochem., 59:583.

Han, M. K., Walbridge, D. G., LaForce, R., Shiber, S., Meadow, N., and Brand, L., 1986, Sulfhydryl studies of Enzyme I of the PTS, Biophys. J., 49:498a.

Han, M. K., Walbridge, D. G., Knutson, J. R., Brand, L., and Roseman, S., 1987a, Nanosecond time-resolved fluorescence kinetic studies of the 5,5'-dithiobis(2-nitrobenzoic acid) reaction with Enzyme I of the phosphoenolpyruvate: glycose phosphotransferase system, Anal. Biochem., 161:479.

Han, M. K., Walbridge, D. G., Knutson, J. R., Hong, S., Roseman, S., and Brand, S., 1987b, Nanosecond time-resolved fluorescence studies of pyrene maleimide labeled Enzyme I of the PTS, Biophys. J., 51:276a.

Han, M. K., Roseman, S., and Brand, L., 1987c, Fluorescence studies of Enzyme I of the PTS: Kinetics of the monomer/dimer association, Proc. in International Biophysics Congress, in press.

Han, M. K., Walbridge, D. G., Knutson, J. R., and Brand, L., 1987d, Nanosecond time-resolved fluorescence: Kinetic studies of macromolecules, Proc. in International Biophysics Congress, in press.

Heitz, J. R., and Brand, L., 1971, Fluorescence changes associated with denaturation of alcohol dehydrogenase, Arch. Biochem. Biophys., 144:286.

Holmgren, A., 1968, Thioredoxin 6. The amino acid sequence of the protein from Escherichia coli B, Eur. J. Biochem., 6:475.

Holmgren, A., 1972, Tryptophan fluorescence study of conformational transitions of the oxidized and reduced form of thioredoxin, J. Biol. Chem., 247:1992.

Holmgren, A., Ohlsson, I., and Grankvist, M.-L., 1978, Radioimmunological and enzymatic determinations in wild type cells and mutants defective in phage T7 DNA replication, J. Biol. Chem., 253:430.

Holmgren, A., and Roberts, G., 1976, Nuclear Magnetic Resonance Studies of Redox-Induced Conformational Changes in Thioredoxin from Escherichia Coli, FEBS Lett., 71:261.

Johnson, M. L., 1983, Evaluation and propagation of confidence intervals in nonlinear, asymmetric variance spaces, Biophys. J., 44:101.

Johnson, M. L., and Frasier, S. G., 1985, Non-linear least-squares analysis, in: "Methods in Enzymology", C. H. W. Hirs, and S. N. Timashef, eds., Vol. 117, Part J. 301, Academic Press, New York.

Kelley, R. F., and Stellwagen, E., 1984, Conformational transitions of thioredoxin in guanidine hydrochloride, Biochemistry, 23:5095.

Kishore, R., Mathew, M. K., and Balaram, P., 1983, A fluorescent peptide model for the thioredoxin active site, FEBS Lett., 159:221.

Knutson, J. R., Walbridge, D. W., and Brand, L., 1982, Decay associated fluorescence spectra and the heterogeneous emission of alcohol dehydrogenase, Biochemistry, 21:4671.

Knutson, J. R., Beechem, J. M., and Brand, L., 1983, Simultaneous analysis of multiple fluorescence decay curves: A global approach, Chem. Phys. Lett., 102:501.

Knutson, J. R., Chen, R. F., Scott, C. S., and Bowman, R. L., 1985, Studies of intrinsic protein fluorescence decay using a mode-locked laser source, Photochem. Photobio., 41:78s.

Knutson, J. R., Davenport, L., and Brand, L., 1986, Anisotropy decay associated fluorescence spectra and analysis of rotational heterogeneity. 1. Theory and applications, Biochemistry, 25:1805.

Knutson, J. R., 1987, Global analysis of fluorescence data: Some extensions, Biophys. J., 51:285a.

Kundig, W., Gosh, S., and Roseman, S., 1964, Phosphate bound to histidine in a protein as an intermediate in a novel phosphotransferase system, Proc. Natl. Acad. Sci. USA, 52:1067.

Kundig, W., and Roseman, S., 1971, Sugar transport. I. Isolation of a phosphotransferase system from Escherichia coli, J. Biol. Chem., 246:1393.

Kukuruzinska, M. A., Harrington, W. F., and Roseman, S., 1982, Sugar transport by the bacterial phosphotransferase system. Studies on the molecular weight and association of Enzyme I, J. Biol. Chem., 257:14470.

Kukuruzinska, M. A., Turner, B. N., Ackers, G. K., and Roseman, S., 1984, Subunit association of Enzyme I of the Salmonella typhimurium phosphoenolpyruvate: glycose phosphotransferase system, J. Biol. Chem., 259:11679.

Loken, M. R., 1973, Determination of rates of excited-state reactions using nanosecond fluorometric techniques, Ph.D. dissertation, The Johns Hopkins University, Baltimore, MD.

McKay, R. H., 1962, Effect of various environments on the intrinsic fluorescence polarization spectra of horse liver alcohol dehydrogenase, Arch. Biochem. Biophys., 135:218.

Meadow, N. D., Kukuruzinska, M. A., and Roseman, S., 1984, The bacterial phosphoenol pyruvate: sugar phosphotransferase system, in: "Enzymes of Biological Membranes", A. Martonosi, Ed., Plenum, New York.

Neyroz, P. N., Brand, L., and Roseman, S., 1987, Sugar transport by the bacterial phosphotransferase system. The intrinsic fluorescence of Enzyme I, J. Biol. Chem., submitted.

Oppenheimer, H. L., Green, R. W., and McKay, R. H., 1967, Function of Zinc in horse liver alcohol dehydrogenase, Arch. Biochem. Biophys., 119:552.

Postma, P. W., and Rosemann, S, 1976, The bacterial phosphoenolpyruvate: Sugar phosphotransferase system, Biochim. Biophys. Acta, 457:213.

Saffen, D. W., Presper, K. A., Doering, T. L., and Roseman, S., 1987, Sugar transport by the bacterial phosphotransferase system. Molecular cloning and structural analysis of the Escherichia coli pts H, pts I, and crr genes, J Biol. Chem., in press.

Schuyler, R., and Isenberg, I., 1971, A monophoton fluorometer with energy discrimination, Rev. Sci. Instrum., 42:813.

Selinger, B. K., and Harris, C. M., 1983, The pile-up problem in pulse fluorometry, in: "Time-resolved fluorescence spectroscopy in Biochemistry and Biology", R. B. Cundall and R. E. Dale, eds., Plenum, New York.

Small, E., and Isenberg, I., 1977, On moment index displacement, J. Chem. Phys., 66:3347.

Stryer, L., Holmgren, A., and Reichard, P., 1967, Thioredoxin. A localized conformational change accompanying reduction of the protein to the sulfhydryl form, Biochemistry, 6:1016.

Turner, B. W., Pettigrew, D. W., and Ackers, G. K., 1981, Measurement and analysis of ligand-linked subunit dissociation equilibria in human hemoglobins, Methods in Enzymol., 76:596.

vandeVen, M., Han, M., Walbridge, D., Knutson, J., Shin, D., Anfinsen, C. B., and Brand, L., 1987, Fluorescence decay studies of thioredoxin: Quenching, oxidation and reduction, Biophys. J., 51:275a.

Walbridge, D. G., Knutson, J. R., and Brand, L., 1982, Fluorescence decay studies of acid denaturation of horse liver alcohol dehydrogenase, Biophys. J., 37:393a.

Walbridge, D. G., Knutson, J. R., and Brand, L., 1987a, Nanosecond time-resolved fluorescence measurements during protein denaturation, Anal. Biochem., 161:467.

Walbridge, D. G., Knutson, J. R., Han, M. K., and Brand, L., 1987b, Nanosecond time-resolved fluorescence: A tool for chemical kinetic studies. Biophys. J., 51:284a.

Ware, W. R., 1971, Transient luminescence measurements, in: "Creation and detection of the excited state, A. A. Lamola, ed., Decker, New York.

Waygood, E. B., 1986, Enzyme I of the phosphoenolpyruvate: Sugar phosphotransferase system has two sites of phosphorylation per dimer, Biochemistry, 25:4085.

Weber, G., 1961, Excited states of Proteins, in: "Light and Life", W. D. McElroy and B. Glass, eds., Johns Hopkins, Baltimore, MD.

Weigel, N., Kukuruzinska, M. A., Nakazawa, A., Waygood, E. B., and Roseman, S., 1984, Sugar transport by the bacterial phosphotransferase system. Phosphoryl transfer reactions catalyzed by Enzyme I of Salmonella typhimurium, J. Biol. Chem., 257:14477.

Yguerabide, J., 1972, Nanosecond fluorescence spectroscopy of macromolecules, in: "Methods in Enzymology", C. H. W. Hirs and S. N. Timashef, eds., Academic Press, New York.

FLUORESCENCE INVESTIGATIONS ON THE ELONGATION FACTOR TU SYSTEM

David M. Jameson and Theodore L. Hazlett

Department of Pharmacology
University of Texas Southwestern Medical Center at Dallas
5323 Harry Hines Blvd.
Dallas, TX 75235 USA

The application of fluorescence spectroscopy to biophysical investigations of protein systems was pioneered by Gregorio Weber nearly four decades ago. In his early work Weber demonstrated the usefulness of intrinsic fluorescence from flavins to study proteins such as diaphorase (Weber, 1948), the applicability of fluorescence probes such as dansyl chloride which could be covalently linked to proteins (Weber, 1952), the utility of probes such as anilinonaphthalene sulfonates which interact non-covalently with proteins (Weber and Laurence, 1954) and the existence and utility of intrinsic protein fluorescence from tyrosine and tryptophan residues (Weber and Teale, 1957). During the past few decades an enormous increase in the number of protein systems studied and in the sophistication of the spectroscopic techniques utilized has occurred; these studies are philosophical and logical continuations of Weber's initial investigations. Indeed, a significant fraction of the research in the last few decades involving fluorescence approaches to biophysical chemistry has come from Professor Weber's laboratory.

In this contribution we shall describe our recent fluorescence investigations on components of the prokaryotic elongation pathway, specifically elongation factor Tu (EF-Tu). We shall recapitulate results obtained on the intrinsic tryptophan emission of EF-Tu and results on a fluorescent analog of GDP; these experiments were carried out in collaboration with Dr. John Eccleston (MRC Laboratory, Mill Hill) and Dr. Enrico Gratton (University of Illinois). We shall then describe preliminary studies on the interaction of EF-Tu with a fluorescamine analogies of GTP (provided by Dr. John Eccleston), a fluorescein derivative of tRNA (provided by Dr. Arthur

61

Johnson; University of Oklahoma) and on ethidium bromide complexes of tRNA.

EF-TU: OVERVIEW

EF-Tu, a 43,000 molecular weight protein, has been purified from E. coli in crystalline form as a complex with GDP or elongation factor Ts (EF-Ts) (Kawakita et al., 1971). Its role in prokaryotic protein biosynthesis includes binding aminoacyl-tRNA (aa-tRNA) and allowing for accurate completion of one peptide elongation cycle. A schematic diagram of the elongation cycle pertaining to EF-Tu is given in Figure 1. EF-Tu binds to the ribosomal 'A' site in the form of a ternary complex consisting of EF-Tu, GTP, and aa-tRNA. Once this ternary complex is properly bound the GTP is hydrolyzed and EF-Tu·GDP is released. The aa-tRNA remains on the ribosome to extend the growing polypeptide chain and, upon interaction with elongation factor G, translocates to the ribosomal 'P' site. A third elongation factor, EF-Ts, catalyzes the exchange of GDP and GTP to form the EF-Tu·GTP complex, capable of binding aa-tRNA to again form the ternary complex. Structural changes of EF-Tu in its interactions with various components of the protein biosynthesis cycle, i.e. guanosine nucleotides, EF-Ts, aa-tRNA, and ribosomes, have been inferred from a wide variety of chemical, spectroscopic, and kinetic evidence (for reviews see Miller and Weissbach, 1977; Parmeggiani and Swart, 1985). Furthermore, particular structural aspects of the ternary complex have been implicated in the mechanism controlling translational accuracy since the number of

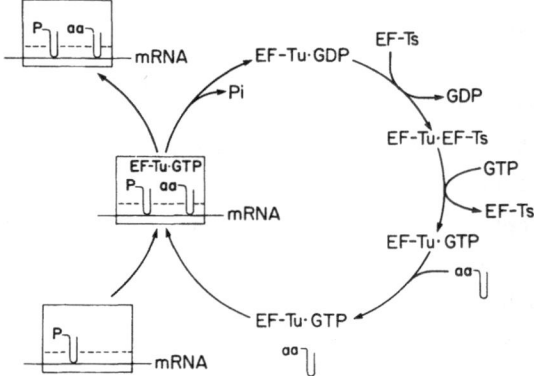

Figure 1. Schematic diagram of the prokaryotic protein elongation cycle.

translation errors is fewer than predicted from codon and anticodon association alone (Grosjean et al., 1978).

The structure of EF-Tu and its various complexes has been investigated with a number of physical techniques including X-ray crystallography, small angle X-ray scattering, and small angle neutron scattering. The partial crystal structure of trypsin-modified EF-Tu·GDP has been published (Jurnak, 1985). Ambiguities in the electron density maps corresponding to the polypeptide backbone prohibit complete analysis: the GDP binding domain, however, was resolved to 2.7 Å. This result is of general interest in light of the homologies recently reported between the GDP binding domain of EF-Tu and regions of the p-21 ras oncogene products (Halliday, 1984; Lochrie et al., 1985). Small angle x-ray studies on EF-Tu·GTP and the ternary aa-tRNA complex demonstrate a significant difference in their respective sizes, as determined from the change in the calculated radii (Osterberg et al., 1981). The data are consistent with the interpretation that the aminoacyl end of tRNA binds directly to EF-Tu·GTP while the anticodon stem protrudes into the surroundings. Though nuclease digestion studies support this hypothesis (Jekowsky et al., 1977), a more recent report involving small angle neutron scattering provides no evidence for a protruding aa-tRNA stem (Antonsson et al., 1986). In analyzing this latter data the authors assume that EF-Tu·GTP undergoes a monomer-dimer equilibrium. Unequivocal evidence for a dimeric form of EF-Tu has not been presented although a pH dependent polymerization of EF-Tu has been reported by one group (Beck, 1979). Our recent fluorescence studies on EF-Tu (Jameson et al., 1987; Eccleston et al., 1987) indicated a monomeric form for EF-Tu·GDP at concentrations up to approximately 20 micromolar, however, preliminary results from our laboratory suggest that, under similar conditions, a EF-Tu·GTP dimer may be present.

Though crystal studies have elucidated the GDP binding site on EF-Tu, the specific residues which form the binding domain for aa-tRNA are still unknown. Numerous techniques have shown that the aminoacyl-tRNA 3'end interacts closely with EF-Tu in the ternary complex. Furthermore, ribonuclease digestion experiments on aa-tRNA in the ternary complex and aa-tRNA alone show that EF-Tu protects the tRNA 3'end from enzymatic attack, and indicates that the amino acid acceptor stem and the TψC arm are protected, as well (Jekowsky et al. 1977; Wikman et al., 1982). Other investigators have attempted to determine specific contact points between aa-tRNA and EF-Tu in order to define residues on EF-Tu within the aa-tRNA binding domain (Duffy et al., 1981; Van Noort et al., 1984). Efficient crosslinking of the reactive aa-tRNA analogue N-bromoacetyl-lys-tRNA with the His-66 residue of EF-Tu has been reported and remains one of the best characterized contact points (Duffy et al., 1981).

Conformational changes in EF-Tu associated with ligand binding have been predicted based on the observation that aa-tRNA binds significantly tighter to EF-Tu·GTP (Kd = 0.1 to 5.0 nM; Abrahamson et al., 1985) than to EF-Tu·GDP (Kd > 10μM; Pingoud et al., 1982). Early evidence for a conformational alteration came from 1-anilino-8-naphthalalenesulfonate (ANS) binding studies in which two ANS sites were reported for EF-Tu·GDP yet three sites for EF-Tu·GTP (Crane and Miller, 1974). The authors concluded that an ANS site is located in the aa-tRNA binding domain and is exposed only upon the formation of the EF-Tu·GTP complex. The hypothesis that binding of GTP to EF-Tu exposes the aa-tRNA binding site is supported by a modification study which identified an accessible sulfhydryl group in EF-Tu·GTP 10-fold more reactive than in EF-Tu·GDP (Arai et al., 1974). Differences between EF-Tu·GTP and EF-Tu·GDP also have been observed with total intrinsic fluorescence, ORD, and tritium exchange (Arai et al., 1975; Arai et al., 1977).

One goal of our research is to utilize spectroscopic techniques to acquire detailed information on the nature of the structural and dynamical alterations which occur in EF-Tu as a consequence of its interactions with relevant ligands. Our initial investigations included the characterization of both intrinsic and extrinsic fluorophores. Specifically, we sought to acquire detailed information on the local environment of each fluorophore, their motional properties, accessibility to solvent and the hydrodynamic properties of the various EF-Tu complexes. Some of our findings are summarized in the following sections.

INTRINSIC FLUORESCENCE OF EF-TU

Steady-State Measurements

EF-Tu contains a single tryptophan residue at position 184 (Arai et al., 1980). The x-ray diffraction studies of crystals of a modified (protease treated) form of EF-Tu·GDP show that this tryptophanyl residue is located in the N-terminal domain, the domain containing the nucleotide binding site, at the end of α-helix F between β-strand 6 and α-helix E (Figure 2 and Jurnak, 1985). EF-Ts, however, has no tryptophan residues

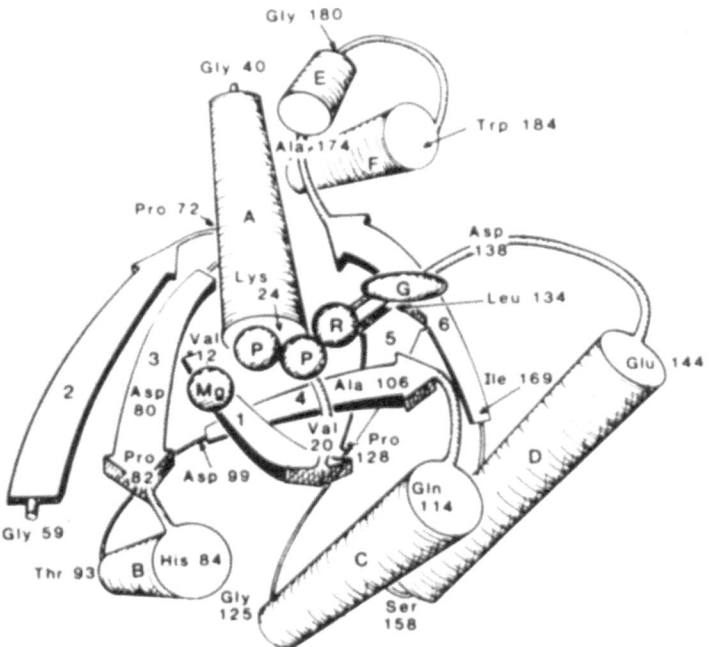

Figure 2. Schematic of GDP binding domain of trypsin modified EF-Tu·GDP showing location of tryptophan 184 (Adapted from Jurnak (1985)).

(An et al., 1981). Hence we are able to observe uniquely the fluorescence properties of the EF-Tu tryptophan residue in both the EF-Tu·GDP and EF-Tu·EF-Ts complexes. Although the presence of ten tyrosine residues in EF-Tu and three in EF-Ts renders the tyrosine emission per se too complex for most quantitative interpretations the possibility of tyrosine to tryptophan energy transfer can provide an additional probe of protein conformation in the complexes.

The corrected emission spectra of EF-Tu·GDP and EF-Tu·EF-Ts are shown in Figure 3. Upon excitation at 280 nm the spectra of both protein complexes are dominated by the tyrosine contribution - this observation is consistent with the earlier results of Arai et al. (1977). Excitation at 300 nm, however, leads exclusively to tryptophan emission, as expected, with emission maxima of 336 nm for both complexes. We should note that the position of these maxima alone does not permit us to draw firm conclusions regarding the degree of solvent exposure of the tryptophan residue. Sufficient exceptions to the general tendency of "bluer" emissions from buried tryptophans exist to warrant caution in drawing such conclusions (Longworth, 1983). The microscopic details of the environment, such as the

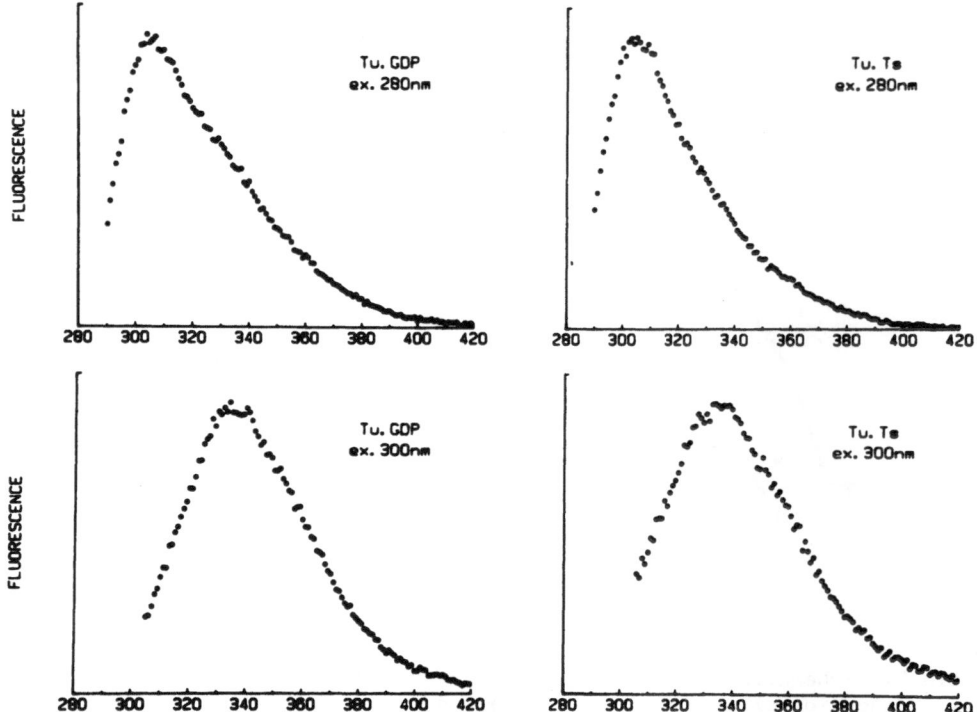

Figure 3. Emission spectra of EF-Tu·GDP and EF-Tu·EF-Ts excited at 280
and 300 nm. Excitation and emission bandpasses were 2 and 4 nm respectively. Reprinted with permission from: Jameson et al. 1987. Copyright by
the American Chemical Society.

extent of the polarizability, the presence of water molecules and specific
interactions of the ground state or excited state transition dipole of the
indole ring with polar groups in the protein will all influence the emission
process.

The accessibility of the tryptophan residue in both EF-Tu·GDP and
EF-Tu·EF-Ts complexes to quenchers, specifically iodide ion and acrylamide,
was investigated. The quenching studies were carried out using 300 nm
excitation to insure that only tryptophan would be viewed. The results,
shown in Figure 4, demonstrate that acrylamide quenches the tryptophan
emission of both complexes substantially better than does iodide ion.
These data suggest that, in both systems, the local charge environment of
the tryptophan residue may be somewhat negative thus inhibiting access of
the negatively charged iodide ion to the fluorophore. Indeed, the primary
sequence of EF-Tu supports this interpretation since the residues on either
side of the tryptophan are glutamic acids (Arai et al., 1980). The quench-

66

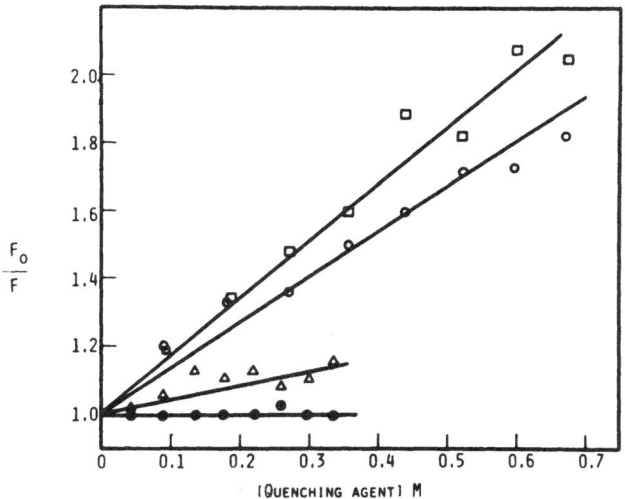

Figure 4. Stern-Volmer plots of the quenching of tryptophan 184 in EF-Tu·GDP and EF-Tu·EF-Ts by acrylamide and iodide. EF-Tu·GDP: acrylamide (o); iodide (•). EF-Tu·EF-Ts: acrylamide (□); iodide (Δ). Reprinted with permission from: Jameson et al. 1987. Copyright by the American Chemical Society.

ing data may be analyzed according to the Stern-Volmer equation (see, for example, Eftink and Ghiron, 1981):

$$F_o/F = 1 + K_{sv} [Q] \tag{1}$$

where F_o and F are the fluorescence intensities in the absence and presence of quencher, respectively, [Q] is the quencher concentration and K_{sv} is the Stern-Volmer quenching constant which equals (in the case of dynamic quenching) the product of the bimolecular quenching constant (k_q) and the unquenched lifetime (τ_o). The K_{sv} values for acrylamide quenching were 1.34 M^{-1} and 1.68 M^{-1} for EF-Tu·GDP and EF-Tu·EF-Ts, respectively. These values fall within the lower range of the acrylamide quenching constants measured by Eftink and Ghiron (1981) for a number of single tryptophan proteins. The fluorescence lifetime results, discussed in detail in a later section, demonstrated that the quenching by acrylamide is indeed dynamic. The bimolecular quenching constants, calculated using the major lifetime components of 4.8 ns and 4.6 ns for EF-Tu·GDP and EF-Tu·EF-Ts, respectively (discussed in detail later), were 2.8x10⁸$M^{-1}s^{-1}$ and 3.7x10⁸$M^{-1}s^{-1}$. These values indicate that, in both complexes, tryptophan is somewhat accessible to the solvent but that formation of the EF-Tu·EF-Ts complex enhances the accessibility.

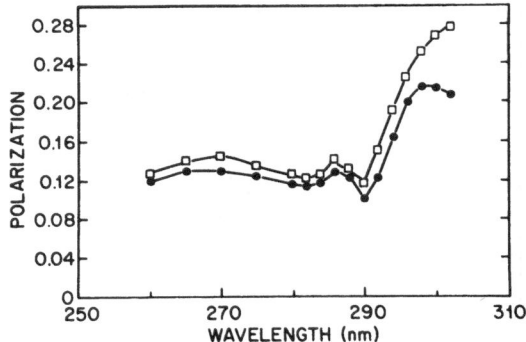

Figure 5. Excitation polarization spectra for EF-Tu·GDP (squares) and EF-Tu·EF-Ts (circles) for emission wavelengths > 345 nm. Reprinted with permission from: Jameson et al. 1987. Copyright by the American Chemical Society.

Excitation polarization spectra for EF-Tu·GDP and EF-Tu·EF-Ts are shown in Figure 5. For these data emission wavelengths greater than 345 nm were observed so that even upon excitation below 295 nm where tyrosine can be excited the observed polarizations may still be attributed to the tryptophan residue. The most striking aspect of these data is the lower polarization values for EF-Tu·EF-Ts compared to EF-Tu·GDP. At excitation wavelengths below about 295 nm both tyrosine and tryptophan residues are excited and the possibility of tyrosine to tryptophan energy transfer (discussed in more detail later) precludes a simple interpretation. With 300 nm illumination, however, the tryptophan residue in each complex is exclusively excited and the decrease in polarization, in conjunction with the similarities of the lifetime data for the two complexes, must be interpreted as due to increased motional properties for the tryptophan residue upon formation of EF-Tu·EF-Ts despite the increased molar volume of the complex. At 300 nm excitation, and 25°C the polarization values were 0.270 and 0.216 for EF-Tu·GDP and EF-Tu·EF-Ts respectively. A starting point for the analysis of steady-state polarization values is the Perrin equation:

$$(1/P - 1/3) = (1/Po - 1/3)(1 + 3\tau/\rho_h) \tag{2}$$

where P is the observed polarization, Po the limiting polarization, τ the fluorescence lifetime, and ρ_h the harmonic mean of the Debye rotational relaxation times of the principle axes of rotation. The Po value for tryptophan upon 300 nm excitation is 0.405 (Valeur and Weber, 1978). Using this value and lifetimes of 4.8 ns and 4.6 ns for EF-Tu·GDP and EF-Tu·EF-Ts, respectively (as discussed later) the Perrin equation gives ρ_h values of 25 and 14 ns for these complexes. These values are significantly lower than one would expect for globular proteins of molecular weights 43,000 (EF-Tu·GDP) and 74,000 (EF-Tu·EF-Ts) and suggest local mobility of the tryptophan residue in addition to the global rotation of the protein complex. Moreover, the smaller ρ_h value for the larger EF-Tu·EF-Ts complex suggests an increase in this local motion upon formation of the dimer protein. The time-resolved studies described in the subsequent section support these conclusions.

Time-resolved Measurements

Multifrequency phase and modulation fluorometry was utilized for lifetime and anisotropy decay determinations (see Gratton et al., 1984a for review). In this approach the intensity of the exciting light is modulated and the phase shift and relative demodulation of the emitted light, with respect to the excitation, are determined. Lifetimes are then calculated according to the equations:

$$\tan[P] = \omega\tau^P$$

$$M = [1+(\omega\tau^M)^2]^{-\frac{1}{2}}$$

(3)

where P is the phase shift and M the demodulation and ω the angular modulation frequency. Two independent lifetime determinations, τ^P and τ^M, are thus obtained. These measurements have usually been carried out on instruments which utilize either acousto-optic (Debye-Sears tanks) or electro-optic (Pockels cells) devices for sinusoidal modulation of the exciting light; both xenon arc lamps and lasers have been utilized as light sources. The recent development of techniques utilizing the harmonic content of pulsed light sources such as synchrotron radiation or mode-locked lasers have greatly extended the UV capabilities and the frequency modulation range of phase fluorometry (Gratton and Barbieri, 1986). Phase and modulation lifetime measurements on intrinsic fluorescence of proteins at high modulation frequencies were first performed using synchrotron radiation (Gratton et al. 1984b) and the preliminary time-resolved measurements on the EF-Tu complexes were obtained using the synchrotron radiation facility in Frascati, Italy on the instrumentation previously described (Gratton et al., 1984c). The final time resolved measurements, however, were performed with a multifrequency phase fluorometer utilizing the harmonic content of a mode-locked synchronously pumped dye laser (Alcala et al., 1985). The pulse train from the dye laser was amplitude modulated by an acousto-optic modulator and then frequency doubled providing excitation in the region 280 nm to 310 nm. The high repetition rate pulsed source gives a set of equally spaced harmonic frequencies. Amplitude modulation of the pulse train permits variation of the modulation frequency quasi-continuously from a few hertz to the gigahertz regime. In this study the frequency set utilized ranged from 1 MHz to 440 MHz. An ISS1-ADC interface (I.S.S., Inc., Champaign, IL) to an IBM-PC and software from I.S.S., Inc. were used for data collection and analysis.

An emitting system characterized by a single exponential decay will yield identical phase and modulation lifetimes irrespective of the modulation frequency. In the case of heterogeneous emitting systems (multiple non-interacting fluorescent species) the phase lifetime will be less than the modulation lifetime and these values will furthermore be dependent upon the modulation frequency, namely decreasing as the modulation frequency increases (Spencer and Weber, 1969).

The measured phase and modulation values may be analyzed assuming either a sum of exponentials or by a new and alternative approach to lifetime heterogeneity analysis which involves the use of continuous lifetime distribution models (Alcala et al., 1987a). In both cases the goodness of the fit may be judged by the value of the reduced chi-square (χ^2) defined as

$$\chi^2 = \Sigma\{[(P_c - P_m)/\sigma^P]^2 + [(M_c - M_m)/\sigma^M]^2\}/(2n - f - 1) \qquad (4)$$

where the sum is carried over the measured values at n modulation frequencies and f is the number of free parameters. The symbols P and M correspond to phase shift and relative demodulation values respectively while the indices c and m indicate the calculated and measured values. σ^P and σ^M are the standard deviations of each phase and modulation measurement, respectively. The calculated values of phase and modulation were obtained using the following equations

$$P = \tan^{-1}[S(\omega)/G(\omega)] \qquad (5)$$

$$M^2 = S(\omega)^2 + G(\omega)^2$$

where the functions $S(\omega)$ and $G(\omega)$ have different expressions depending on the fitting model utilized. Details of the analysis methods employed are

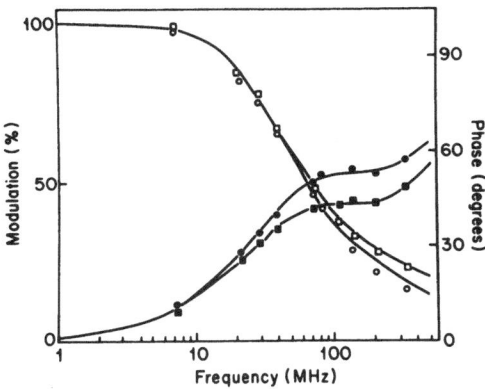

Figure 6. Multifrequency phase (closed symbols) and modulation (open symbols) data for EF-Tu·GDP (squares) and EF-Tu·EF-Ts (circles). Reprinted with permission from: Jameson et al. 1987. Copyright by the American Chemical Society.

given in our recently published report on the EF-Tu system (Jameson et al., 1987).

The phase and modulation lifetime data for EF-Tu·GDP and EF-Tu·EF-Ts were obtained with 300 nm excitation using cut-on filters on the emission side chosen to pass wavelengths longer than 340 nm. For these measurements the exciting light was polarized at 35° relative to the vertical laboratory axis and observed without an emission polarizer to eliminate possible polarization effects upon the lifetime determinations. The frequency dependence of the phase angles and relative modulation ratios for both protein systems are shown in Figure 6. The two-component analysis of these data yielded lifetimes of 4.9 ns and 0.34 ns with fractional intensities of 0.77 and 0.23, respectively, for EF-Tu·GDP and 4.7 ns and 0.36 ns with fractional intensities of 0.86 and 0.14, respectively, for EF-Tu·EF-Ts; the calculated curves corresponding to these components are shown in Figure 6. The χ^2 values were 13 and 22, respectively, for EF-Tu·GDP and EF-Tu·EF-Ts while three-component fits yielded lower χ^2 values of 4 and 20, respectively, demonstrating that the decay is more complex than two discrete exponentials. The continuous distribution analysis gave lower or comparable χ^2 values than were obtained with the discrete component method; the results on both protein complexes obtained using a bimodal Lorentzian distribution are shown in Figure 7A,B. Clearly, the decay law associated with the single tryptophan in both protein complexes is quite complex. Such complex decays for proteins containing single tryptophan residues are, in fact,

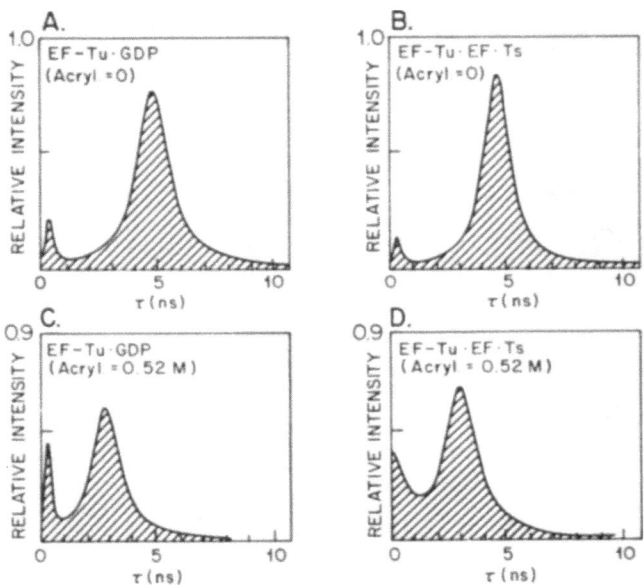

Figure 7. Continuous distribution lifetime plots for EF-Tu·GDP (A,C) and EF-Tu·EF-Ts (B,D) in the absence (A,B) and presence (C,D) of 0.52 M acrylamide. Reprinted with permission from: Jameson et al. 1987. Copyright by the American Chemical Society.

generally observed (Beechem and Brand, 1984). At the present time insufficient information exists on the precise influences of protein structure and dynamics on the excited state lifetimes of tryptophanyl residues to warrant a detailed modeling of the lifetime results in terms of distinct conformational states of the Tu systems. Theoretical approaches to such modeling for proteins in general are being developed, however (Alcala et al., 1987b).

In the dynamic polarization measurements the sample is illuminated by light polarized parallel to the vertical laboratory axis with intensity modulated at variable frequencies. The phase delay, $\Delta\Phi$, between the perpendicular and parallel polarization components of the emission can then be directly determined as well as the ratio of their AC components, Y. For an isotropic rotation one obtains the expressions (Weber, 1977):

$$\Delta\Phi = \tan^{-1}\{(3\omega rR)/[(k^2+\omega^2)(1+r-2r^2)+R(R+2k+kr)]\}$$

$$Y^2 = \{[k+6R/(1-r)]^2 + \omega^2\}/\{[k+6R/(1+2r)]^2+\omega^2\} \tag{6}$$

r is the limiting anisotropy, R the rotational diffusion coefficient and k

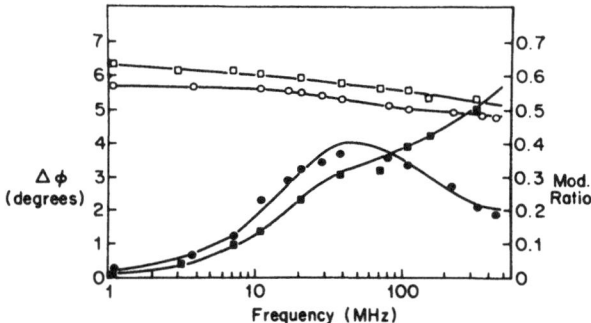

Figure 8. Multifrequency differential phase (closed symbols) and modulation ratio (open symbols) data for EF-Tu·GDP (circles) and EF-Tu·EF-Ts (squares). Solid lines were calculated using the following parameters: EF-Tu·GDP, $\tau_1=4.8$ ns, $\tau_2=0.31$ ns, $f_1=0.79$, $\rho_1=63$ ns, $r_1=0.23$, $\rho_2=1.5$ ns, $r_2=0.05$; EF-Tu·EF-Ts, $\tau_1=4.6$ ns, $\tau_2=0.31$ ns, $f_1=0.82$, $\rho_1=84$ ns, $r_1=0.19$, $\rho_2=0.3$ ns and $r_2=0.12$. Reprinted with permission from: Jameson et al. 1987. Copyright by the American Chemical Society.

the radiative decay rate $(1/\tau)$. For a residue in a protein, the expression for the differential phase and modulation ratio contains more terms. Firstly, a protein is generally not an isotropic rotator. However, assuming that the protein can be considered as an ellipsoid and that the ratio of the long axis to the short axis is not large, the rotational relaxation times about the different axes will not differ greatly and our values should correspond approximately to weighted averages of these relaxation times. Also, the relative weight of the rotational times along the different axes depends on the orientation of the fluorophore's absorption and emission dipoles with respect to the long and short axes of the protein (see for example, Beechem et al., 1986). Since we are interested in distinguishing internal motions from the tumbling of the entire protein molecule we will consider only large differences in the rotational relaxation times. The following expression for time-dependent anisotropy was used for the rotational analysis:

$$r(t) = r_o \sum_j g_j e^{-t/\theta_j} \tag{7}$$

where r_o is the time zero anisotropy, θ the rotational correlation time, g the associated amplitudes and the index j can have a value of 1 or 2. This expression describes the rotation of the entire protein (global rotation) plus the internal motion (local rotation) of the tryptophan residue. Equation 7 was converted to the frequency domain using the method outlined by Weber (1977) and then a fit was performed using a non-linear least squares analysis (Lakowicz et al. 1985). The dynamic polarization results on the EF-Tu·GDP and EF-Tu·EF-Ts systems are shown in Figure 8. The solid lines in the figure correspond to the best fit using two rotational motions, specifically for EF-Tu·GDP the rotational relaxation times were 63 and 1.5 ns with relative amplitudes (anisotropies) of 0.23 and 0.05, respectively, while for EF-Tu·EF-Ts the rotational relaxation times were 84 and 0.3 ns with relative amplitudes of 0.19 and 0.12, respectively. In both cases the longer relaxation times are in the range expected for globular proteins of their respective molecular weights. As we shall discuss in the subsequent section a larger rotational relaxation time of 88 ns was determined for EF-Tu·GDP using a fluorescent GDP analog. This larger value is more consistent with the ellipsoidal model for EF-Tu proposed by Sjoberg and Elias (1978) than the 63 ns value determined using the intrinsic tryptophan emission. The relaxation time obtained using the tryptophan emission, however, assumes a unique lifetime value while the actual tryptophan lifetime data are better described using a distribution as previously noted. The presence of the rapidly rotating component also complicated the analysis. We thus feel that the 88 ns value, obtained as described in the next section, is probably a better reflection of the overall hydrodynamic properties of EF-Tu. What remains clear, however, is that formation of the EF-Tu·EF-Ts dimer leads to a substantial increase in the local mobility of the tryptophan residue. The relaxation time of 0.3 ns is exceptionally rapid considering that free tryptophan in buffer at 20°C has a rotational relaxation time of about 0.25 ns (Alcala et al., 1985).

FLUORESCENT GDP ANALOG

Guanosine nucleotide binding proteins often show high specificity for the purine ring. EF-Tu, for example, binds the fluorescent analog 2-azido-6-oxy-purine riboside (Weigand and Kaleja, 1976) 90 fold weaker than GDP and exhibits no detectable binding of 2-aminopurine riboside 5'-diphosphate (Eccleston, 1981). The three dimensional structure of the nucleotide binding domain of EF-Tu·GDP demonstrates, however, that the 2'3' cis diol of GDP projects out of the protein (Jurnak, 1985) and hence suggests that modification of the nucleotide at these ribose positions, e.g., by incorporation of a fluorescent moiety, may not significantly affect the nucleotide's binding properties.

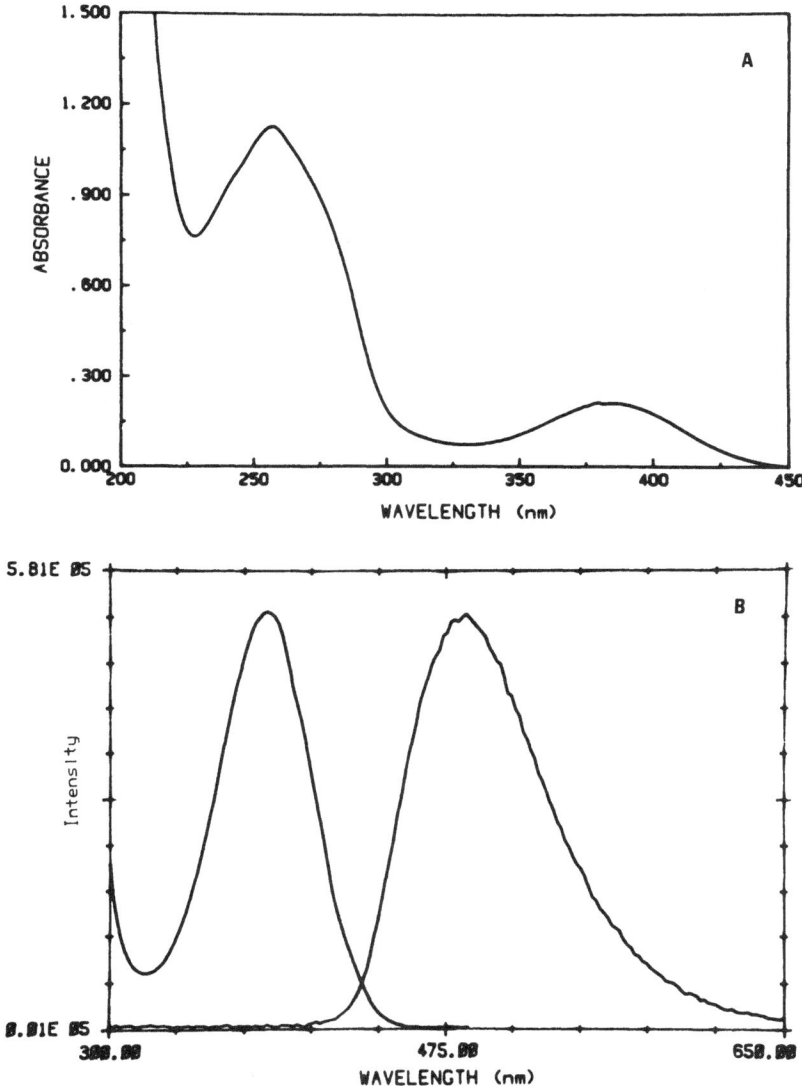

Figure 9. Spectroscopic properties of fluorescamine-GDP. (A) Absorption spectrum; 36 μM fluorescamine-GDP (B) Corrected excitation spectrum (emission 480 nm) and uncorrected emission spectrum (excitation 390 nm). Reprinted with permission from: Jameson et al. 1987. Copyright by the American Chemical Society.

A suitable precursor to such analogs is 2'-amino 2'-deoxy guanosine since the reactive 2'-amino group facilitates the introduction of various fluorescent probes. Probes can then be selected on the basis of the appropriateness of their fluorescent properties such as lifetime, spectral maxima and environmental sensitivities. Our collaborator, Dr. John Eccleston, has synthesized several fluorescent nucleotide analogs including fluorescamine and anthraniloyl derivatives. We have recently published a study characterizing the spectroscopic properties of the fluorescamine analog and its interaction with EF-Tu (Eccleston et al., 1987) and shall review its properties here. The original description of the spectroscopic properties of several fluorescamine-amine adducts in various solvents was reported by DeBernardo, et al., 1974. We should also note that the stratagem of modifying guanosine nucleotides on the ribose moiety was independently adopted by Faulhammer et al., 1986, who incorporated a nitroxide spin label on the 3' position of 3'-amino 3'-deoxy guanosine and studied the interaction of this analogue with EF-Tu. We will comment on their findings later.

The synthetic route to the synthesis of the fluorescamine deriva- tive of 2'-amino-2'-deoxyguanosine 5'-diphosphate has been published (Eccleston et al.,1987). The absorption spectrum of fluorescamine-GDP in buffer shown in Figure 9A has maxima at 259 nm and 380 nm. The 259 nm band evidently derives from both the guanine and fluorescamine moities since the fluorescamine-ethylamine derivative also absorbs at 260 nm. The extinction coefficient for fluorescamine-GDP was determined to be 2400 M^{-1} cm^{-1} at 380 nm; this value was determined by measurement of the labile phosphate content (Mahoney and Yount, 1984) of a fluores- camine-GDP solution while taking account of the solution's absorbance at 380 nm.

The corrected excitation spectrum and uncorrected emission spectrum for fluorescamine-GDP are shown in Figure 9B; the excitation and emission maxima are 385 nm and 484 nm respectively. The absorption and emission properties of fluorescamine-GDP are slightly modified upon binding to EF-Tu (Eccleston et al., 1987). Upon displacement from EF-Tu (by excess GDP) the emission intensity of fluorescamine-GDP increases about two- fold (with excitation at 390 nm). By recording the emission intensity from an EF-Tu·fluorescamine-GDP solution as a function of time after addition of excess GDP we were able to determine a dissociation rate constant of 5.0×10^{-3} s^{-1}.

The fluorescence lifetimes of fluorescamine-GDP, EF-Tu·fluoresca- mine-GDP and fluorescamine-ethylamine were determined using the multi- frequency phase and modulation fluorometry technique previously described (Eccleston, et al. 1987). In all cases the data were well-fit by a single-exponential decay (χ^2 values were near one). Lifetime values of 7.74 ns, 11.03 ns and 1.45 ns were obtained for fluores- camine-GDP, EF-Tu·fluorescamine-GDP and fluorescamine-ethylamine,

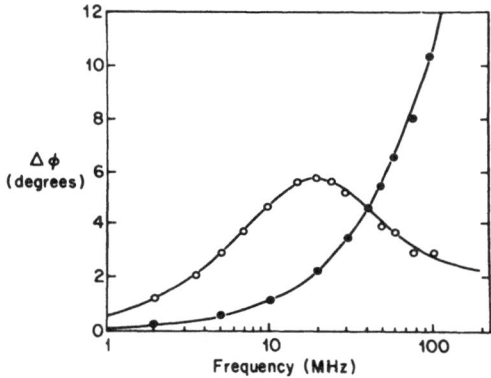

Figure 10. Differential phase measurements of fluorescamine-GDP and EF-Tu·fluorescamine-GDP. The starting protein solution contained 20 μM EF-Tu·fluorescamine-GDP in 50 mM Tris-HCl, 10 mM MgCl$_2$, and 0.5 mM DTE, pH 7.6 at 20C. Data correspond to EF-Tu·fluorescamine-GDP (o) and displaced (by 100 μM GDP) fluorescamine·GDP (•). The solid lines represent the best-fit differential phase (ΔΦ) values for ρ = 1.1 ns, τ = 7.74 ns, and r$_0$ = 0.31 (free fluorescamine·GDP) and ρ$_1$ = 88 ns, τ$_1$ = 11.03 ns (f$_1$=0.95), ρ$_2$ = 1.1 ns, τ$_2$ = 7.74 ns and r$_0$ = 0.31 for EF-Tu·fluorescamine-GDP. Reprinted with permission from: Jameson et al. 1987. Copyright by the American Chemical Society.

respectively. The lifetime observed for the displaced fluorescamine-GDP was the same as that observed for fluorescamine-GDP placed directly in buffer. Dynamic polarization measurements on the EF-Tu·fluorescamine-GDP complex were also carried out. The results, shown in Figure 10, represent differential phase values for fluorescamine-GDP bound to elongation factor Tu and then displaced by excess GDP. The best fit for the bound case was obtained using a two-component analysis corresponding to fractional contributions of 0.95 for bound fluorescamine-GDP (τ=11.03 ns) and 0.05 for free fluorescamine-GDP (τ=7.74 ns). The Debye rotational relaxation times obtained for the bound and free analogues were 88 ns and 1.1 ns, respectively. The rotational relaxation time of a rigid, spherical protein with a molecular weight of 43,000 is approximately 50 ns (Eccleston et al., 1987). Our value of 88 ns is thus consistent with an asymmetric protein but the data, at present, does not allow us to distinguish unambiguously between the published models such as those proposed by Sjoberg and Elias (1978) or Morikawa et al. (1978). The rotational correlation time of 2 ns (which corresponds to a Debye rotational relaxation time of 6 ns) reported by Faulhammer et al., (1986) for complexes of elongation factor Tu with spin label analogues of GDP and GTP is dramatically lower than our value and is not consistent with the relaxation times expected for this protein complexes. The reasons for this discrepancy are not clear.

INVESTIGATIONS ON THE TERNARY COMPLEX AND ITS COMPONENTS

Of the various EF-Tu complexes only the ternary complex initiates an elongation step when bound to the ribosomal 'A' site. As such, the structure and hydrodynamics of this complex, as well as those of its individual components EF-Tu·GTP and tRNA, are of general interest. Furthermore the large intracellular GTP/GDP ratio suggests that the ternary and the EF-Tu·GTP complexes are the primary forms of EF-Tu in vivo. We have begun to investigate these complexes by utilizing the fluorescent probes fluorescamine-GTP, fluorescein-labeled tRNA, and ethidium bromide. Preliminary data will be presented in the following discussions. Data were taken primarily at 5°C to minimize the acyl hydrolysis of the aa-tRNA and the residual EF-Tu GTPase activity (Block and Pingoud, 1981).

In addition to fluorescamine-GDP mentioned above, Dr. Eccleston has generously provided to us the complementary fluorescamine-GTP derivative for comparison. Our aim is to elucidate the conformational differences between the EF-Tu·GDP and ·GTP complexes which allow for the more favorable binding of aa-tRNA. Conformational alterations should be greatest within the tRNA and GTP binding domains and may well be reflected in the fluorescamine label on the nucleotide ribose ring. Time-resolved lifetime and polarization analyses were performed as described earlier in this chapter.

Lifetimes at 5°C for the fluorescamine derivatives of ethylamine, GDP, and GTP were found to be 2.3 ns, 8.6 ns, and 8.8 ns, respectively. Multifrequency phase and modulation data on EF-Tu-bound fluorescamine-GTP were best fit to a lifetime between 11 and 12 ns. Present uncertainties in the concentration of free fluorophore and the resultant broad chi-squared minimum in the analysis hindered a more precise determination. Additional information from the dynamic polarization measurements indicated that only 1 or 2 percent of a quickly rotating species ie., free fluorescamine-GTP, could be present. Differential phase results were analyzed for a single rotational relaxation time. No significant improvement in the fit was observed if a second relaxation time was included in the analysis. The relaxation time at 5°C was 195 ns. This value is equivalent to 123 ns at 20°C (adjusting for temperature and viscosity differences) and is significantly longer than the 88 ns value (at 20°C) reported for the EF-Tu·fluorescamine-GDP complex (Eccleston et al., 1987). These results suggest that EF-Tu·GTP may,

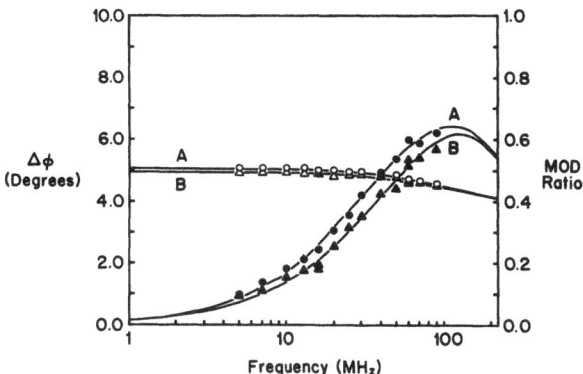

Figure 11. Multifrequency phase (closed symbols) and modulation (open symbols) data for [A] Phe-tRNA^{Phe}-F^8 (circles) and [B] Phe-tRNA^{Phe}-F^8·EF-Tu·GTP (triangles) taken at 5°C. Excitation wavelength was 488 nm; emission measured through a Schott OG530 cut-on filter. Data set [A] was analyzed using τ = 4.3 ns and best fit to a single rotating species with two relaxation times: ρ_1 = 103 ns and ρ_2 = 4.8 ns with amplitudes, r_o, of .247 and .092, respectively. Data set [B] contains Phe-tRNA^{Phe}-F^8 bound to EF-Tu·GTP (80%) best fit to two relaxation times ρ_1 = 206 ns and ρ_2 = 4.7 ns with amplitudes of 0.244 and 0.098, respectively. Solid lines represent the best fit analyses.

under appropriate conditions, undergo self-association. Indeed, other workers have also suggested that EF-Tu may form a dimer (Antonsson et al., 1986) or much higher order complexes (Beck, 1979; Weiser et al., 1982). However, not all our work is consistent on this point and additional effort will be required to substantiate these hypotheses.

We have begun to investigate the fluorescein-labeled tRNA (Phe-tRNA^{Phe}-F^8) provided by Dr. Arthur Johnson, University of Oklahoma, Norman, Oklahoma. Bound to EF-Tu·GTP or free in solution, Phe-tRNA^{Phe}-F^8 has a single lifetime component of 4.27 ± 0.05 ns. Differential phase and modulation results at 5°C for Phe-tRNA^{Phe}-F^8, free and bound, are shown in Figure 11. Time-resolved measurements on free tRNA were best fit (χ^2 = 1.1) by including one slow and one fast rotational relaxation time. We attribute the 103 ns value to global rotation of the tRNA and the 4.8 ns value to local motion of the fluorophore. Indeed, the 103 ns global rotation is equivalent to a 65 ns relaxation time at 20°C, in the range of relaxation times for tRNA reported in the literature (Tao et al., 1970; Ehrenberg et al., 1979; Thomas et al., 1984; Ferguson and Yang, 1986). We obtained similar results for non-acylated tRNA^{Phe}-F^8.

In contrast, Phe-tRNA^{Phe}-F^8 bound to EF-Tu·GTP exhibits rotational

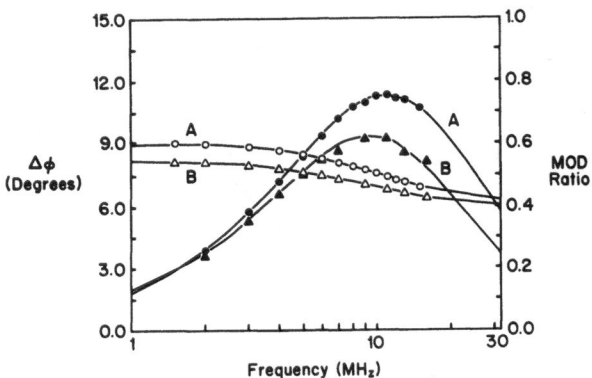

Figure 12. Multifrequency phase (closed symbols) and modulation (open symbols) data for Ethidium bromide plus tRNA at 20°C (circles) and 5°C (triangles). Excitation wavelength was 514 nm; the fluorescence was measured through Schott OG550 and OG570 filters. Solid lines represent two-component, best fit analyses for 99% bound and 1% free ethidium bromide. Best fit values for bound ethidium of ρ = 88.2 ns and r_0 = 0.340, at 20°C, and ρ = 129 ns and r_0 = 0.346, at 5°C, were obtained. Lifetimes used for the free and bound ethidium were 1.78 ns and 25.6 ns at 20°C, and 1.86 ns and 26.0 ns at 5°C.

relaxation times of 206 ns and 4.7 ns (Figure 11). The 206 ns value is reasonable for the rotational relaxation time of the ternary complex at 5°C considering the relaxation times of the components, i.e., 140 ns for EF-Tu (equivalent to 88 ns value determined at 20°C) and 103 ns for the tRNA. Hence, the sum of the rotational relaxation times of the individual components, 243 ns, is in good agreement with the observed value, 206 ns, considering the possibility of hydration effects and structurally related shape changes upon complex formation. If the ternary complex were spherical we would expect its rotational relaxation time to be near 165 ns (at 5°C). The observed value is somewhat larger and indicates, as one might expect, that the ternary complex is irregularly shaped although details of the anisotropic motions were not resolvable. Unfortunately, the short lifetime of fluorescein relative to the rotational relaxation times of the complexes limits the precision and resolving power of this probe for slow, global rotations.

The fast local motions of the fluorescein on the free and bound Phe-tRNA^{Phe}-F^8 were 4.8 ns and 4.7 ns, respectively. Thus, upon binding to EF-Tu·GTP the local freedom of the probe has remained unchanged. The fluorescein is covalently attached to the s^{4}U-8 position placing it between the D-stem and the acceptor stem (Johnson et al., 1982). Our data, then, suggests that the D-stem domain is not highly restricted in the ternary complex and does not associate directly with EF-Tu·GTP. This conclusion supports those from ribonuclease digestion experiments which indicate that the D-stem and D-loop regions are not protected in the presence of EF-Tu·GTP (Jekowsky et al., 1977), and those from quenching studies on Phe-tRNA[Phe]-F[8] which indicate that the solvent accessibility of the fluorescein also is unaffected by the presence of EF-Tu·GTP (Adkins et al., 1983).

Ethidium bromide has a single strong binding site to tRNA and has been used to study the rotational diffusion of tRNA in solution (Tao et al., 1970; Thomas et al., 1984). Upon binding to tRNA the quantum yield of this fluorophore increases 14-fold with a concomitant increase in lifetime from 1.8 ns to approximately 26 ns (Tao et al., 1970 and Ferguson et al., 1986; Hazlett and Jameson, unpublished). The relatively long lifetime of the bound probe facilitates its utility in the resolution of global rotations of macromolecular complexes such as tRNA free in solution and bound to EF-Tu·GTP. Figure 12 shows the differential phase and modulation results for ethidium bromide bound to tRNA at 20°C (A) and 5°C (B). Non-linear least-squares fits for two rotating species, free and bound probe, are shown. The relaxation time of the free probe was determined independently and fixed in the rotational analyses. The rotational relaxation times attributed to the bound probe, 88 ns (20°C) and 129 (5°C), are within the range reported for tRNA but slightly longer than what we have observed with Phe-tRNA[Phe]-F[8]. This difference may reflect differences in the orientations of the probes relative to the tRNA rotational axes. The lack of discernible fast, local fluorophore rotation is not surprising since ethidium bromide is thought to intercalate into tRNA and have little local freedom (Jones et al., 1978; Thomas et al., 1984). Work is now in progress to utilize ethidium bromide to investigate the hydrodynamics and local environment of the aa-tRNA in the ternary complex.

SUMMARY

The detailed examination of the fluorescence properties of several intrinsic and extrinsic probes of the elongation factor Tu system is well under way. Our results to date have demonstrated that the intrinsic tryptophan fluorescence of EF-Tu may be used to monitor conformational alterations in the protein's nucleotide binding domain. Our work with extrinsic probes such as fluorescamine - guanine nucleotides, fluorescein and ethidium bromide have demonstrated the utility of such fluoro-phores in studies of the hydrodynamic properties of the various macro-molecular complexes. These experiments and the more refined measure-ments, including energy transfer studies, currently being carried out should, in conjunction with the forthcoming crystallographic structures, enhance our understanding of the structural and dynamic alterations which occur as the components of this system interact in the prokaryotic protein biosynthesis cycle.

ACKNOWLEDGEMENTS

Our research was supported in part by National Science Foundation Grant DMB 8706440.

REFERENCES

Abrahamson, J. K., Laue, T. M., Miller, D. L., and Johnson, A. E., 1985, Direct Determination of the Association Constant Between Elongation Factor Tu-GTP and Aminoacyl-tRNA Using Fluorescence, Biochemistry, 24:692.

Adkins, H. J., Miller, D. L., and Johnson, A. E., 1983, Changes in Aminoacyl Transfer Ribonucleic Acid Conformation upon Association with Elongation Factor Tu - Guanosine 5'-Triphosphate. Fluorescence Studies of Ternary Complex Conformation and Topology, Biochemistry, 22:1208

Alcala, J. R., Gratton, E., and Jameson, D. M., 1985, A Multifrequency Phase Fluorometer Using the Harmonic Content of a Mode-Locked Laser, Anal. Instrum., 14:225.

Alcala, J. R., Gratton, E. and Prendergast, F. G., 1987a, Resolvability of Fluorescence Lifetime Distributions Using Phase Fluorometry, Biophysical J., 51:587.

Alcala, J. R., Gratton, E. and Prendergast, F. G., 1987b, Fluorescence Lifetime Distributions in Proteins, Biophysical J., 51:597.

An, G., Bendiak, D. S., Mamelak, L. A., and Friesen, J. D., 1981, Organization and Nucleotide Sequence of a New Ribosomal Operon in Escherichia coli Containing the Genes for Ribosomal Protein S2 and Elongation Factor Ts, Nucleic Acids Res., 9:4163.

Antonsson, B., Leberman, R., and Jacrot, B., 1986, Small-Angle Neutron Scattering Study of the Ternary Complex Formed Between Bacterial

Elongation Factor Tu, Guanosine 5'-Triphosphate, and Valyl-tRNA^{Val}, Biochemistry, 25:3655.

Arai, K., Kawakita, M., Nakamura, S., Ishikawa, I., and Kaziro, Y., 1974, Studies on the Polypeptide Elongation Factors From E. coli. VI. Characterization of Sulfhydryl Groups in EF-Tu and Ef-Ts, J. Biochem. (Tokyo), 76:523.

Arai, K., Arai, T., Kawakita, M., and Kaziro, Y., 1975, Conformational Transitions of Polypeptide Chain Elongation Factor Tu. I. Studies with Hydrophobic Probes, J. Biochem. (Tokyo), 77:1095.

Arai, K., Arai, T., Kawakita, M., and Kaziro, Y., 1977, Further Studies on the Properties of the Polypeptide Chain Elongation Factors Tu and Ts: Hydrogen-Tritium Exchange, Optical Rotatory Dispersion, and Intrinsic Fluorescence, J. Biochem. (Tokyo), 81:1335.

Arai, K., Clark, B. F .C., Duffy, L., Jones, M. D., Kaziro, Y., Laursen, R. A., L'Italién, J., Miller, D. L., Nagarkatti, S., Nakamura, S., Nielsen, K. M., Petersen, T. E., Takahashi, T., and Wade, M., 1980, Primary Structure of Elongation Factor Tu from Escherichia coli, Proc. Natl. Acad. Sci. (U.S.A.), 77:1326.

Beck, B. D., 1979, Polymerization of the Bacterial Elongation Factor for Protein Synthesis, EF-Tu, Eur. J. Biochem., 97:495.

Beechem, J. R. and Brand, L., 1984, Time Resolved Fluorescence Decay in Proteins, Annu. Rev. Biochem., 54:43.

Beechem, J. R., Knutson, J. R. and Brand, L., 1986, Global Analysis of Multiple Dye Fluorescence Anisotropy Experiments on Proteins, Biochem. Soc. Trans., 14:832.

Block, W. and Pingoud A., 1981, The Identification and Analysis of Nucleotides Bound to the Elongation Factor Tu from Escherichia coli, Biochemistry, 114:112.

Crane, L. J., and Miller, D. L., 1974, Guanosine Triphosphate and Guanosine Diphosphate as Conformation-Determining Molecules. Differential Interaction of a Fluorescent Probe with the Guanosine Nucleotide Complexes of Bacterial Elongation Factor Tu, Biochemistry, 13:933.

DeBernardo, S., Weigele, M., Toome, V., Manhart, K., Leimgruber, W., Böhlen, P., Stein, S., and Udenfriend, S., 1974, Studies on the Reaction of Fluorescamine with Primary Amines, Arch. Biochem. Biophys., 163:390.

Duffy, L. K., Gerber, L., Johnson, A. E., and Miller, D. L., 1981, Identification of a Histidine Residue Near the Aminoacyl Transfer Ribonucleic Acid Binding Site of Elongation Factor Tu, Biochemistry, 20:4663.

Eccleston, J. F., Gratton, E., and Jameson, D. M., 1987, Interaction of
a Fluorescent Analogue of GDP with Elongation Factor Tu: Steady-
State and Time-Resolved Fluorescence Studies, <u>Biochemistry</u>, 26:3902.

Eftink, M. R., and Ghiron, C. A., 1981, Fluorescence Quenching Studies
with Proteins, Review, <u>Anal. Biochem.</u>, 114:199.

Ehrenberg, M., Rigler, R., and Wintermeyer, W., 1979, On the Structure
and Dynamics of Yeast Phenylalanine-accepting Transfer Ribonucleic
Acid in Solution, <u>Biochemistry</u>, 18:4588.

Gratton, E. and Barbieri, B., 1986, Multifrequency Phase Fluorometry
Using Pulsed Sources: Theory and Applications, <u>Spectroscopy</u>, 1:28.

Gratton, E., Jameson, D. M., and Hall, R. D., 1984a, Multifrequency
Phase Fluorometry, <u>Ann. Rev. Biophys. Bioeng.</u>, 13:105.

Gratton, E., Jameson, D. M., Rosato, N. and Weber, G., 1984b,
Multifrequency Cross-Correlation Phase Fluorometer Using
Synchrotron Radiation, <u>Rev. Sci. Instrum.</u>, 55:486.

Gratton, E., Jameson, D. M., Rosato, N. and Weber, G., 1984c,
Multifrequency Cross-Correlation Phase Fluorometry Using
Synchrotron Radiation, <u>Biophysical J.</u>, 45:321a.

Grosjean, H. J., deHenau, S., and Crothers, D. M., 1978, On the Physical
Basis for Ambiguity in Genetic Coding Interactions, <u>Proc. Natl.
Acad. Sci. (U.S.A.)</u>, 75:610.

Faulhammer, H. G., Denninger, G., Hartl, P. J., Azhayev, A. V.,
Schwoerer, M., and Sprinzl, M., 1986, Spin-Labelled Analogues of
GDP and GTP as Site-Specific Reporter Groups for Guanosine
Nucleotide-Binding Proteins, <u>Biochim. Biophys. Acta</u>, 884:182.

Ferguson, B. Q., and Yang, D. C. H., 1986, Localization of
Non-covalently Bound Ethidium in Free and Methionyl-tRNA Synthetase
bound tRNA[fMet] by Singlet-Singlet Energy Transfer, <u>Biochemistry</u>,
25:5298.

Halliday, K., 1984, Regional Homology in GTP-Binding Proto-Oncogene
Products and Elongation Factors, <u>J. Cyclic Nucleotide Prot. Phosphoryl. Res.</u>, 9:435.

Jameson, D. M., and Gratton, E., 1983, Analysis of Heterogeneous
Emissions by Multifrequency Phase and Modulation Fluorometry, <u>in</u>:
"New Directions in Molecular Luminescence", D. Eastwood, ed.,
American Society for Testing and Materials, Philadelphia.

Jameson, D. M., Gratton, E., and Eccleston, J. F., 1987, Intrinsic
Fluorescence of Elongation Factor Tu in its Complexes with GDP and
Elongation Factor Ts, <u>Biochemistry</u>, 26:3894.

Jekowsky, E., Miller, D. L., and Schimmel, P. R., 1977, Isolation,
Characterization and Structural Implications of a Nuclease-digested

Complex of Aminoacyl Transfer RNA and <u>Escherichia</u> <u>coli</u> Elongation
Factor Tu, <u>J. Mol. Biol</u>, 114:451.

Johnson, A. E., Adkins, H. J., Matthews, E. A., and Cantor, C. R., 1982,
Distance Moved by Transfer RNA During Translocation from A site to
P site on the Ribosome, <u>J. Mol. Biol.</u>, 156:113.

Jones, C. R., Bolton, P. H., and Kearns, D. R., 1978, Ethidium Bromide
Binding to Transfer RNA: Transfer RNA as a Model System for
Studying Drug-RNA Interactions, <u>Biochemistry</u>, 17:601.

Jurnak, F., 1985, Structure of the GDP Domain of EF-Tu and Location of
the Amino Acids Homologous to <u>ras</u> Oncogene Proteins, <u>Science
(Washington D.C.)</u>, 230:32.

Kawakita, M., Akai, K., and Kaziro, Y., 1971, Identification of the Two
^{3}H-GTP Binding Proteins in <u>E. coli</u> Supernatant as Tu-Ts and Tu
Factors, <u>Biochem. Biophys. Res. Commun.</u>, 42:475.

Lakowicz, J. R., Cherek, H., Maliwal, B. P., and Gratton, E., 1985,
Time-Resolved Fluorescence Anisotropies of Diphenylhexatriene and
Perylene in Solvents and Lipid Bilayers Obtained from Multifre-
quency Phase-Modulation Fluorometry, <u>Biochemistry</u>, 24:376.

Lochrie, M. A., Hurley, J. B., and Simon, M. I., 1985, Sequence of the
Alpha Subunit of Photoreceptor G Protein: Homologies Between
Transducin, <u>ras</u>, and Elongation Factors, <u>Science (Washington,
D.C.)</u>, 228:96.

Longworth, J. W., 1983, Intrinsic Fluorescence of Proteins, <u>in</u>: "Time-
Resolved, Fluorescence Spectroscopy in Biochemistry and Biology,"
R. B. Cundall and R. E. Dale, eds., Plenum Press, New York.

Mahoney, C. W., and Yount, R. G., 1984, Purification of Micromolar
Quantities of Nucleotide Analogs by Reverse-Phase High-Performance
Liquid Chromatography Using a Volatile Buffer at Neutral pH, <u>Anal.
Biochem.</u>, 138:246.

Miller, D. L., and Weissbach, H., 1977, Factors Involved in the Transfer
of Aminoacyl-tRNA to the Ribosome, <u>in</u>: "Molecular Mechanisms of
Protein Biosynthesis, H. Weissbach and S. Pestka, eds., Academic
Press, New York.

Morikawa, K., LaCour, T. F. M., Nyborg, J., Rasmussen, K. M., Miller,
D. L.and Clark, B. F. C., 1978, High Resolution X-ray
Crystallographic Analysis of a Modified Form of the Elongation
Factor Tu: Guanosine Diphosphate Complex, <u>J. Mol. Biol.</u>, 125:325.

Osterberg, R., Sojoberg, B., Ligaarden, R., and Elias, P., 1981, A
Small-Angle X-ray Scattering Study of the Complex Formation Between
Elongation Factor Tu·GTP and Valyl-tRNA$_1^{Val}$ from <u>Escherichia</u> <u>coli</u>,
<u>Eur. J. Biochem.</u>, 117:155.

Parmeggiani, A., and Swart, C. W. M., 1985, Mechanism of Action of Kirromycin-Like Antibiotics, Ann. Rev. Microbiol., 39:557.

Pingoud, A., Block, W., Wittenhofer, A., Wolf, H., and Fischer, E., 1982, The Elongation Factor Tu Binds Aminoacyl-tRNA in the Presence of GDP, J. Biol. Chem., 257:11261.

Sjöberg, B., and Elias, P., 1978, A Small Angle X-ray Study of the Elongation Factory Complex Tu·GDP From Escherichia coli, Biochim. Biophys. Acta, 519:507.

Spencer, R. D., and Weber, G., 1969, Measurements of Subnanosecond Fluorescence Lifetimes with a Cross-Correlation Phase Fluorometer, Ann. N. Y. Acad. Sci., 158:361.

Tao, T., Nelson, J. H., and Cantor, C. R., 1970, Conformational Studies on Transfer Ribonucleic Acid. Fluorescence lifetime and Nanosecond Depolarization Measurements on Bound Ethidium Bromide, Biochemistry, 23:5407.

Thomas, J. C., Schurr, J. M., and Hare, D. R., 1984, Rotational Dynamics of Transfer Ribonucleic Acid: Effect of Ionic Strength and Concentration, Biochemistry, 23:5407.

Valeur, B., and Weber, G., 1977, Resolution of the Fluorescence Excitation Spectrum of Indole Into the ^1L_a and ^1L_b Excitation Bands, Photochem. Photobiol., 25:441.

Weber, G., 1948, Ph.D. Dissertation. St. John's College, University of Cambridge.

Weber, G., 1952, Polarization of the Fluorescence of Macromolecules; Theory and Experimental Method, Biochem. J., 51:145.

Weber, G., and Laurence, D. J. R., 1954, Fluorescent Indicators of Adsorption in Solution and on the Solid Phase, Biochem. J., 56:31.

Weber, G. and Teale, F. W. J., 1957, Ultraviolet Fluorescence of the Aromatic Amino Acids, Biochem. J., 65:476.

Weber, G., 1977, Theory of Differential Phase Fluorometry: Detection of Anisotropic Molecular Rotations, J. Chem. Phys., 66:4081.

Weigand, G., and Kaleja, R., 1976, Fluorescent Guanosine-Nucleotide Analogs Suitable for Photoaffinity-Labeling Experiments, Eur. J. Biochem., 65:473.

Weiser, J., Mikulik, K., Zizka, Z., Stastna, J., Janada, I., and Jiranova, A., 1982, Isolation and Characterization of Streptomyces aureofaciens Protein - Synthesis Elongation Factor Tu in an aggregated State, Eur. J. Biochem., 129:127.

Wikman, F. P., Siboska, G. E., Peterson, H. U., and Clark, B. F. C., 1982, The Site of Interaction of Aminoacyl-tRNA with Elongation Factor Tu, EMBO J., 9:1095.

EFFECTS ON INTRINSIC FLUORESCENCE INDUCED BY ACTIVE SITE MODIFICATIONS OF RHODANESE

Alessandro Finazzi Agrò,* Nicola Rosato*, Rodolfo Berni° and Carlo Cannella°

*Department of Experimental Medicine and Biochemical Sciences
Universite of Rome "Tor Vergata" and CNR Center for Molecular
Biology, Rome, Italy
°Institutes of Biochemistry and Molecular Biology
University of Parma, 43100 Parma, Italy

Rhodanese is a mitochondrial enzyme of unknown physiological function catalyzing "in vitro" the transport of zero valence sulfur from an appropriate donor like thiosulfate to a nucleophilic acceptor. The enzymic reactions occur via a double-displacement with the intermediacy of a sulfur-loaded form of enzyme where the zero valence sulphur is bound to an active site cysteine (CYS 247) in a persulfide linkage (Westley,1973).

It has been previously shown that the sulfur-loaded enzyme (ES-SH) has a 30% lower fluorescence than the sulfur-free enzyme (ESH) though the other fluorescence parameters were identical (Finazzi Agrò et al., 1972). The removal of the transferred sulfur allows the CYS 247 to be chemically modified in several ways. It is alkylated by treatment with iodoacetate (Horowitz and Criscimagna, 1982) or linked in a mixed disulfide bridge by addition of thioglycolate. The reactions of CYS 247 are accompanied by a fluorescence quenching, and in the case of alkylation with iodoacetate also by a red shift (~5 nm). Upon reaction with H_2O_2 rhodanese is inactivated due to the formation of an intramolecular disulfide bridge (Canella and Berni, 1983). Also in this case a fluorescence quenching and a red shift are apparent (Table 1). It is interesting to note that most thiolic compounds are not able to react with the sulfur free form of rhodanese even at 10-20 mM concentration, i.e. 1000 fold in excess over enzyme.

The reactions of CYS 247 producing a red shift are also accompanied by a spectral perturbation typical of the exposition of a previously

Table 1. Quenching of Sulfur-Free Rhodanese Fluorescence by Active Site Modifications

Reagent	F/FO	Red Shift	Inactivation
iodoacetate	0.7	5 nm	irr.
thioglycolate	0.7	-	rev.
3-mercaptopyruvate	0.8	-	rev.
H_2O_2	0.8	5 nm	rev.
2-mercaptoethanol			
cysteine			
3-mercaptopropionate		no effect	
thiolactate			
red. glutathione			

buried tryptophyl side chain(s) to the solvent (Figure 1). We have measured the fluorescence lifetime of some of these rhodanese derivatives with the single photon counting technique. All derivatives studied give decay curves that can be fitted by a sum of two exponentials (Table 2). The average lifetime is decreased in the ESSH form by the same percent as the quantum yield suggesting that a dynamic quenching mechanism is operating. This finding supports the hypothesis of a non-radiative energy transfer from tryptophans to the sulphide group (Finazzi Agrò et al., 1972). Interestingly the longer lifetime appears to be the more affected (Table 2. In the other cases, the dynamic fluorescence measurements show a slight decrease in the average lifetime which does not account for the quenching. Instead, a significant decrease in the fractional fluorescence emitting with longer lifetime is apparent. These observations might

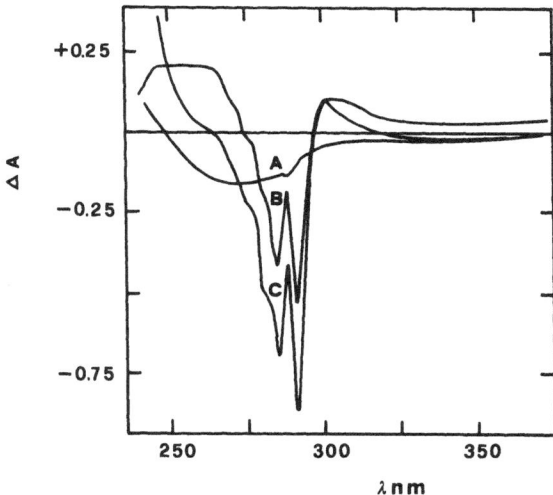

Figure 1. Spectroscopic changes of rhodanese induced by hydrogen peroxide or iodoacetate. Difference absorption spectra. The base line was made with both sample and reference cuvettes containing 26 μM sulfur-loaded enzyme (ES-SH) in 50 mM phosphate buffer, pH 7.6. Curve A = ESH vs ES-SH, upon addition of 30 μM cyanide to the sample. To the sulfur-free enzyme was added a stoichiometric amount of hydrogen peroxide (curve B) or 50 μM iodoacetate (curve C). The spectra were taken after 30 and 50 min respectively, at 20 C using 1-cm light-path cuvettes. Reagents were added to the sample in microliter amounts and a corresponding amount of buffer was added to the reference.

indicate that the tryptophans emitting with longer lifetimes are the buried ones, i.e., those emitting at lower wavelength. Upon reaction with iodo-acetate or with H_2O_2 a fraction of them becomes exposed to the solvent giving rise to a red-shifted fluorescence. This fraction might also change its lifetime to shorter values.

Table 2. Fluorescence Quantum Yields and Lifetimes of Rhodanese Derivatives

derivative	relative quantum yield X 100	$\langle\tau\rangle_{av}$	%	τ_1	α_1	τ_2	α_2	χ^2
ESH	100	2.6	100	1.7	0.45	3.4	0.54	1.4
ESSH	77	2.0	76	1.3	0.54	2.8	0.46	1.6
E + iodoacetate	68	2.4	90	1.6	0.72	4.5	0.28	1.3
E + hydrogen peroxide	74	2.4	90	1.5	0.72	4.6	0.28	1.1

89

REFERENCES

Cannella, C. and Berni, R., 1983, Interaction of Rhodanese with Intermediates of Oxygen Reduction, <u>FEBS Lett.</u>, 162:180.

Finazzi Agrò, A., Federici, G., Giovagnoli, C., Cannella, C., and Cavallini, D., 1972, Effect of Sulfur Binding on Rhodanese Fluorescence, <u>Eur. J. Biochem.</u>, 28:89.

Horowitz, P., and Criscimagna, N., 1982, The Specificity of Active-Site Alkylation by Iodoacetic Acid in the Enzyme Thiosulfate Sulfurtransferase, <u>Biochim. Biophys. Acta</u>, 702:173.

Westley, J., 1973, Rhodanese, <u>Adv. Enzymol. Relat. Areas Mol. Biol.</u>, 39:327.

LIFETIME MEASUREMENTS OF COLOR CENTERS BY FLUOROMETRY WITH SYNCHROTRON RADIATION

Mario Piacentini*

Istituto di Struttura della Materia del C.N.R.
Via E. Fermi 38, 100044 Frascati, Italy
*Present Address: Dipartimento di Energetica, Universita La
Sapienza di Roma, Via E. Scarpa 14, I00161 Roma, Italy

INTRODUCTION

The measurements of emission lifetimes by means of time-resolved luminescence have been of great importance in the understanding of properties of electronic excited states of pure crystals as well as of defects in insulating crystals, as has been so clearly explained by Bassani in his talk at this conference. The availability of a perfectly stable, pulsed light source such as synchrotron radiation available from storage rings makes the measurements of lifetimes using the phase shift method competitive with other techniques utilizing the short light pulses available from lasers and fast electronics. In addition, synchrotron radiation has the advantage of tunability of the exciting radiation over a broad range of energies.

In the next section the most important features of synchrotron radiation will be summarized and in the subsequent section the experimental set-up available at the Italian synchrotron radiation facility PULS using the radiation emitted from the ADONE storage ring of the INFN National Laboratories at Frascati will be presented. The measurements performed on several colors centers in NaF crystals will be discussed at the end of this chapter.

SYNCHROTRON RADIATION

We call synchrotron radiation the electromagnetic radiation emitted by electrons moving along circular paths at relativistic speeds. The main features of synchrotron radiation can be summarized as follows (Kunz, 1979; Winich and Doniach, 1980): i) it has a smooth, continuous spectral distribution extending over a broad range from the infrared to the hard X-rays; ii) it is highly collimated; iii) it has a high intensity; iv) it is polarized; v) it is a pulsed source.

The total power P (in KW) irradiated from the electrons orbiting in a storage ring is given by:

$$P = 88.5 \ E^4 I/R \qquad (1)$$

where E (in GeV) is the electrons energy, I (in amperes) is the electrons' current and R (in meters) is the radius of curvature of the electrons' orbit. For ADONE E = 1.5 GeV, R = 5 m, I = 50 mA, and P = 4.5 KW. Instantaneously the radiation is emitted within a very narrow cone of aperture ψ = 1/E mrad with the axis tangent to the electrons orbit. During the electrons motion, this well collimated beam of radiation sweeps the observation area and appears as a narrow streak of light. The spectral properties of synchrotron radiation depend on a characteristic wavelength, λ_c, which is a parameter characteristic for a given storage ring:

$$\lambda_c (A) = 5.59 \ R/E^3. \qquad (2)$$

The photon flux emitted by the electrons going along an arc of 1 mrad of orbit is given by:

$$N(\lambda) = 2.23 \times 10^{14} IEG(\frac{\lambda}{\lambda_c}) \ \text{photons/(s mrad 1\% bandwidth)} \qquad (3)$$

where the function, $G(\frac{\lambda}{\lambda_c})$,

reproduced in Figure 1, does not depend on the parameters of a particular storage ring. The maximum of the spectral distribution occurs at $\lambda \doteq 4\lambda_c$. For $\lambda << \lambda_c$ the spectral distribution decays very fast (exponentially) and depends strongly on the particular storage ring. For this reason, λ_c is considered also as a cut-off wavelength. Instead, in the long wavelength region ($\lambda >> \lambda_c$) the spectral distribution decays slowly as

$$N(\lambda) = 9.35 \times 10^{13} I(\frac{R}{\lambda})^{1/3} \ \text{photons/(s mrad 1\% bandwidth)} \qquad (4)$$

and depends very little on the storage ring characteristics.

Synchrotron radiation is elliptically polarized with the major axis of the ellipse lying in the orbit plane of the electrons. The two polarization components (parallel and perpendicular to the orbit plane, respectively) have different spectral behaviors. For $\lambda \leq \lambda_c$ the radiation is polarized almost linearly in the orbit plane.

Since the electrons circulate grouped in almost monoenergetic bunches inside the storage ring, a flash of radiation reaches the observer whenever an electron bunch moves along the portion of observed orbit. In the case of a single bunch running inside the storage ring, all the pulses have the same intensity and are equally separated. For ADONE the pulses are almost gaussian in shape, with $\sigma \cong 2$ ns (σ is the gaussian halfwidth) and are separated by 351 ns.

At a given storage ring, the spectral properties of the synchrotron radiation available at the bending magnets can be boosted using special devices mounted along straight sections, where the electrons are forced to perform one to several oscillations around their straight orbit using a series of magnets of opposite polarity. Wigglers are constructed with few magnetic poles, typically one to three. The magnetic field is higher than that at the bending magnets. Consequently, R and λ_c are shorter and the spectral distribution is pushed towards the higher energies. In the case of undulators, the electrons are forced to perform a large number of oscillations (20-100) and the radiation is emitted almost entirely at a single wavelength

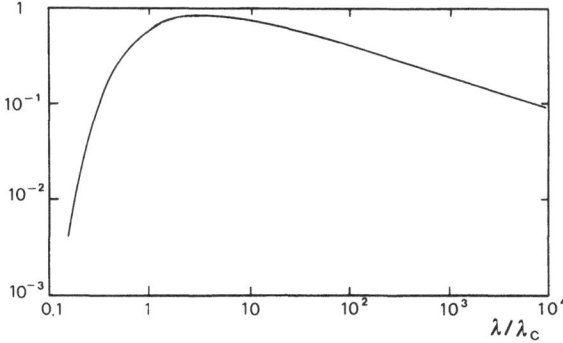

Figure 1. Universal function $G(\frac{\lambda}{\lambda_c})$ or the spectral distribution of synchrotron radiation.

93

$$\lambda_u \cong 1.3 \times 10^{-7} \frac{\lambda_o}{E^2}, \qquad\qquad (5)$$

where λ_o is the undulator period, or in a few odd harmonics of λ_u.

PHASE FLUOROMETRY WITH SYNCHROTRON RADIATION

The measurement of lifetimes using the phase and modulation technique is based on a very simple effect (Spencer and Weber, 1969). If a sample, with a single exponential decay time τ is excited with a radiation beam, the intensity of which is sinusoidally modulated with an angular frequency, ω:

$$I_o(t) = I_o(1 + M_o \cos\omega t), \qquad\qquad (6)$$

also the luminescence signal is modulated at the same frequency, according to

$$I_L(t) = I_L(1 + M_L \cos(\omega t + \Phi)) \qquad\qquad (7)$$

($M_{o,L}$ are called the modulation amplitudes). The phase shift Φ and the modulation ratio $M - M_L/M_o$ between the two modulation amplitudes are given by:

$$M = (1 + \omega^2\tau^2)^{-\frac{1}{2}}; \quad \tan \Phi = \omega\tau. \qquad\qquad (8)$$

By measuring either one of the two quantities M or Φ, τ is obtained immediately. Assuming for example that the sensitivity for measuring Φ is 10 mrad at $\omega = 10^8$ rad/s, the shortest lifetime that can be measured is 100 ps. For measuring even shorter lifetimes, it is necessary either to employ higher frequencies or to improve the sensitivity for Φ. The latter can be obtained by converting the signal modulated at high frequency into a signal modulated at low frequency using the cross correlation method devised several years ago by Spencer and Weber (1969).

The relations of eq. 7 are simple only in the case of a single exponential decay. In the presence of several lifetimes τ_i, the corresponding relations become quite complicated. In this case it is necessary to measure M and Φ over a large number of frequencies and to employ a best-fit procedure for obtaining the lifetimes (Jameson and Gratton, 1983; Jameson et al., 1984).

Since synchrotron radiation is a pulsed, repetitive, highly stable source:

$$I(t) = \frac{A}{\sigma\sqrt{2\pi}} \sum_{m=-\infty}^{\infty} \exp\left[\frac{-1}{2}\left(\frac{t-mT_0}{\sigma}\right)^2\right], \qquad (9)$$

it contains intrinsically a large number of harmonics of the fundamental frequency $\omega_0 = 1/T_0$ of the storage ring:

$$I(1) = A\nu_0\left[1+2\sum_{n=1}^{\infty}\exp\left[-\frac{1}{2}(n\omega_0\sigma)^2\right]\cos n\omega_0 t\right] \qquad (10)$$

where $\nu_0 = 2\pi\gamma_0$ with high intensity up to ~$1/\sigma$. For ADONE, $\sigma=2$ ns, $\omega_0=2.86$ MHz and the harmonic content extends up to about 80 MHz, as indicated in Figure 2. Thus, synchrotron radiation is an ideal multifrequency source to be used for lifetime measurements with the phase fluorometer (Gratton and Lopez-Delgado, 1980) and for this reason we have constructed such an apparatus at the PULS facility, sketched in Figure 3 (Antonangeli, et al., 1983).

The synchrotron radiation emitted by the electrons orbiting in the ADONE storage ring is first monochromatized and then it is split into a reference beam, detected by the photomultiplier PM1, and the exciting beam falling onto the sample. The emission is monochromatized before reaching a second photomultiplier PM2 identical to PM1. A first frequency synthesizer FS1 generates the radio frequency signal driving the storage ring ($\nu_0=8568500$ Hz). A second frequency synthesizer FS2 is used to generate all the harmonic frequencies $m\nu_0$ (m=1,2,3...), all of them shifted by a small amount $\Delta\nu=24$ Hz. For the correct operation of the phase fluorometer, it is essential that the two frequencies generated by FS1 and FS2 maintain a constant phase relation between each other. For this reason, the two frequency synthesizers are driven by the same crystal oscillator.

FS2 supplies a bias voltage that, applied to the last dynode of PM1 and PM2, modulates their gain at the frequency $m\nu_0+\Delta\nu$. The output voltage is proportional to the detected light intensity times the detector gain:

$$V(t)\propto I(t)\cdot g(t) =$$

$$[I_0 + \sum_{n=1}^{\infty} I_n \cos(n\omega_0 t+\Phi_n)]\cdot[1+g_0\cos(m\omega_0+\Delta\omega)t] = \qquad (11)$$

$$I_o + [\sum_{n=1}^{\infty} I_n \cos(n\omega_o t + \Phi_n)] + I_o g_o \cos(m\omega_o + \Delta\omega)t + \sum_{\substack{n=1 \\ n \neq m}}^{\infty} I_n g_o \cdot$$

$$[\cos(((n+m)\omega_o + \Delta\omega)t + \Phi_n) + \cos(((n-m)\omega_o - \Delta\omega)t + \Phi_n)] +$$

$$I_m g_o \cos((2m\omega_o + \Delta\omega)t + \Phi_m) + I_m g_o (\cos(\Delta\omega t \Phi_m))$$

where I_n and Φ_n are the modulation amplitude and the phase shift of the nth harmonic of the luminescence signal. Thus V(t) contains terms with high frequencies, multiples of ω_o. There is also a d.c. term, I_o, that gives the average intensity. The last term at low frequency $\Delta\omega$ carries the same phase modulation information as the m-th harmonic. All the high frequency terms can be easily filtered out and the phase and modulation of the luminescence signal can be measured with great accuracy from the low frequency component. In a practical run, for each harmonic frequency $m\omega_o$, Φ_m and M_m are measured with respect to the reference channel firstly for the sample and secondly for a reference scatterer placed at the same position. In this way the delays introduced by the different optical paths in the sample and reference channels are taken into account. The sample and the reference scatterer are mounted onto the cold finger of a rotating liquid nitrogen cryostat.

EXPERIMENTAL RESULTS

We have used the phase fluorometer constructed at the synchrotron radiation facility PULS described above for studying the F and F-aggregate

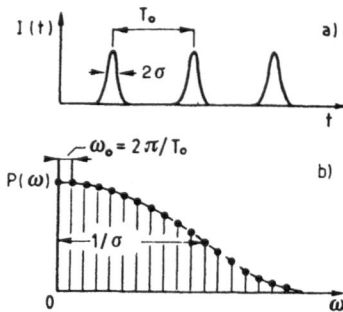

Figure 2.   a) The synchrotron radiation pulsed time structure. b) The harmonic content of synchrotron radiation. For ADONE T = 351 ns, σ = 2 ns, ν_o = 2.86 MHz, $1/\sigma = 5 \times 10^8$ rad/s.

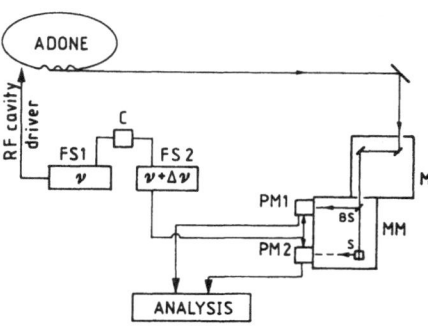

Figure 3. Schematic diagram of the phase fluorometer realized at the vacuum ultraviolet beam line of the Italian synchrotron radiation facility PULS. M = monochromator, MM = measuring module, BS = beam splitter, S = sample, PM1 and PM2 = photomultipliers, FS1 and FS2 = frequency synthesizers driven by the same crystal oscillator C.

centers in alkali halides (Grassano et al., 1986). As it is well known, an F center corresponds to an electron trapped at a negative ion vacancy in alkali halides (Farge and Fontana, 1979). It may happen that two or more vacancies are located at neighbor sites and F-aggregate centers, mainly F_2 and F_3^*, form. The absorption and emission bands of these color centers have been studied in detail, but their lifetimes, in particular those of the aggregates, are known only in part. In Figure 4 we show the absorption

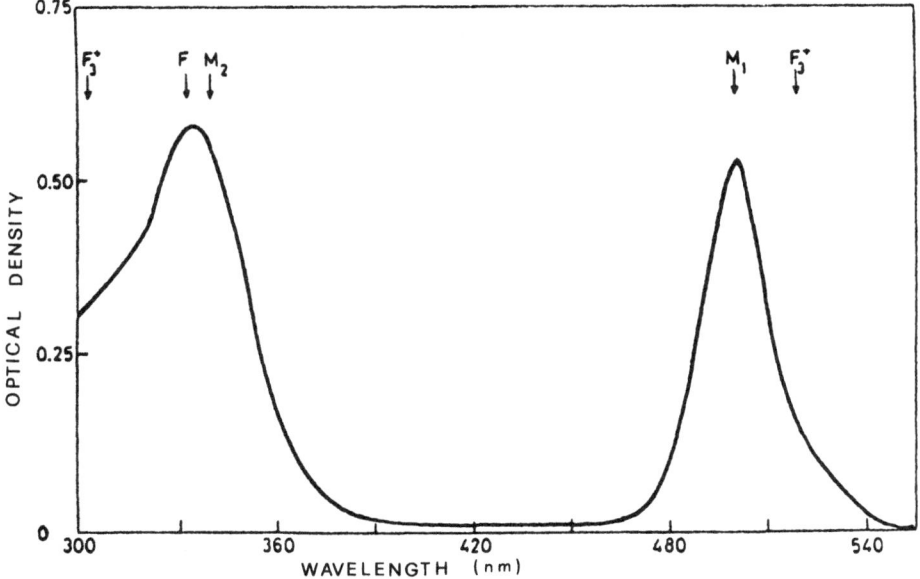

Figure 4. Absorption bands of a γ-irradiated NaF crystal measured at liquid nitrogen temperature.

Table 1. Wavelengths (in nm) of the Absorption and Emission Peaks of the Bands of Colors Centers in NaF (Measured at 77 K) and Lifetimes (in ns) of the Emission Bands Excited in the Fundamental Absorption Bands

Center	Absorption	Emission	Lifetime
F	333	750	51 ± 2
F_2 { M_2, M_2'	340	650	10 ± 0.2
M1	498		
F_3^+ {	303	575	8.55 ± 0.05
	512		

peaks of the F center and several aggregates in NaF. In Table 1 we report their energies and lifetimes. To our best knowledge, the lifetime of the F_3^+ emission was never reported before our work. The shorter lifetime of the F_3^+ center with respect to the F_2 center that we have found for NaF is in contrast with respect to other alkali halides and indicates a smaller electron-lattice coupling in the aggregate center (Grassano et al., 1986).

A very interesting problem in the study of defects and color centers is the existence of radiative and non-radiative transitions between the higher excited states of the same center. For example, the F center shows only the emission from the lowest excited state to the ground state even

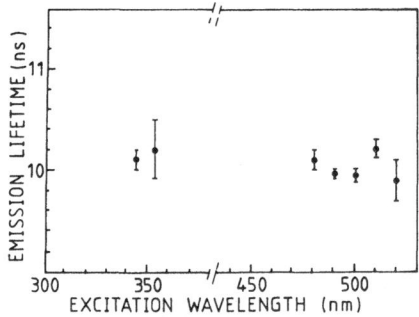

Figure 5. Emission lifetimes of the first two excited states of the F_2 center in γ-irradiated NaF crystals as a function of the exciting photon wavelength measured at liquid nitrogen temperature.

when it is excited to higher levels. An accurate measurement of the rise time of the emission or of the delay between the emission and the excitation can give interesting information on the configurational relaxation of different excited states.

We have tried (Grassano et al. 1986) to determine the delay of the emissions excited in the different bands of the aggregate centers of NaF, by measuring the apparent lifetime of the luminescence of the F_2 center excited at various wavelengths around 500 nm (fundamental F absorption) and around 340 nm (second excited state absorption). The results are shown in Figure 5. The accuracy of the measurements with the present set-up, especially those taken in the weak emission excited in the ultraviolet band, is not sufficient to show, unambiguously, delays in the higher energy excitation.

In conclusion, we have described the main characteristics of synchrotron radiation and the phase fluorometer constructed at the Italian synchrotron radiation facility PULS, using the pulsed nature of synchrotron radiation as a multifrequency light source. We have also discussed, as an example, the application of such a device to the study of colors centers in NaF.

REFERENCES

Antonangeli, F., Bassani, F., Campolungo, F, Finazzi-Agro, A., Grassano, U. M., Gratton, E., Jameson, D. M., Piacentini, M., Rosato, N., Savoia, A., Weber, G., and Zema, N., 1983, A Multifrequency Cross-Correlation Phase Fluorometer with Picosecond Resolution Using Synchrotron Radiation, Report LNF-83/68(R).

Farge, Y. and Fontana, M. P., 1979, Electronic and Vibrational Properties of Point Defects in Ionic Crystals, North Holland, Amsterdam.

Grassano, U. M., Piacentini, M., and Zema, N., 1986, Lifetime Measurements of Colour Centers by a Multifrequency Phase Fluorometer, Nuovo Cimento, 7D:379.

Gratton, E., and Lopez Delgado, R., 1980, Measuring Fluorescence Decay Times by Phase-Shift and Modulation Techniques Using the High Harmonic Content of Pulsed Light Sources, Nuovo Cimento, B56:110.

Gratton, E., Jameson, D. M., Rosato, N., and Weber, G., 1984, Multifrequency Cross-Correlation Phase Fluorometer Using Synchrotron Radiation, Rev. Sci. Instrum., 55:486.

Jameson, D. M., and Gratton, E., 1983, Analysis of Heterogeneous Emissions by Multifrequency Phase and Modulation Fluorometry, in: "New Directions in Molecular Luminescence", ed., D. Eastwood and L. Cline-Love, ASTM, Philadelphia.

Jameson, D. M., Gratton, E., and Hall, R. D., 1984, The Measurement and Analysis of Heterogeneous Emissions by Multifrequency Phase and Modulation Fluorometry, Appl. Spectr. Rev., 20:55.

Kunz, C., ed., 1979, Synchrotron Radiation Techniques and Applications, Springer Verlag, Berlin.

Spencer, R. D., and Weber, G., 1969, Measurements of Subnanosecond Fluorescence Lifetimes With a Cross-Correlation Phase Fluorometer, Ann. N. Y. Acad. Sci., 158:361.

Winich, H. and Doniach, S., eds., 1980, Synchrotron Radiation Research, Plenum Press, New York.

SLAB GEL ELECTROPHORESIS PERFORMED AT HIGH HYDROSTATIC PRESSURE: A NEW APPROACH FOR THE STUDY OF OLIGOMERIC PROTEINS

Alejandro A. Paladini
Ingebi-Conicet, Facultad de Ciencias Exactas y Naturales
Universidad de Buenos Aires, Argentina

Jerson L. Silva and Gregorio Weber
Department of Biochemistry, School of Chemical Sciences
University of Illinois, Urbana, IL 61801 USA

In the last few years several studies have demonstrated the reversible dissociation of oligomeric proteins under pressure (for review see: Heremans, 1982; Weber and Drickamer, 1983). The dissociating effects of pressure are generally ascribed to the intersubunit surfaces and the restrictive effect of the covalent bond architecture of the protein. The pressure dissociation has been detected either by indirect or direct methods. Indirect methods, such as hybridization studies (Jaenicke and Koberstein, 1971) and activity measurements (Penniston, 1971; Schade et al., 1980; Seifert et al., 1985) have been employed with some success. These indirect methods are valuable in revealing qualitatively the involvement of protein dissociation in the observed effect of pressure. However, the stoichiometry of dissociation and the thermodynamic parameters (volume change and dissociation constant) cannot be unambiguously determined. Methods which directly reveal the state of association are more suitable for a thermodynamic approach to the pressure dissociation, although they are not immune to experimental problems.

An early direct detection of protein dissociation was performed by ultracentrifugation studies (Josephs and Harrington, 1967). The main drawback of this method is that pressures generated in the ultracentrifuge are limited to the order of 100 bars, permitting only the detection of dissociation in those systems which present large change in volume upon dissociation, and at protein concentrations for which the equilibrium is already poised at atmospheric pressure. For pressures in the range of 1 to 4000 bars, optical methods have been preferred. Light scattering has been

101

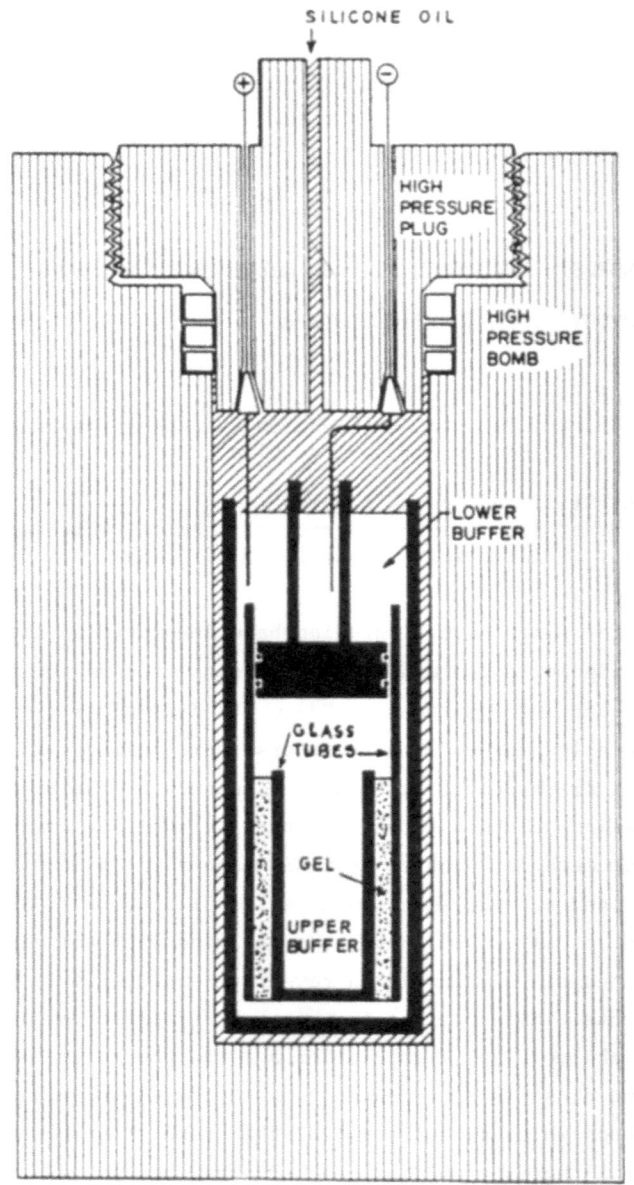

Figure 1. Side view of the high pressure bomb with the slab gel apparatus in running position. The high pressure bomb is 16 inches long and 6 inches outer diameter. The dimensions of the inner chamber are 1 inch in diameter and 8 inches depth. Seals are made around the plug with the use of three rings: two made of brass and a lead one in between. The pressure transmitting fluid is silicone oil.

employed when large aggregates are dissociated into many small particles (Payens and Heremans, 1969; Engelborghs et al., 1976) and fluorescence

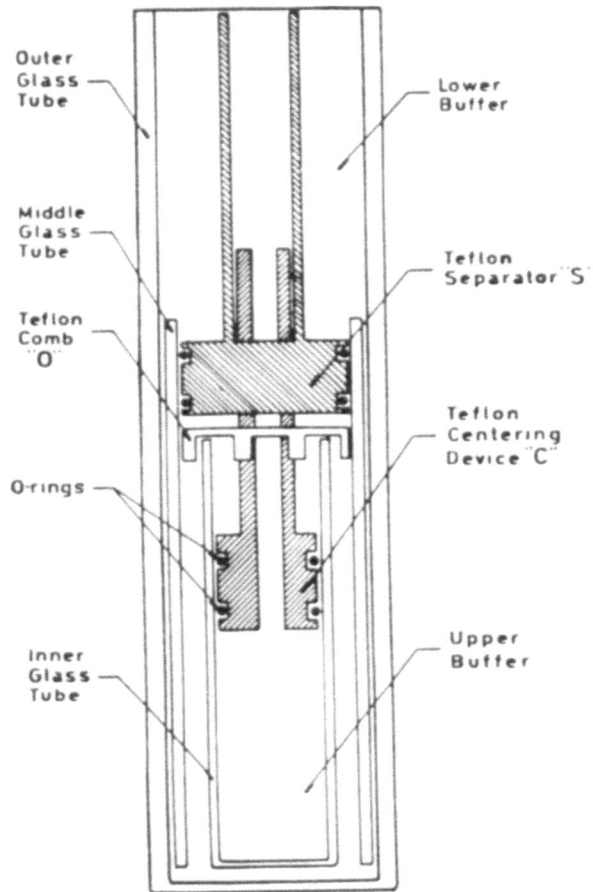

Figure 2. Side view of the complete slab gel electrophoresis apparatus. The dimensions of the glass tubes are such that the setup fits inside the high pressure inner chamber. The difference between the outer radius of the inner tube and the inner radius of the middle tube determines the slab gel thickness. Teflon centering device "C" and the comb "O" are removed during the high pressure electrophoresis runs.

polarization and energy distribution of fluorescence emission have been used to study the pressure dissociation of small oligomers, mostly dimers and tetramers (Paladini and Weber, 1981a; King and Weber, 1986; Silva et al., 1986; Royer et al., 1986; Verjovski-Almeida et al., 1986).

In order to demonstrate the pressure-dissociation process directly by a mass transport method, we have used the electrophoretic migration of oligomeric proteins when subjected to hydrostatic pressures of 1-2 Kbars. If the subunits have different mobilities dissociation should be easily detectable and even permit the subunit isolation in pure condition. The

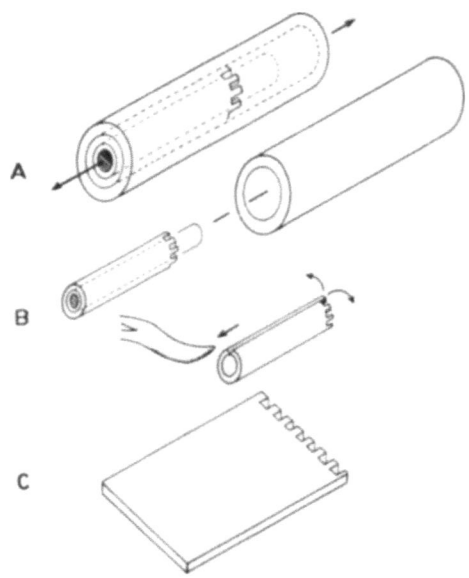

Figure 3. Diagram showing the slab gel handling protocol. Pulling the inner tube from the middle one (A) exposes the gel (B). A longitudinal cut is then performed and the slab gel is obtained and ready to be either stained or optically scanned (C).

apparatus developed by us allows the use of a polyacrylamide slab gel with a capacity up to 12 wells, therefore permitting the study of several samples in one experiment. The experimental set up and the results obtained with a dimeric protein (tryptophan synthase β_2 subunit) are described in the present work.

The complete gel electrophoresis apparatus is shown in figure 1: it consists of three concentric glass tubes resting inside a cylindrical steel reservoir (pressure bomb), designed to resist up to 3 Kbars of pressure. These allow the gel preparation and loading of the samples. The polyacrylamide gel is polymerized between the inner tubes and two buffer reservoirs of 15 ml each are formed between the outer and the middle tubes and between the middle and the inner tubes respectively (figure 2). The number and shape of the sample wells is achieved with the insertion of a cylindrical teflon comb machined to fit snugly between the tubes containing the gel solution. The gel thickness can be varied between the practical limits of 0.5 to 3 mm by using an inner tube of the appropriate external diameter. The maximum length of the slab gel is 80 mm.

A teflon centering plug C is also shown in figure 2. This plug, in conjunction with the teflon comb 0, works as a tube centering device to

Table 1. Mobility of Single-Chain Proteins in Poly-Acrylamide Gel Electrophoresis in the Presence or Absence of SDS

Condition	Protein	Pressure (Kbar)	Mobility
SDS-PAGE	BSA	0.001	0.267
SDS-PAGE	BSA	2.000	0.273
SDS-PAGE	Myoglobin	0.001	0.792
SDS-PAGE	Myoglobin	2.000	0.785
PAGE	BSA	0.001	0.825
PAGE	BSA	2.000	0.646
PAGE	Myoglobin	0.001	0.178
PAGE	Myoglobin	2.000	0.140

achieve uniform gel thickness. Once the gel is formed, the teflon plugs C and S and the Comb O are removed from the system, making the wells accessible for sample loading. After replacement of the plug S and filling the reservoirs with buffer, the device is carefully placed inside the pressure bomb. The rest of the volume of the bomb is filled with silicone oil and the closing plug, with the electrodes, is torqued down. The system is now ready for a pressure run.

The cylindrical symmetry of this apparatus simplifies the electrode connections, because they simply extend from the closing plug of the pressure bomb into the buffer reservoirs as can be seen in figure 1. The electrodes are made of platinum wire and they are connected through the closing plug of the pressure bomb to the power supply. Pressure leaks in the electrode holes are prevented using a conic (9°) piece of steel welded to each electrode and a conic teflon O-ring.

The pressure bomb is made of tool steel machined to a cylindrical

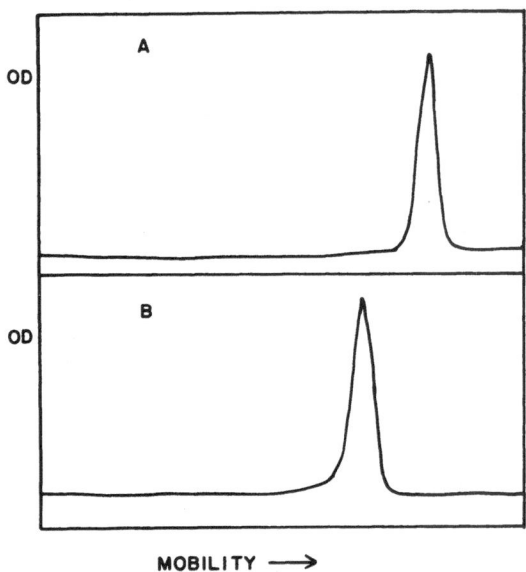

Figure 4. Densitometric scans of electrophoresis runs of bovine serum albumin at atmospheric pressure (A) and at 2.0 Kbars (B). 15 μl of 0.2 mg/ml BSA was applied in each condition.

shape and mounted in a water bath for temperature control during the electrophoretic runs. The rest of the equipment consists of a pressure generator, a Bourdon gauge and an electrophoresis power supply. As a general reference to the high pressure setup, see Paladini and Weber (1981b).

Dow Corning silicone oil #200 of 100 cstokes at 25°C was used both to avoid electric shorts between the lower buffer and the bomb wall and as the pressurizing fluid. After a run the glass tubes are removed from the bomb and the cylindrical gel between them is carefully extracted. The inner wall of the middle tube should be lightly lubricated with silicone grease prior to gel polymerization. A single cut is then made along the axis of the hollow cylindrical gel to obtain a slab as can be seen in figure 3. Either staining or direct optical scanning and elution of the protein bands is possible using standard electrophoresis techniques.

Polyacrylamide gels with 6% acrylamide and 0.11% of bis-acrylamide as the cross-linking agent were used for the "native" runs (without SDS). SDS-PAGE were performed with 7.5% acrylamide and 0.2% bis-acrylamide. Gel preparation was done in the standard way. The running buffer consisted of 1 mM EDTA, 0.5 mM DTE, 24 mM TRIS, 192 mM glycine, pH 8.3. For the SDS-PAGE, 0.2% SDS was added to the running and gel buffers.

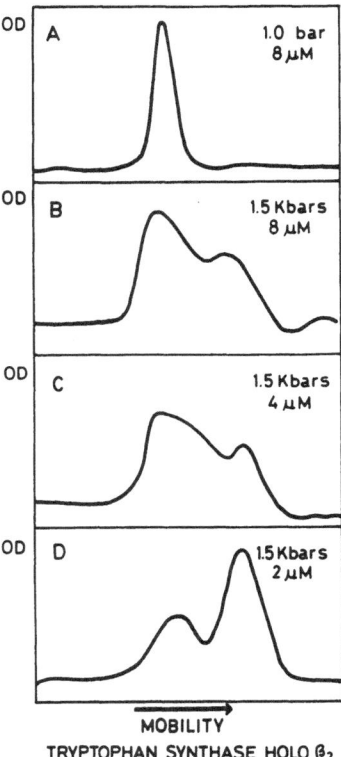

Figure 5. Comparison of the densitometric profiles of electrophoresis of holo β2 tryptophan synthase subunit at atmospheric pressure (A) and under 1.5 Kbars (B,C,D). Concentrations of the enzyme were as follows: (A) 8 μM; (B) 8 μM; (C) 4 μM; (D) 2 μM.

High pressure electrophoresis in the presence of SDS was carried out to investigate whether the hydrostatic pressure affects the sieving structure of the polyacrylamide gel. Table 1 shows that the mobilities of single polypeptide-SDS complexes at 1 bar and 2.0 Kbars, were identical, indicating that the gel matrix pore size was not affected, however, the electrophoresis under pressure requires longer times (about 1.4 fold) in comparison to those at atmospheric pressure for the same voltage and current conditions.

The high pressure electrophoresis of globular single chain proteins (myoglobin and BSA) in the absence of SDS were also explored. Both proteins ran as single peaks at atmospheric conditions and under pressure. Table 1 shows that the mobilities were smaller under pressure. The ratio of the mobilities under 2.0 Kbars and at atmospheric pressure were the same for both proteins (0.78), suggesting that the mobility decrease was caused

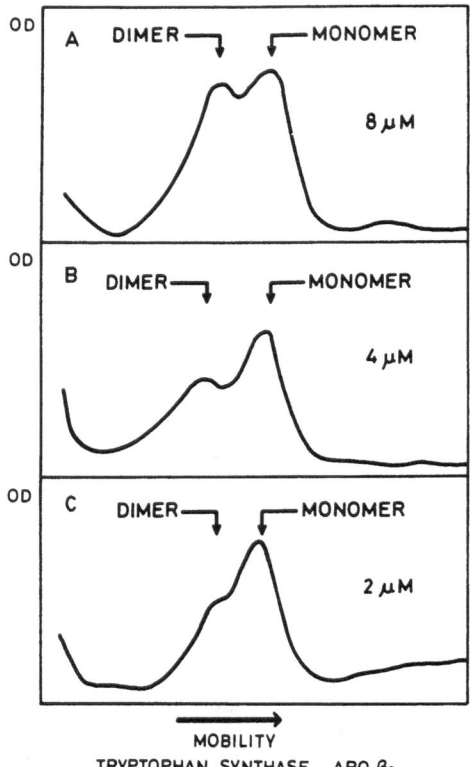

Figure 6. Concentration dependence of the pressure dissociation of apo β2 tryptophan synthase subunit as shown by high pressure electrophoresis (1.5 Kbars). Enzyme concentrations: (A) 8 µM; (B) 4 µM; (C) 2 µM.

by an external effect. This general reduced mobility can be explained by a pH decrease induced by pressure. It has long been known that the ionization constant (Ka) of weak acids, such as glycine, undergoes a large increase with pressure (Zipp and Kauzmann (1973) and Neuman et al. (1973)). For pressures in the order of Kbars, the increase of Ka of glycine might decrease the pH of the buffer enough to alter the net charge of proteins. Figure 4 compares the electrophoresis under pressure of bovine serum albumin (2 Kbars) to the control run at atmospheric pressure. No spreading or asymmetries appeared in the BSA band that was run under pressure. If 2 Kbars pressure had elicited any significant effect on the tertiary structure, it would have been apparent either by the appearance of another peak or by a distortion of the protein peak.

In order to investigate the effect of pressure on the electrophoretic behavior of oligomeric proteins, a simple dimer protein was used. The

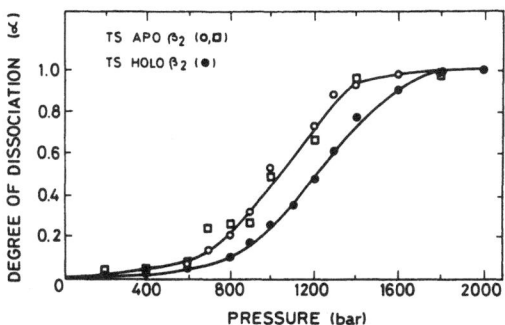

Figure 7. Plot of degree of dissociation of apo β2 (o, □) and holo β2 tryptophan synthase (●) versus pressure at 2 μM concentration. Measurements were made by center of mass displacement of intrinsic fluorescence (●, o) and by fluorescence polarization of the 2,5 DNS conjugate (□).

tryptophan synthase β_2 subunit was chosen due to the availability of a detailed spectroscopic study under pressure of this protein (Silva et al., 1986). Either BSA or myoglobin were run together with the β_2 subunit in order to correct the mobilities measured under pressure. Figure 5 shows the densitometric profiles of the electrophoresis of holo β_2 subunit (pyridoxal-phosphate bound enzyme) performed either under high or atmospheric pressure. Different protein concentrations of holo β_2 dimer were loaded to a 6% polyacrylamide gel. The proteins were run in a 2.5% stacking gel at atmospheric pressure, then the system was transferred to the pressure bomb and electrophoresis was carried out at 1.5 Kbars for 3 hours. The coomassie blue stained gel revealed the presence of a second peak for the gel that was run at high pressure (figure 5B,C,D), not present in the control gel obtained at atmospheric pressure (figure 5A). It is noteworthy that as the protein concentration was decreased, the fastest of the two clearly separated components, increased (figures 5 and 6). The concentration dependence indicates that the extra peak corresponds to a monomer and permits to explain unequivocally the electrophoretic pattern as the result of a dissociation process.

The effect of pressure on the holo β_2 subunit of the tryptophan without pyridoxal-P bound (apo β_2). The gel densitometric scan (figure 6) shows that the ligand free protein (apo β_2) was more readily dissociated by pressure than the holoprotein. The electrophoretic pattern is more complex in the holoprotein than in the apoprotein. A complex pattern would be expected if subunit dissociation is accompanied by increased dissociation of the prosthetic group. These results are in agreement with activity measurements performed after release of pressure (Seifert et al., 1985) and with the spectroscopic data obtained under pressure (Silva et al., 1986). These last results are shown in figure 7 where changes in average energy of tryptophan fluorescence emission (given by displacement of center of spectral mass) can be seen while the pressure was increased for the two forms of the β_2 subunit. The apparatus and conditions of these experiments have been recently described (Silva et al., 1986).

Size analysis of the β_2 subunit eluted from the monomeric position of a non-stained gel that was run under pressure was performed by HPLC. The dimeric structure was fully recovered in agreement with fluorescence polarization measurements and similar HPLC experiments performed after release of pressure applied to the protein in solution (Silva et al., 1986).

In addition to the experiments already described, it can be mentioned that one of us (J.L.S.) studying the effects of pressure upon RNA polymerase by the technique described here was able to show the dissociation of this oligomer at pressures around 2 Kbars. (Data not shown.)

In conclusion, we have demonstrated the feasibility of an apparatus to run slab gel electrophoresis under pressure and its application as a new and simple tool for the study of the aggregation of oligomeric proteins that together with ultracentrifugation provides a direct method to prove dissociation under pressure by actual separation of the oligomers. Electrophoresis has the advantage, over ultracentrifugation, in that the pressures can be one order of magnitude greater and the separation of the components can be made complete. Furthermore it does not involve large pressure gradients which can introduce important artifactual effects. Although, spectroscopic methods are more suitable to provide information about thermodynamic parameters of protein dissociation (volume change and dissociation constant), the interpretation of the spectroscopic changes is not always straightforward since it is based on the assumption that there are only two states, respectively stable at the highest and lowest pressures. This assumption can be tested by an electrophoresis run of different protein concentrations under pressure. The electrophoresis separation of pressure dissociated forms of apo and holo β_2 tryptophan synthase (figures 5 and 6) corroborates spectroscopic studies done under pressure (Silva et al., 1986), with respect to the protein concentration and pressure range in which the dissociation takes place. For pressure studies of larger aggregates and hetero-oligomers, it might be mandatory to use electrophoresis under pressure in order to determine the dissociation pathway.

Hawley and Mitchell (1975) used high pressure electrophoresis to study the denaturation of a single-polypeptide protein (chymotrypsinogen) by pressures of the order of 5-8 Kbars using a tube electrophoresis method. Their apparatus allowed to run only one sample at a time. This limitation forced them to take very elaborate and laborious precautions to achieve reproducibility among different runs. The slab gel system described by us can easily be adapted to study protein unfolding by placing the system in pressure bombs rated to support higher pressures.

In addition to studies on the energetics of protein association, our device can be used to separate and isolate the constituent peptide chains of oligomeric proteins. We demonstrated that the tryptophan synthase β_2 dimer could be reconstituted from eluted monomer. In fact, since this protein is made of equal subunits the reconstitution probably occurred as soon as the pressure was released. In contrast, in a hetero-oligomer, the separated polypeptides being different, could be isolated. We have shown that the electrophoretic run of single chain proteins submitted to pressures below 2.0 Kbars occurs without significant effects upon the conformation. Numerous observations indicate that the conformation of single polypeptide chains is not conspicuously affected by pressures smaller than about 3 Kbars (Heremans, 1982; Weber and Drickamer, 1983). Therefore, the electrophoresis of oligomeric proteins at pressures under 2 Kbars offers a general method by which separation of the polypeptide chains can be performed without unfolding. Since its original description by Weber and Osborn (Weber and Osborn, 1969), the electrophoresis in the presence of SDS has been of great value as an analytical tool. However, its applicability as a preparative method is limited since the recovery of a peptide chain denatured by SDS is burdensome, and in many cases, unfolding is practically irreversible. In contrast, recovery after high pressure electrophoresis only involves the separate elution of the protein components from the gel. Although it is too early to claim general applicability, electrophoresis under pressure appears to furnish an important tool to study dissociation processes and to separate the different constituent polypeptides of an oligomeric complex.

REFERENCES

Engelborghs, Y., Heremans, K. A., DeMaeyer, L., and Hoebeke, J., 1976, Effect of Temperature and Pressure on Polymerization Equilibrium of Neuronal Microtubules, Nature, 256:686.

Heremans, K. A., 1982, High Pressure Effects on Proteins and Other Biomolecules, Ann. Rev. Biophys. and Bioeng., 11:1.

Hawley, S. A., and Mitchell, R. M., 1975, An Electrophoretic Study of Reversible Protein Denaturation: Chymotrypsinogen at High Pressures, Biochemistry, 14:3257.

Hogberg-Raibaud, A., and Goldberg, M. E., 1977, Isolation and Characterization of Independently Folding Regions of the β Chain of Escherichia coli Tryptophan Synthetase, Biochemistry, 16:4014.

Jaenicke, R., and Koberstein, R., 1971, High Pressure Dissociation of Lactic Dehydrogenase, FEBS Lett., 17:351.

Josephs, R., and Harrington, W. F., 1967, An Unusual Pressure Dependence for a Reversibly Associating Protein System; Sedimentation Studies on Myosin, Proc. Natl. Acad. Sci. U.S.A., 58:1587.

King, L., and Weber, G., 1986, Conformational Drift of Dissociated Lactate Dehydrogenases, Biochemistry, 25:3632.

Miles, E. W., and Moriguchi, 1977, Tryptophan Synthase of Escherichia coli Removal of Pyridoxal 5'-Phosphate and Separation of the α and β$_2$ Subunits, J. Biol. Chem., 252:6594.

Neuman, R. C., Kauzmann, W., and Zipp, A., 1973, Pressure Dependence of Weak Acid Ionization in Aqueous Buffers, J. Phys. Chem., 77:2687.

Paladini, A.A., and Weber, G., 1981a, Pressure-Induced Reversible Dissociation of Enolase, Biochemistry, 20:2587.

Paladini, A. A., and Weber, G., 1981b, Absolute Measurements of Fluorescence Polarization at High Pressures, Rev. Sci. Instrum., 52:419.

Payens, T.A.J., and Heremans, K.A.H., 1969, Effect of Pressure on the Temperature-Dependent Association of B-Casein, Biopolymers, 8:335.

Penniston, J. T., 1971, High Hydrostatic Pressure and Enzymic Activity: Inhibition of Multimeric Enzymes by Dissociation, Arch. Biochem. Biophys., 142:322.

Royer, C. A., Weber, G., Daly, T. J., and Matthews, K. S., 1986, Dissociation of the Lactose Repressor Protein Tetramer Using High Pressure, Biochemistry, 25:8308.

Schade, B. C., Rudolph, R., Ludemann, H. D., and Jaenicke, R., 1980, Reversible High-Pressure Dissociation of Lactic Dehydrogenase from Pig, Biochemistry, 19:1121.

Seifert, T., Bartholmes, P., and Jaenicke, R., 1985, Influence of Cofactor

Pyridoxal 5'-Phosphate on Reversible High-Pressure Denaturation of Isolated β_2 Dimer of Tryptophan Synthase Bienzyme Complex from Escherichia coli, Biochemistry, 24:339.

Silva, J. L., Miles, E. W., and Weber, G., 1986, Pressure Dissociation and Conformational Drift of the β Dimer of Tryptophan Synthase, Biochemistry, 25:5780.

Verjovski-Almeida, S., Kurtenbach, E., Amorim, A. F., and Weber, G., 1986, Pressure-Induced Dissociation of Solubilized Sarcoplasmic Reticulum ATPase, J. Biol. Chem., 261:9872.

Weber, G., and Drickamer, H. G., 1983, The Effect of High Pressure Upon Proteins and other Biomolecules, Quarterly Review of Biophysics, 16:89.

Weber, K., and Osborn, M., 1969, The Reliability of Molecular Weight Determinations by Dodecyl Sulfate-Polyacrylamide Gel Electrophoresis, J. Biol. Chem., 244:4406.

Zipp, A., and Kauzmann, W., 1973, Pressure Denaturation of Metmyoglobin, Biochemistry, 12:4217.

LIPID TRANSFER REACTIONS: FLUORESCENCE STUDIES

William W. Mantulin

Laboratory for Fluorescence Dynamics
Departments of Physics, Physiology and Biophysics, and
Biochemistry
University of Illinois at Urbana-Champaign
Urbana, IL 61801 USA

INTRODUCTION

In cells the transport of lipid molecules from sources of biosynthesis or ingestion to the various membranes and organelles represents an important component of metabolism (Pagano and Sleight, 1985). The analogous problems of protein synthesis, membrane assembly, turnover, and secretion are also being actively studied by cell biologists. The important sites of lipid trafficking include the plasma compartment of blood and the cells of the vascular and extravascular compartment. The lipid composition of these sites includes cholesterol, cholesteryl esters, triglycerides, fatty acids, and phospholipids. Figure 1 schematically illustrates several pathways of lipid transfer in a membrane; these include lateral diffusion of the lipid in the membrane plane, spontaneous desorption into the surrounding aqueous compartment with subsequent uptake by an adjacent membrane surface, and bilayer translocation, which is commonly referred to as flip-flop. This latter mechanism is often accelerated by enzymes termed "translocases". Other lipid transfer mechanisms not pictured include transient collisional complexes of lipid surfaces with lipid exchange occurring at the contact point, protein mediated transfer between lipid surfaces involving specific lipid-carrier proteins (e.g., phospholipid exchange protein), enzymatic degradation of lipid molecules and subsequent release of soluble lipid fragments from the membrane, and finally, transport of large lipid aggregates, such as lipoproteins, across the cell membrane via endocytosis. Receptors, which are membrane proteins, regulate the binding and internalization of the protein.

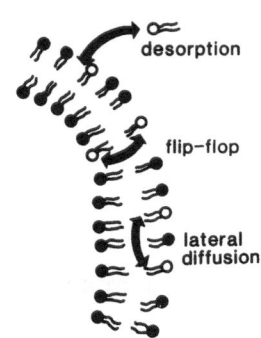

Figure 1. A model bilayer with three types of lipid transfer represented: Spontaneous desorption from the bilayer, flip-flop (bilayer translocation) and lateral diffusion within the bilayer leaflet.

Since much of the experimental work in lipid transfer reactions involves conditions of very low solubility, the high sensitivity of fluorescence methods has proven useful for these studies. This report will focus on the kinetics and mechanisms of lipid transfer, the consequences of transfer, and the role that fluorescence studies have played in unraveling the lipid transfer process. There are multiple routes for lipid trafficking, and it is beyond the scope of this review to consider them all in depth; instead, we will concentrate on intermembrane transfer reactions, especially of phospholipids, and direct the reader to other sources for additional information on transfer reactions. Excellent monographs detailing topics in lipid transfer are available: cellular cholesterol exchange (Phillips et al., 1987), cellular lipid transport (Pagano and Sleight, 1985), transfer in lipid bilayers (Lange, 1986), protein mediated lipid transfer (Wirtz, 1982), and receptor mediated endocytosis (Gianturco and Bradley, 1987; Brown and Goldstein, 1986).

NATIVE AND MODEL SYSTEMS

To understand the mechanisms of lipid transfer *in vivo*, it is important to identify and characterize both native and model systems suitable for such studies. In general, lipids exhibit limited aqueous solubility and, thus, tend to separate into complex assemblies, which often contain protein. Cell membranes and plasma lipoproteins are two examples of native macromolecular assemblies containing both lipids and proteins. In simple terms, a cell membrane is an insoluble bilayer array containing a variety

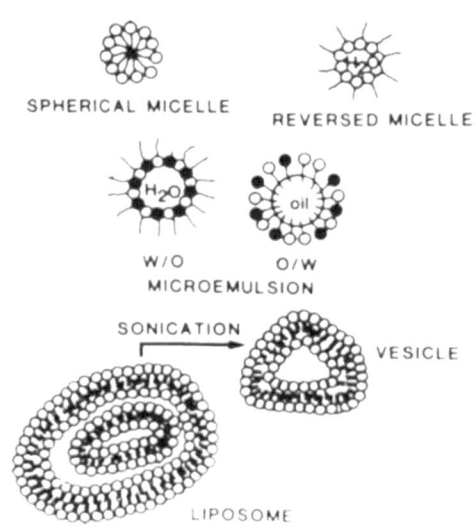

Figure 2. Models of lipid array organization.

of lipids and proteins. Plasma lipoproteins are also composed of a mixture
of lipids and proteins, however, these water-soluble structures are smaller
and their surface is highly curved. Lipoproteins share a common architecture, namely, a surface monolayer of protein and polar lipids surrounding a
core of neutral lipids. They are operationally ranked by density: high
density (HDL), low density (LDL), and very low density (VLDL). The protein
and lipid compositions of various lipoproteins differ and the denser
lipoproteins are smaller. Both native cell membranes and plasma lipoproteins are complex macromolecular assemblies, whose structural properties
and composition do not readily yield to experimental manipulation. For
this reason, several models of artificial membranes and recombinant lipoproteins have been developed. Figure 2 schematically represents surfactant
micelles and microemulsion structures. These are useful models for studying
lipid aggregation phenomena. Figure 2 also illustrates the most commonly
used artificial membranes: multi-lamellar bilayer liposomes of phosphatidylcholine, which convert to single bilayer vesicles upon ultrasonic irradiation (Huang, 1969). Detergent solubilization techniques permit incorporation of protein into the vesicle bilayer (Parente and Lentz, 1984), thereby
producing a structure more characteristic of native membranes. Recombinant
models of lipoproteins are prepared by either spontaneous association of
apolipoproteins and lipids (Pownall et al., 1978) or by detergent solubilization and subsequent detergent removal through dialysis (Matz and Jonas,
1982). Figure 3 presents a bilayer disk model of recombinant HDL, which is

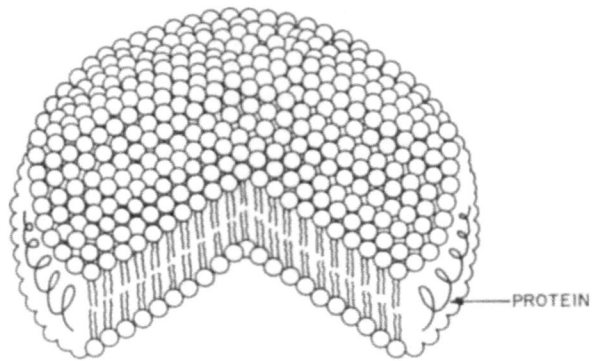

Figure 3. A bilayer disk model for recombinant high density lipoprotein, with helical protein on the circumference. The diameter is approximately 200 Å.

approximately 200 Å in diameter. Current evidence suggests that the bilayer disk is surrounded by a circumferential band of protein (Atkinson et al., 1980). The advantages of these artificial systems include their structural similarities to native lipid aggregates and the flexibility afforded the researcher in manipulating their chemical composition and size.

KINETIC MODEL FOR SPONTANEOUS LIPID TRANSFER

For some time a controversy in the literature has sought to discriminate between the "transient collisional complex" and "aqueous diffusion" models for spontaneous lipid transfer. In the transient collisional complex model, a reversible transient fusion between donor and acceptor membranes occurs, allowing for lateral diffusion of lipids between the membranes (Ferrell et al., 1985). Subsequently, the membranes separate with the resultant lipid redistribution. The aqueous diffusion model proposes spontaneous desorption of a lipid monomer from the donor membrane into the aqueous phase and its subsequent, rapid uptake by the acceptor membrane. At first glance, the collisional model holds intuitive appeal, especially in light of the limited aqueous solubility of most lipids, however, theoretical and experimental evidence shows that it is rarely applicable. Studies of surfactant transfer in micelles (Nakagawa, 1974) provided a quantitative treatment for the aqueous phase diffusion model. Nichols and Pagano (1981, 1982) extended the model for diffusion-mediated

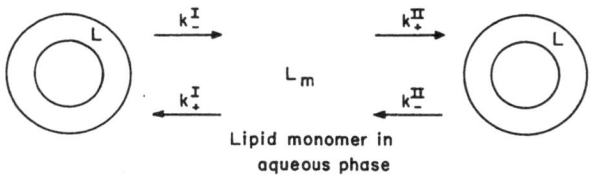

DONOR VESICLE (I)   ACCEPTOR VESICLE (II)

Figure 4. The model for lipid (L) transfer between donor vesicles (I) of phospholipid A and acceptor vesicles (II) of phospholipid B. The lipid monomer (L_m) diffuses through the aqueous phase.

transfer by considering the effects of membrane structure on desorption and uptake rates (Figure 4). Assuming that the desorption rate of lipid (L) from a membrane is proportional to its concentration on the membrane surface, whereas the adsorption rate is proportional to the product of the free monomer (L_m) and the membrane surface area, mass action kinetics yield the following rate equations for lipid transfer between donor (I) and acceptor (II) vesicles

$$\frac{d[L]_I}{dt} = \kappa_+^I [L]_m \{ S_L [L]_I + S_A [A]_I \} - \kappa_-^I [L]_I \tag{1}$$

$$\frac{d[L]_{II}}{dt} = \kappa_+^{II} [L]_m \{ S_L [L]_{II} + S_B [B]_{II} \} - \kappa_-^{II} [L]_{II} \tag{2}$$

κ_+ and κ_- correspond to the adsorption and desorption rate constants. $[L]_m, [A]_I$ and $[B]_{II}$ are, respectively, the concentrations of free lipid monomer, phospholipid A in donor vesicle I, and phospholipid B in acceptor vesicle II. S represents the surface area per mole of the corresponding lipid.

At steady-state equilibrium

$$\frac{d[L]_I}{dt} + \frac{d[L]_{II}}{dt} = 0 \tag{3}$$

In transfer experiments, the initial concentration of L in the acceptor vesicle is zero ($[L]_{II} = 0$), thereby simplifying equation (2). After algebraic manipulation, the initial rate of lipid transfer is given by

$$\nu_o = \frac{d[L]_{II}}{dt} = \frac{\kappa_+^{II}\kappa_-^{I}[L]_I[B]_{II}}{\left\{\frac{S_L}{S_B}|L|_I + \frac{S_A}{S_B}|A|_I\right\} + \kappa_+^{II}|B|_{II}} \tag{4}$$

This formidable expression can be simplified if the acceptor vesicle is present in substantial excess (at least 5-10 fold). Then,

$$\kappa_+^{II}[B] \gg \left\{\frac{S_L}{S_B}|L|_I + \frac{S_A}{S_B}|A|_I\right\} + \kappa_+^{II}|B|_{II} \tag{5}$$

and equation (4) reduces to

$$\nu_o = \kappa_-^{I}[L]_I \tag{6}$$

Equation (4) also reduces to equation (6) under roughly equal concentration of donor and acceptor vesicles, if $\kappa_+^{II} \gg \kappa_+^{I}$; thereby, underscoring the potential influence of membrane structure on the transfer rates vis-a-vis the adsorption process. The formulation arising from the events depicted in Figure 4 assume a rapid steady-state equilibrium. Storch and Kleinfeld (1986) present a solution independent of this assumption and they demonstrate that, for reasonable values of κ_+ and κ_-, the assumption is valid. Furthermore, the solution of Storch and Kleinfeld (1986) explicitly includes flip-flop reactions and further demonstrates that intermembrane aqueous diffusion transfer originates from the outer bilayer leaflet and not the inner layer.

A similar derivation of the equations for a transient collisional model would reveal a dependence on both donor and acceptor vesicle concentrations. This quadratic dependence ($[A]_I[B]_{II}$) has been reviewed by Lange (1986). Examination of equation (6) provides several experimentally useful criteria for discriminating between the transient collisional and the aqueous diffusion models. In the case of aqueous diffusion, the kinetics of transfer should be first order, independent of donor or acceptor membrane concentration, independent of acceptor membrane structure, and proportional to the aqueous solubility of the transferring lipid monomer. The transient collision model can not account for different transfer rates of various lipids, since diffusion during fusion is roughly equivalent for many lipids. If fusion does not occur during the transient collision, then lipid transfer reverts to the aqueous diffusion scenario (Ferrell et al., 1985).

FLUORESCENCE TRANSFER ASSAYS

Due, in part, to the poor solubility of lipids, the two most common
assays used in lipid transfer experiments are the radioactive label lipid
tracer and the fluorescent lipid analog. In the former, trace amounts of
radioactivity labeled lipid are incorporated into the donor membrane, which
is mixed with the acceptor membrane. At specified time intervals, aliquots
are removed from the mixture. The transfer reaction is quenched, the donor
and acceptor populations are separated, and the radioactivity level in each
group establishes the kinetics of transfer (Lange, 1986). The disadvantages
of this approach lie in the laborious separation of donor and acceptor
populations and the absence of real-time information about the course of
the reaction. In contrast, fluorescence transfer assays utilizing a
fluorescently labeled lipid analog provide immediate kinetic information
without the necessity of separating donor and acceptor membrane populations. The disadvantage of fluorescent lipid analogs is the uncertainty in
how well the analogs mimic endogenous lipid behavior. Control experiments
to classify this point compare fluorescence and radioactive tracer data,
vary the chemical identity of the extrinsic fluorophore, and compare
homologous series of fluorescently labeled lipids. The fluorescence
transfer assays fall into two broad categories: those utilizing pyrene
photophysics, and others involving quenching or energy transfer of fluorescence as a consequence of transfer. In either category, a synthetic
fluorescent lipid analog is assumed to retain the fluorophores spectroscopic properties. The fluorescence transfer assay of Nichols and Pagano
(1981, 1982, 1983), for example, uses a 7-nitro-2,1,3-benzoxadiazol-4-υω
acyl chain labeled phospholipid (NBD-PC), whose fluorescence signal is
diminished by resonance energy transfer to a "non-transferable" energy
acceptor, N-Rh-PE (head group labeled lissamine, rhodamine B sulfonyl-
phosphatidylethanolamine). For the transfer reaction measurement, the
NBD-PC is in the donor vesicle and N-Rh-PE may occupy either the donor or
acceptor vesicle (Figure 4). If N-Rh-PE resides in the donor vesicle, then
after vesicle mixing the efficiency of resonance energy transfer decreases
during NBD-PC transfer and the time course of NBD-PC fluorescence intensity
changes (increase) reflects lipid transfer. Conversely, N-Rh-PE may occupy
the acceptor vesicle and diminish NBD-PC fluorescence intensity by quenching
the signal after uptake of NBD-PC by the acceptor vesicles. Storch and
Kleinfeld (1986) use a conceptually similar resonance energy transfer
assay, although with different fluorophores, to monitor inter- and trans-
bilayer movement of long-chained, free fatty acids in bilayer vesicle
systems. The results of both research groups confirm that lipid transfer

occurs through diffusion of monomeric lipid in the aqueous phase (Figure 4).

Probably the most widely encountered fluorescence based lipid transfer assay exploits the photophysics of pyrene labeled lipid analogs (Roseman and Thompson, 1980; Doody et al., 1980; Mantulin and Pownall, 1983; Pownall et al., 1987 and references therein). Pyrene fluorescence is exceedingly long lived (~100 nsec in solution) (Birks, 1970) and, as a consequence, collisional interactions of ground and excited state pyrene (Py) molecules result in fluorescent excimers (*excited dimers*). Thus, the emission spectrum of pyrene is characterized by fluorescence originating from monomeric and excimeric species according to the following scheme:

$Py + h\nu \longrightarrow Py^*$	Light absorption
$Py^* \longrightarrow Py + h\nu'$	Monomer fluorescence
$Py^* \longrightarrow Py$	Monomer radiationless transition
$Py^* + Py \longrightarrow (Py - Py)^*$	Excimer formation
$(Py - Py^*) \longrightarrow (Py - Py) + h\nu''$	Excimer fluorescence
$(Py - Py)^* \longrightarrow Py^* + Py$	Excimer decomposition
$(Py - Py)^* \longrightarrow (Py - Py)$	Excimer radiationless transition
$(Py - Py) \longrightarrow 2Py$	Dimer decomposition

From analysis of the rate equations and diffusion theory, it follows that the ratio of excimer to fluorescence intensity is proportional to the microscopic concentration of pyrene [Py].

$$I_E/I_M = [P_y]TK/\eta \qquad (7)$$

where T, η, and K are the temperature, viscosity, and a constant, respectively. This relation has been useful in evaluating lateral diffusion rates of membrane lipids (Chong and Thompson, 1985; Mantulin et al., 1981). Figure 5 schematically demonstrates the consequences of lipid transfer in the pyrene excimer assay. When the pyrene lipid analog is compartmentalized in a lipid array its emission spectrum consists of a structured monomer fluorescence (360 to 420 nm) and a broad excimer band (centered at 475 nm), reflecting the microscopic concentration of pyrene. Mixing with an unlabeled acceptor membrane population increases the available lipid compartments for pyrene solubilization, resulting in the spectrum at the right of Figure 5. Since pyrene is poorly soluble in aqueous solution, the ratio of pyrene to membrane lipid concentrations determines the microscopic concentration of pyrene and its resulting excimer intensity. The time

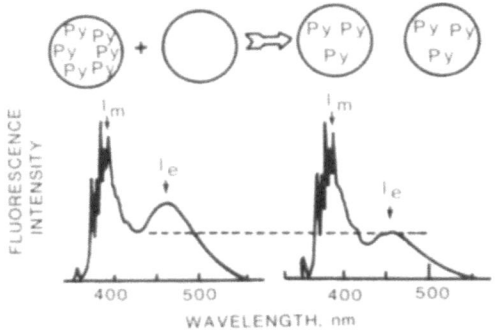

Figure 5. Schematic representation of the fluorescence transfer assay.

course of excimer intensity decay represents the kinetics of pyrene lipid analog transfer rate.

TRANSFER BETWEEN LIPID SURFACES

The early experiments of Charlton et al. (1976), which measured the transfer of pyrene between high density lipoproteins (HDL), establish several important criteria of the excimer assay. First, the microscopic concentration of pyrene in HDL is proportional to the excimer intensity and, thus, dilution results in transfer. Second, the kinetics of transfer are rapid (~3 msec) and first order. Third, the transfer rate is invariant with HDL concentration. Fourth, the transfer rate is invariant with respect to the composition or size of the acceptor species, including a control experiment with dilution transfer into the aqueous phase (i.e., no acceptor species). Fifth, desorption into the aqueous phase is the rate limiting step with subsequent uptake by acceptor species being more rapid (see Figure 4). These observations have been confirmed for a large number of different pyrene-lipid analogs: phospholipids (Massey et al., 1982; Roseman and Thompson, 1980), cholesterol (Kao et al., 1986), fatty acids (Wolkowicz et al., 1984; Doody et al., 1980), sphingomycelins (Pownall et al., 1982), diglycerides (Charlton and Smith, 1982), cerebrosides (Via et al., 1985) and other biological lipophiles (Pownall et al., 1983).

The most striking observation of these kinetic studies is that the dominant structural determinant is the hydrophobic content of the aliphatic portion of the lipid; the presence or identity of a polar head group is less pronounced. Massey et al. (1982a) measured the kinetics of transfer

123

for a homologous series of pyrene-labeled phosphatidylcholine and found a ten-fold decrease in transfer rate for the addition of each methylene unit (from 14 to 18 carbons in the aliphatic chain).

$$Py-[CH_2]_8-C(=O)-O-CH_2 ... R-O-CH_2 ... H_2C-O-P(=O)(O^\ominus)-O-CH_2-CH_2-N^\oplus(CH_3)_3$$

R = 14	1-Myristoyl-2[9'(3'-pyrenyl)nonanoyl] phosphatidylcholine (MPNPC)
R = 16	1-Palmitoyl- (PPNPC)
R = 18	1-Stearoyl- (SPNPC)
R = 18:1	2-Oleyl- (OPNPC)
R = 18:2	1-Linoleyl- (LPNPC)

In contrast, the introduction of unsaturation in the aliphatic chain (OPNPC or LPNPC) increases the transfer rate four-fold. The activation energy for transfer is high ($E_a \sim 20\text{-}24$ kcal/mol) and corresponds to an increase of +700 cal/methylene or -800 cal/double bond. To establish the effect of lipid head group on transfer rates, Massey et al. (1982b) compared to the fluorescent (MPN) glycerides: phosphatidylcholine (PC), phosphatidylserine (PS), phosphatidylglycerol (PG), phosphatidylethanolamine (PE), phosphatidic acid (PA), and diglyceride (DG). The rates changed in the following sequence DG<PE<PC~PG~PA~PS, although the effects were much smaller than those introduced by increasing hydrocarbon chain length.

Aniansson et al. (1976) proposed a model of lipid transfer rates based on lipid aqueous solubility. In amphiphilic molecules with a varying hydrophobic-hydrophilic balance, the correlation between the free energy of transfer (ΔG_t) from a hydrophobic environment to the aqueous phase and the activation free energy ($\Delta G\ddagger$) (from absolute rate theory) is schematically presented in Figure 6. The variation of $\Delta G\ddagger$ with the number of methylene units (Massey et al., 1982a; Pownall et al., 1983) correlates very well with the reported values for ΔG_t (Tanford, 1980). The activation energy barrier to transfer increases with decreasing aqueous solubility (Figure 6). For example, molecules with micromolar aqueous solubility such as pyrene, transfer readily (milliseconds), whereas more hydrophobic molecules like cholesterol (Phillips et al., 1987) or long-chain phospholipids

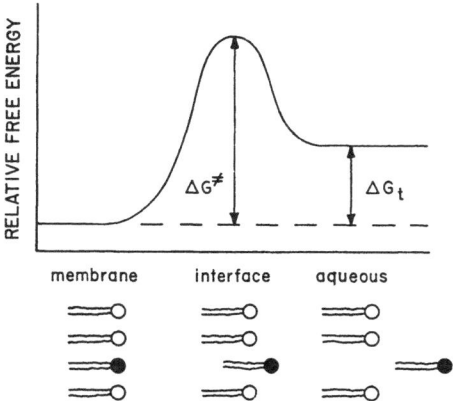

Figure 6. The activated state for lipid transfer from a membrane to the aqueous phase.

(Massey et al., 1982a,b) face a significant energy barrier to spontaneous desorption. In the activated state, the transferring lipid occupies an aqueous cavity at the lipid-water interface and is influenced by interactions with its lipid neighbors (Figure 6). Consistent with this model, $\Delta G\ddagger$ reflects the physical state of the donor lipid surface (Figure 4) and transfer is significantly slower from gel-phase, as compared to fluid liquid-crystalline membranes (McLean and Phillips, 1984). Modifying the lipid-water interface by pH changes (Doody et al., 1980) or through addition of chaotropic salts (Massey et al., 1982a), significantly and predictably changes the rates of lipid transfer.

A useful algorithm linking transfer rates of homologous lipids and their retention times by reverse-phase chromatography predicts transfer rates for other lipids in the series (Massey et al., 1984). Spontaneous desorption rates of homologous pyrene-phospholipid derivatives decreased with decreasing aqueous solubility. Partitioning of the lipids between the hydrocarbon stationary phase of the column and the mobile phase yielded a linear relationship between the logarithms of the retention time and the transfer rate. Extended retention times correlate with slow transfer. This algorithm can be a reliable indicator of whether the spontaneous transfer and/or cellular redistribution of a given lipid occurs on a physiologically important time scale.

PRESSURE PERTURBATION OF TRANSFER

In contrast to variations in temperature, which simultaneously changes

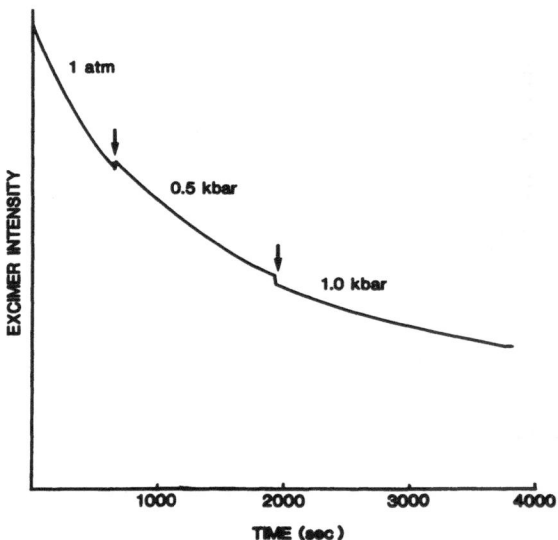

Figure 7. The effect of hydrostatic pressure on the time course of MPNPC transfer between recombinant HDL.

the thermal energy and volume of a system, pressure manipulation isolates effects due to volume. Since pressure provides a means for continuously and reversibly perturbing biological systems (Weber and Drickamer, 1983), we have conducted studies of pressure perturbation of lipid transfer reactions (Mantulin et al., 1984). Figure 7 shows the transfer kinetics of the pyrene-phospholipid analog MPNPC between recombinant, bilayer disk, HDL. At atmospheric pressure, the transfer reaction is first-order as predicted by equation 6. Subsequent application of 0.5 and 1.0 kilobar of hydrostatic pressure (Figure 7) slows the reaction, which, however, remains first-order. This result indicates that under pressure perturbation the mechanism of spontaneous transfer remains lipid monomer diffusion through the aqueous phase. The basic equation relating reaction velocity to pressure is:

$$k = k_o exp[-P\Delta V\ddagger/(RT)] \tag{8}$$

where k and k_o are the reaction rates at pressure (P) and 1 atm, respectively, and R is the gas constant. At a given temperature T, the activation volume $\Delta V\ddagger$, is calculated from the slope of the plot of $ln\ k$ against pressure (Figure 8). The activation volume is positive ($\Delta V\ddagger \sim 12.5$ ml/mol), which means the activation volume is larger than the initial

126

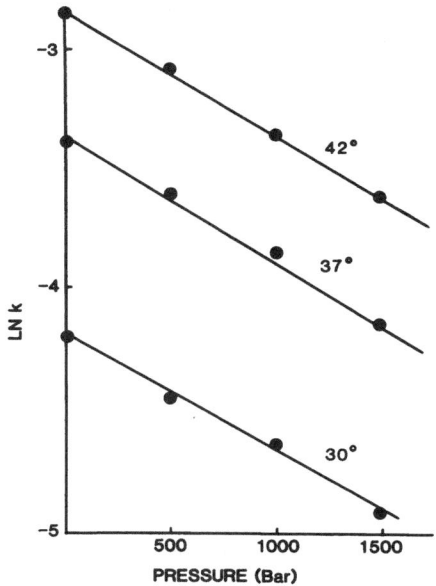

Figure 8. The pressure dependence of the transfer rate for MPNPC at 30, 37, and 42°C.

volume. In this pressure domain, neither the host lipid (POPC), nor the protein of the recombinant HDL undergoes a phase transition. Pressure does, however, compress the host lipid and the positive $\Delta V\ddagger$, in part, reflects the work necessary to overcome this effect as predicted by the transition state model of Figure 6.

To further explore how pressure perturbs the lipid-water interface and retards the transfer rate, we compared transfer rates with two different measures of fluidity in the bilayer disk of recombinant HDL. By measuring the rotational depolarization of a lipophilic fluorescence probe, diphenyl-hexatriene (DPH), Chong and Weber (1983) demonstrated the compensatory effects of temperature and pressure on lipid bilayer fluidity. Under different conditions of temperature and pressure the same DPH fluorescence polarization value is obtained, reflecting the same state of fluidity in the bulk lipid bilayer. We measure the same compensatory relation of temperature and pressure in recombinant HDL. However, comparing MPNPC transfer rates at given (T,P) couples with fluidity values (DPH polarization) at those same (T,P) conditions reveals a non-equivalence (Figure 9). The same fluidity of the donor HDL does not result in the same transfer rate. The excimer to monomer intensity ratio of pyrene (equation 7), which

127

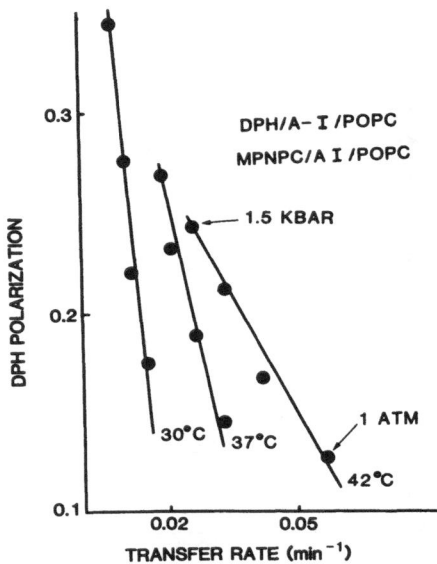

Figure 9. Non-equivalence of membrane fluidity, as assessed by DPH polarization, and MPNPC transfer rates between recombinant HDL at various temperatures (30, 37, and 42°C) and pressures (1 atm and 0.5, 1.0, and 1.5 kbar).

reflects lateral diffusion in the bulk lipid, provides a different assay for membrane fluidity. Measuring this ratio at various (T,P) couples for MPNPC in recombinant HDL reveals that common fluidity values exist under different compensating T-P conditions. However, comparison of excimer/-monomer ratios with transfer rates (Figure 10) again reveals a non-equivalence. This observation lends itself to the conclusion that the lipid desorption model of Figure 6, is not dependent on the state of fluidity of the bulk lipid in any simple, linear fashion.

CONCLUSIONS AND PERSPECTIVES

The application of fluorescent lipids to the study of lipid transfer reactions both *in vitro* and *in vivo* has provided valuable correlations between structure and function. New developments in fluorescent microscopy, with digital image enhancement (Benson et al., 1984; Jacobson, et al., this volume), should produce considerable progress in studying recycling of lipids between the plasma membrane and cell interior, translocation of lipids between intracellular compartments and observations of the metabolic fate of a given lipid in a living cell.

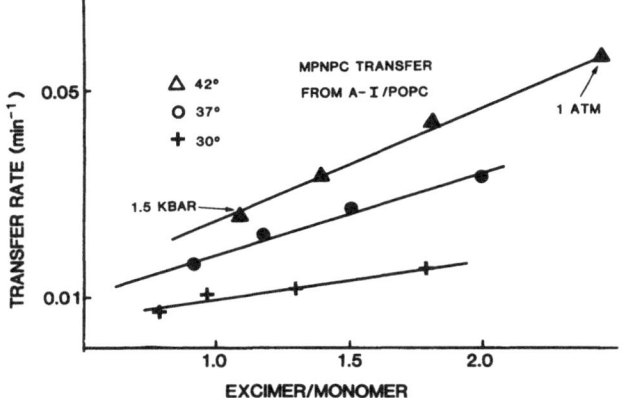

Figure 10. Non-equivalence of membrane fluidity, as assessed by the excimer to monomer ratio of MPNPC, and MPNPC transfer rates between recombinant HDL at various temperatures (30, 37, and 42°C) and pressures (1 atm and 0.5, 1.0, and 1.5 kbar).

ACKNOWLEDGEMENTS

This work was supported by NIH Grant RR03155 and a grant from the American Heart Association, Illinois Affiliate. The assistance of Julia Butzow in manuscript preparation and Nancy Beechem in artwork is acknowledged.

REFERENCES

Aniansson, E.A.G., Wall, S. N., Almgren, M., Hoffmann, H., Kielmann, I., Ulbricht, W., Zana, R., Lang, J., and Tondre, 1976, Theory of the Kinetics of Micellar Equilibria and Quantitative Interpretation of Chemical Relaxation Studies of Micellar Solutions of Ionic Surfactants, J. Phys. Chem., 80:905.

Atkinson, D., Small, D. M., and Shipley, G. G., 1980, X-ray and Neutron Scattering Studies of Plasma Lipoproteins, Annals of the New York Acad. Sci., 348:284.

Birks, J. B., 1970, Photophysics of Aromatic Molecules, Chapter 7, in: "Excimers", Wiley-Interscience, New York.

Brown, M. S., and Goldstein, J. L., 1986, A Receptor-mediated Pathway for Cholesterol Homeostasis, Science, 232:34.

Charlton, S. C., and Smith, L. C., 1982, Kinetics of Transfer of Pyrene and

rac-1-Oleyl-2-[4-(3-pyrenyl)butanoyl]glycerol between Human Plasma
Proteins, <u>Biochemistry</u>, 21:4023.

Charlton, S. C., Olson, J. S., Hong, K-Y., Pownall, H. J., Louie, D. D.,
and Smith, L. C., 1976, Stopped Flow Kinetics of Pyrene Transfer
between Human High Density Lipoproteins, <u>J. Biol. Chem.</u>, 251:7952.

Chong, P.L-G., and Thompson, T. E., 1985, Oxygen Quenching of Pyrene--Lipid
Fluorescence in Phosphatidylcholine Vesicles: A Probe for Membrane
Organization, <u>Biophys. J.</u>, 47:613.

Chong, P.L-G., and Weber, G., 1983, Pressure Dependence of 1,6-Diphenyl-
1,3,5-hexatriene Fluorescence in Single-component Phosphatidylcholine
Liposomes, <u>Biochemistry</u>, 22:5544.

Doody, M. C., Pownall, H. J., Kao, Y. J., and Smith, L. C., 1980, Mechanism
and Kinetics of Transfer of a Fluorescent Fatty Acid between Single-
walled Phosphatidylcholine Vesicles, <u>Biochemistry</u>, 80:108.

Ferrell, J. E. Jr., Lee, K-J., and Heustis, W. H., 1985, Lipid Transfer
between Phosphatidylcholine Vesicles and Human Erythrocytes:
Exponential Decrease in Rate with Increasing Acyl Chain Length,
<u>Biochemistry</u>, 24:2857.

Gianturco, S. H., and Bradley, W. A., 1987, Lipoprotein Receptors, <u>in</u>:
"Plasma Lipoproteins", A. M. Gotto, Jr., ed., Elsevier Publishing
Company, New York.

Huang, C., 1969, Studies on Phosphatidylcholine Vesicles, <u>Biochemistry</u>,
8:344.

Jonas, A., 1987, Lecithin Cholesterol Acyltransferase, <u>in</u>: "Plasma Lipoproteins", A. M. Gotto, Jr., ed., Elsevier Publishing Company, New York.

Kao, Y. J., Doody, M. C., and Smith, L. C., 1986, Transfer of Cholesterol
and a Fluorescent Cholesterol Analog, 3'-pyrenylmethyl-23,24-*dinor*-5-
cholen-22-oate-3β-01, between Human Plasma High Density Lipoproteins,
<u>J. Lipid Res.</u>, 27:781.

Lange, Y., 1986, Molecular Dynamics of Bilayer Lipids, <u>in</u>: "The Physical
Chemistry of Lipids", D. M. Small, ed., Plenum Press, New York.

Mantulin, W. W., Gotto, A. M., Jr., and Pownall, H. J., 1984, The Effect of
Hydrostatic Pressure on the Transfer of a Fluorescent Phosphatidylcho-
line between Apolipoprotein—Phospholipid Recombinants, <u>J. Amer. Chem.
Soc.</u>, 106:3317.

Mantulin, W. W., Massey, J. B., Gotto, A. M., Jr., and Pownall, H. J.,
1981, Reassembled Model Lipoproteins, <u>J. Biol. Chem.</u>, 256:10815.

Mantulin, W. W., and Pownall, H. J., 1983, Plasma Lipoproteins: Fluorescence as a Probe of Structure and Dynamics, <u>in</u>: "Excited States of
Biopolymers", R. F. Steiner, ed., Plenum Press, New York.

Massey, J. B., Gotto, A. M., Jr., and Pownall, H. J., 1982a, Kinetics and

Mechanism of the Spontaneous Transfer of Fluorescent Phosphatidylcholines between Apolipoprotein—Phospholipid Recombinants, Biochemistry, 21:3630.

Massey, J. B., Gotto, A. M., Jr., and Pownall, H. J., 1982b, Kinetics and Mechanisms of the Spontaneous Transfer of Fluorescent Phospholipids between Apolipoprotein—Phospholipid Recombinants: Effect of the Polar Headgroup, J. Biol. Chem., 257:5444.

Massey, J. B., Hickson, D., She, H. S., Sparrow, J. T., Via, D. P., Gotto, A. M., Jr., and Pownall, H. J., 1984, Measurement and Prediction of the Rates of Spontaneous Transfer of Phospholipids Between Plasma Lipoproteins, Biochim. Biophys. Acta, 794:274.

McLean, L. R., and Phillips, M. C., 1984, Kinetics of Phosphatidylcholine and Lysophosphatidylcholine Exchange between Unilamellar Vesicles, Biochemistry, 23:4624.

Nakagawa, T., 1974, Critical Examination of Published Data Concerning the Rate of Micelle Dissociation and Proposal of a New Interpretation, Colloid and Polymer Sci., 252:55.

Nichols, J. W., and Pagano, R. E., 1983, Resonance Energy Transfer Assay of Protein-mediated Lipid Transfer between Vesicles, J. Biol. Chem., 258:5368.

Nichols, J. W., and Pagano, R. E., 1982, Use of Resonance Energy Transfer to Study the Kinetics of Amphiphile Transfer between Vesicles, Biochemistry, 21:1720.

Nichols, J. W., and Pagano, R. E., 1981, Kinetics of Soluble Lipid Monomer Diffusion between Vesicles, Biochemistry, 20:2783.

Pagano, R. E., and Sleight, R. G., 1985, Defining Lipid Transport Pathways in Animal Cells, Science, 229:1051.

Parente, R. A., and Lentz, B. R., 1984, Phase Behavior of Large Unilamellar Vesicles Composed of Synthetic Phospholipids, Biochemistry, 23:2353.

Phillips, M. C., Johnson, W. J., and Rothblat, G. H., 1987, Mechanisms and Consequences of Cellular Cholesterol Exchange and Transfer, Biochim. Biophys. Acta, 906:223.

Pownall, H. J., Hickson, D., Gotto, A. M., Jr., and Massey, J. B., 1982, Kinetics of Spontaneous and Plasma-stimulated Sphingomyelin Transfer, Biochim. Biophys. Acta, 712:169.

Pownall, H. J., Hickson, D. L., and Smith, L. C., 1983, Transport of Biological Lipophiles: Effect of Lipophile Structure, J. Amer. Chem. Soc., 105:2440.

Pownall, H. J., Homan, R., and Massey, J. B., 1987, Pyrene-labeled Amphiphiles: Dynamic and Structural Probes of Membranes and Lipoproteins, SPIE, 743:137.

Pownall, H. J., Massey, J. B., Kusserow, S. K., and Gosso, A. M., Jr., 1978, Kinetics of Lipid—Protein Interactions: Interaction of Apolipoprotein A—I from Human Plasma High Density Lipoproteins with Phosphatidylcholines, Biochemistry, 17:1183.

Roseman, M. A., and Thompson, T. E., 1980, Mechanism of the Spontaneous Transfer of Phospholipids between Bilayers, Biochemistry, 19:439.

Storch, J., and Kleinfeld, A. M., 1986, Transfer of Long-chain Fluorescent Free Fatty Acids between Unilamellar Vesicles, Biochemistry, 25:1717.

Tall, A., Swenson, T., Hesler, C., and Granot, E., 1987, Mechanisms of Facilitated Lipid Transfer Mediated by Plasma Lipid Transfer Proteins, in: "Plasma Lipoproteins", A. M. Gotto, Jr., ed., Elsevier Publishing Company, New York.

Tanford, C., 1980, in: "The Hydrophobic Effect", 2nd Edition, Wiley Interscience, New York.

Weber, G., and Drickamer, H. G., 1983, The Effect of High Pressure Upon Proteins and Other Biomolecules, Quart. Rev. Biophys., 16:89.

Wolkowicz, P. E., Pownall, H. J., Pauly, D. F., and McMillin-Wood, J. B., 1984, Pyrenedodecanoylcarnitine and Pyrenedodecanoyl Coenzyme A: Kinetics and Thermodynamics of Their Intermembrane Transfer, Biochemistry, 23:6426.

THOUGHTS AND HINTS ON POSSIBLE VIOLATION OF PARITY IN THE BIOLOGICAL MATTER

Meir Shinitzky

Department of Membrane Research
The Weizmann Institute of Science
Rehovot, Israel

My numerous visits to Urbana persistently followed an enchanting pattern which began with a hearty welcome and chauffeuring by Gregorio to his apartment where I stayed. In the evenings, after the ritual shopping at the local supermarket, where we carefully selected sirloin steaks, broccoli and other seasonal vegetables, each of us turned to his expertise in the preparation of supper. In contrast to the meticulously prepared dinner, we consumed the food somewhat detached from each other without much talking. A long night of spontaneous intellectual encounter was awaiting us, it seemed as if we were preparing ourselves.

In one of these wonderful nights, in the serene atmosphere indulged by the tender notes of Schumann's Kinderscenen, we pondered into the mysterious selectivities of asymmetry observed in nature. The chirality of biological molecules came up immediately which triggered a stream of ideas and arguments that we could hardly control. Since that night I became possessed by thoughts on this fascinating subject and both Gregorio and I have begun searching for experimental approaches which may allow us to address specific issues. The following article, which is dedicated to my eminent teacher and collaborator, Prof. Gregorio Weber, describes my starting points in this virgin area.

DISTRIBUTION OF CHIRAL STRUCTURES

Parity in the molecular level is confined, almost exclusively, to spatial asymmetry - i.e. chirality. Two chiral isomers are of opposite "handedness" which renders their structure mirror image to each other. From all conventional aspects the two chiral isomers are identical in chemical and physical properties and have the same intrinsic probability to be synthesized from a non-chiral precursor. In contrast to this basic axiom, nature around us persistently displays preferences, which in some instances reaches complete exclusivity, for one parity over the other (for a comprehensive and artistic account on parity see Gardner, 1979). The most striking example for this selectivity is obviously the chiral biological molecules like the amino acids, nucleotides and sugars. It is currently believed that these molecules originated sporadically and independently at various places on earth and therefore the reason for the sharp selectivity of one type of symmetry through evaluation is still obscure. A somewhat trivial account for it could be that the first biological molecules were synthesized on chiral templates (and how were they originated?) or induced by a physical signal with handedness (e.g. photochemical reaction preceded by absorption of circularly polarized light). These possibilities can explain how handedness of biological molecules might have originated but cannot explain the exclusive preference of one type of chirality over the other.

The discrimination of one 3-dimensional asymmetric configuration over its mirror image extends even to large assemblies of non-chiral molecules. Quartz crystals, for example, are polymers of silicon oxide which can be arranged in either left handed or right handed helical configuration, generally termed as L and D quartz. Yet, the abundance of L-quartz crystals in nature is significantly greater than that of D-quartz. As another example, the common spiral sea shells which are composed of inorganic material (predominantly calcium phosphate) are all, without exception, of a right handed helical structure. The handedness here might be rationalized by the fact that the shell is assembled by a living organism where, on the microscopic scale, the building material is excreted from asymmetric structures. This explanation is obviously unsatisfactory.

In a recent conference entitled "Chiral symmetry breakings in physics, chemistry and biology" (Rouen 17-19 June, 1986) the up-to-date views on the origin of the imbalance in distribution of chiral symmetry were presented (reviewed in Konderpudi and Nelson 1983). A few possibilities related to the violation of parity in β-decay and in the handedness of electron spin have been suggested to be involved in such symmetry breaking. However, in all of these, the statistical difference in the evolution of one optical isomer over the other differs by only a tiny fraction. Only a putative autocatalytic effect could amplify this difference to the level found in nature (MacDermott, 1986). One such autocatalytic mechanism, which has not yet been considered, is that an assembly of identical chiral elements forms a macroscopic chiralic structure where the marginal difference between the individual isomers is amplified to a level which could be decisive as to their eventual distribution during evolution. This possibility and its far-reaching implications are discussed below.

CHIRAL MOVEMENTS AND SIGNALS

Many autonomous rotational movements in living organisms are also of a selective asymmetric mode, while their mirror image is virtually excluded. These include the helicoidal movement of the bacterium flagella or sperm tail and the right handed rotation about the polar lobe of eggs from sea mollusks immediately after fertilization. To some extent such asymmetric rotations could be rationalized by the torque induced by the underlying superhelical muscle fibers or even by vectorial liberation of thermal energy. Again, it seems that a full account for the mechanism underlying such asymmetric movements should be based on a fundamental property of an assembly of chiral molecules like that of the biological tissue.

Translational propagation of molecules or ions in a biological matrix may also be endowed with a chiralic twist. This would introduce a series of informational degrees of freedom which have not yet been explored or even conceived. If a conductance of a nerve impulse, for example, would propagate in a chiralic helicoid mode its handedness and the magnitude of the turns might store a subtle information which could not be resolved by any ordinary electrophysical means. In the central nervous system such chiral electrophysiological signals might provide a spectrum of sensualities (e.g. intellectual or emotional) which are hard to express in pure physical terms.

OPTICAL ACTIVITY OF AMINO ACIDS

Optical rotation and its various spectral techniques (circular dichroism or optical rotatory dispersion) are the conventional means for analyzing chiral molecules in the liquid state or in solution. The magnitude of optical rotation of the polarized Na line serves as the ultimate analytical value for verifying the purity of an optical isomer. This parameter can be measured with high accuracy and is dependent on concentration, solvent and temperature. When these factors are kept constant one should expect that two enantiomers would yield an identical value of optical rotation with an opposite sign. Surprisingly, the absolute values of optical rotation of D and L amino acids appearing in handbooks or commercial catalogs do not always match. As an extreme example, according to Aldrich catalog the $[\alpha]_D^{26}$ value of histidine hypochloride monohydrate of high purity measured at 2% solution in HCl is -15° for the D isomer and +8.0° for the L isomer - a difference which is very hard to reconcile.

Differences of a smaller magnitude in specific rotation between D and L amino acids are quite common.

SEC-BUTANOL

The D and L isomers of sec-butanol are of the simplest cases of chiral molecules and can be obtained from commercial sources in a high purity (around 99%). We have used D and L-sec-butanol for searching some deviations from symmetrical behavior in optical activity. The purity of the L-isomer, as analyzed by gas chromatography was 99.1% while that of the D-isomer was 97.9%. The measured optical rotation of the neat liquid was +8.6° for the D-isomer and -9.8° for the L-isomer (Na line, 23°, 10 cm optical path). At the UV range the circular dichroism spectra of D and L-sec-butanol were markedly different in amplitude, as shown in Figure 1. Upon dilution with water this difference diminished (see insert in Figure 1). One trivial cause for this deviation could be ascribed to a molecular association between an impurity and butanol to yield a high increase (in the case of the D-isomer) or decrease (in the case of the L-isomer) in apparent elipticity. The difference is partially reduced upon dilution due to dissociation of such complexes. Yet, this type of interaction is expected to contribute a similar deviation in optical rotation at the

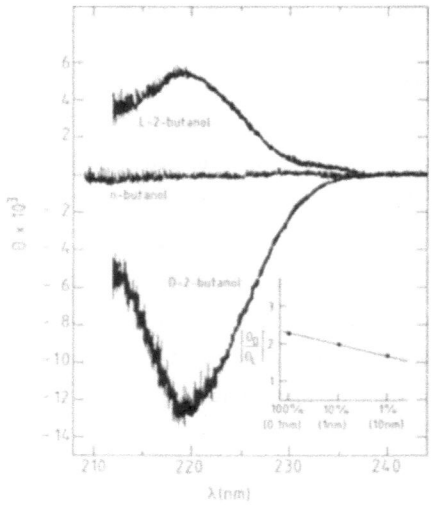

Figure 1. Circular dichroism spectra at 23° (presented as the apparent ellipticity, θ, in 0.1 mm optical path) of L-2-butanol, D-2-butanol and n-butanol (taken as a baseline). The insert represents the absolute value of the ellipticity ratios at 219 nm of the neat liquids, their 10% solution (recorded in 1 mm optical path) and 1% solution (recorded in 10 mm optical path) in water.

visible range, which is not the case. Since the experiment was carried out as a comparison between the two isomers under the same conditions, all other trivial possibilities (e.g. solvent effect) are self excluded.

If the results presented in Figure 1 are indeed real, they indicate a fundamental violation of parity in D or L-sec-butanol which emerges at high concentrations. It seems that this deviation in symmetry is apparent only at high concentration but it could also be that the deviation from symmetry is already inherent in the isolated molecules and is only markedly amplified upon increase in concentration, a possibility which is hard to verify by the conventional optical means.

DISCUSSION AND SPECULATION

From the standpoint of thermodynamics and molecular energetics two enantiomers are identical with respect to all possible intra- and inter-molecular interactions. This is clearly reflected in the identical thermo-dynamic parameters of the neat isomers (e.g. melting point, boiling point and heat of vaporation) and of the isomers in solution (e.g. absorption spectra, heat of solvation and rate of chemical reactions). The marginal difference which may be induced by the tiny preference of electron spin momentum (MacDermott, 1986) is much too small for being of any direct implication. Therefore, if there is any inherent difference between two enantiomers in either the isolated form or in an assembly, it would imply a basic structural feature which has not yet been discovered. The conven-tional theories of classical and quantum mechanics do not offer any trait or peculiarity in the symmetry distribution of chiralic centers. Therefore, at the current state of the art we are bound to speculate on possible extension of existing axioms in an attempt to account for the observations presented here.

At the atomic and sub-atomic level particles (e.g. electrons) "communicate" via their eigen-function. Fermi-Dirac particles (e.g. particles with a spin of 1/2) possess an anti-symmetric wave function and an assembly of such particles is presented by a cumulative wave function which is also anti-symmetric, and each particle there has a distinct identity. Any exchange between the particles inside the domain of the assembly is "sensed" through the change in sign of the eigen-function. Furthermore, in a collection of such assemblies each of them should preserve a distinct identity. Now, let us assume that a chiralic center is

presented by a virtual particle ("chiron") of either one of two parities, L
and D. In other words, the center of the 3-dimensional asymmetric distribution of electron cloud or atomic bonds can be presented by the respective L
or D-chiron. Because of their opposite parity the chirons should comply
with the Fermi-Dirac statistics. Accordingly, in a collection of assemblies of one kind of chirons each of them should preserve its intrinsic
identity. The question that should be considered is whether this power
of preservation of identity is different for an assembly of D-chirons, and
whether this could lead to any statistical difference between such assemblies. The analogous question for electrons would be a comparison between
an assembly of electrons with spin +1/2 to an assembly of electrons with
spin -1/2. However, this situation is physically irrelevant since in
practice, the preservation of spin identity has only a statistical meaning.
Nevertheless, the solution of this basic question could shed some light on
the hypothesis suggested above.

Turning to the biological matter, proteins, nucleic acids and other
biopolymers might be considered as assemblies of chirons. As such, these
polymers are, according to the above hypothesis, expected to have some
basic "advantage" over their mirror image (e.g. a protein made of D-amino
acids). This "advantage" is hard to define in pure physical terms but it
predicts an intrinsic unique feature which may contribute to their apparent
biological and physiological properties. Furthermore, this advantage
implies that there is a basic difference in properties between polymers of
D or L chirons which might extend to a whole organism, so that a D and L
worlds are not exactly a mirror image of each other.

REFERENCES

Gardner, M., 1979, The Ambidextrous Universe, Scribner's Sons, New York
Konderpudi, D. K., and Nelson, G. W., 1983, Chiral Symmetry Breaking in
     Nonequilibrium Systems, Phys. Rev. Letters, 50:1023.
MacDermott, A. J., 1986, Distinguishing True Chirality from its Accidental
     Imitators, Nature, 323:16.

DIGITIZED FLUORESCENCE MICROSCOPY IN CELL BIOLOGY APPLICATIONS

Ken Jacobson,[*+] Akira Ishihara[*], Bruce Holifield[*],
and James DiGuiseppi

Laboratories for Cell Biology[*], Department of Anatomy
and Cancer Research Center[+]
School of Medicine
University of North Carolina at Chapel Hill
Chapel Hill, NC 27514 USA

INTRODUCTION

Fluorescence microscopy is a widely used technique for studying the
location and amount of certain components on or within cells. Specific
cellular components can be visualized by linking fluorescent dyes to
exogenous macromolecules such as antibodies (immunofluorescence) and
ligands, or by producing fluorescent analogs of small molecules normally
found within cells. In the past, these approaches have involved fixation
of cells and the use of high intensity excitation to image the specimen and
produce photographs. More recently, low light level video cameras coupled
with frame rate video digitizers have been employed to capture cellular
images from the fluorescence microscope. The current and growing appeal of
digitized fluorescence microscopy (DFM) in cell biology lies in its ability
to quantitatively image the distribution of single classes of macromolecules, molecules and ions in single living cells. In this sense, it is
really a new form of optical microscopy for living cells in which the image
produced gives directly the distribution of the component probed for. In
other forms of live cell light microscopy, we must infer, from other information, the composition of the structure visualized in the image, since we
have no a priori knowledge of the molecular structure producing contrast.

DFM allows one to achieve maximum sensitivity thereby minimizing the
exposure of the sample to the exciting light. This is desirable for at
least two reasons: first, "fading" or photobleaching of the image which is
directly related to excitation intensity and which gradually erodes the

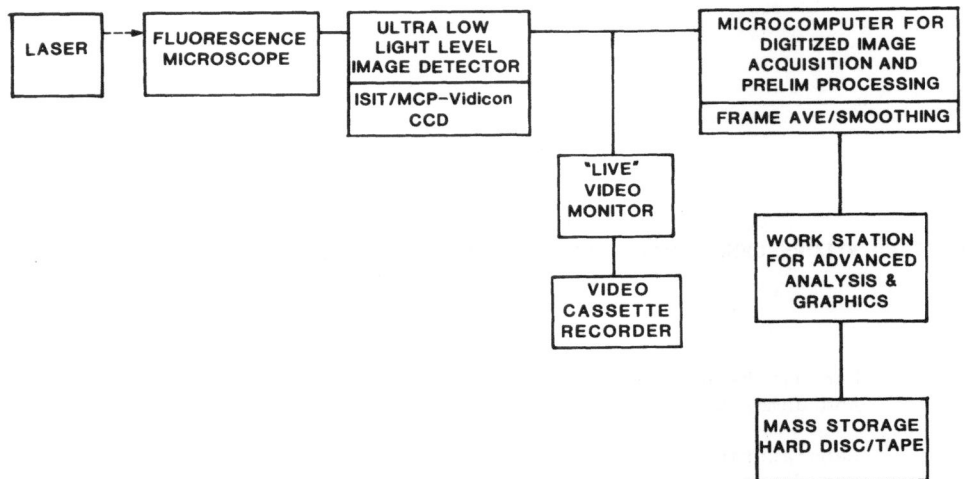

Figure 1. DIGITIZED FLUORESCENCE MICROSCOPY
Block diagram of digitized fluorescence microscopy instrument. Laser at
left is optional and can be used for image photobleaching (video FRAP)
studies.

image intensity is minimized; second, any phototoxic effects which may
alter or abrogate the living cell activity under study are reduced.

INSTRUMENTATION

Most existing DFM systems are approximately described by the very
simplified block diagram shown in Figure 1. Laser excitation for producing
fluorescence images as for photobleaching experiments is optional. More
detailed schematic diagrams are given in recent publications (see, for
example, MacGregor, et al. (1984); DiGuiseppi, et al. (1985); Arndt-Jovin
et al. (1985); Fay et al. (1986). Basically, the systems combine an
extremely sensitive video camera with a frame rate video digitizer and
associated circuitry capable of performing some preliminary processing such
as frame (time) averaging or spatial smoothing. This circuitry is generally located within an on-line host microcomputer. In such preliminary
processing operations, the image signal to noise ratio is improved at the
expense of either temporal or spatial resolution, respectively. Images may
then be passed to a more powerful computer for advanced image analysis
and/or sophisticated graphic display. Finally, the systems must have a
mass storage system to archive images.

The front end of the system, the camera, is the most crucial component as it converts the flux of photons emitted by the fluorescing specimen into an image for subsequent processing. Thus, its characteristics in terms of sensitivity, spatial fidelity, and temporal response are extremely important. A number of cameras are being employed (see Arndt-Jovin et al, (1985) for review). At the lower end of the price range, SIT (silicon intensifier target) tubes and ISIT (intensifier silicon intensifier target) tubes are available. Also available are various combinations of first (electrostatic) or second (microchannel plate) generation image intensifiers coupled to vidicons, newvicons or charge coupled devices CCD). Cooled CCD cameras having chips with very low read noise promise to provide excellent low light level images in terms of a linear response, lack of spatial distortion and wide dynamic range.

Although the dark adapted eye is a very good detector, an approximate psychometric experiment performed in the Video Center of the Department of Anatomy at the University of North Carolina shows the advantage of a low light level imaging device in recognizing extremely faint scenes. In this experiment, a number of individuals were asked whether they could detect various dim light levels or a weakly illuminated test specimen (stage micrometer). The subjects were partially dark adapted since they waited in the darkened microscope laboratory for 5 minutes before the test. It is apparent from the results shown in Figure 2 that while the human eye-brain combination, under these conditions, can perceive very low illumination levels (Figure 2, top panel), it cannot compete with the ISIT camera coupled with frame averaging in recognizing objects illuminated at very low levels. In an approximate sense, this experiment demonstrates the advantage of such a detector-computer combination in the work to be described. Furthermore, even allowing for the sensitivity of the fully dark adapted human eye, photographically recording one, let alone a sequence of, such low light level images becomes a formidable technical task.

SURVEY OF SOME CURRENT APPLICATIONS

Several recent examples reflect the variety of problems which can be approached by these techniques (for review, see Arndt-Jovin et al. 1985). The range of these applications is schematically illustrated in Figure 3. Important information on the three-dimensional disposition of structures within the cell is offered by performing a through-focus series of optical sections of cells fixed and stained with a fluorescent probe for the

141

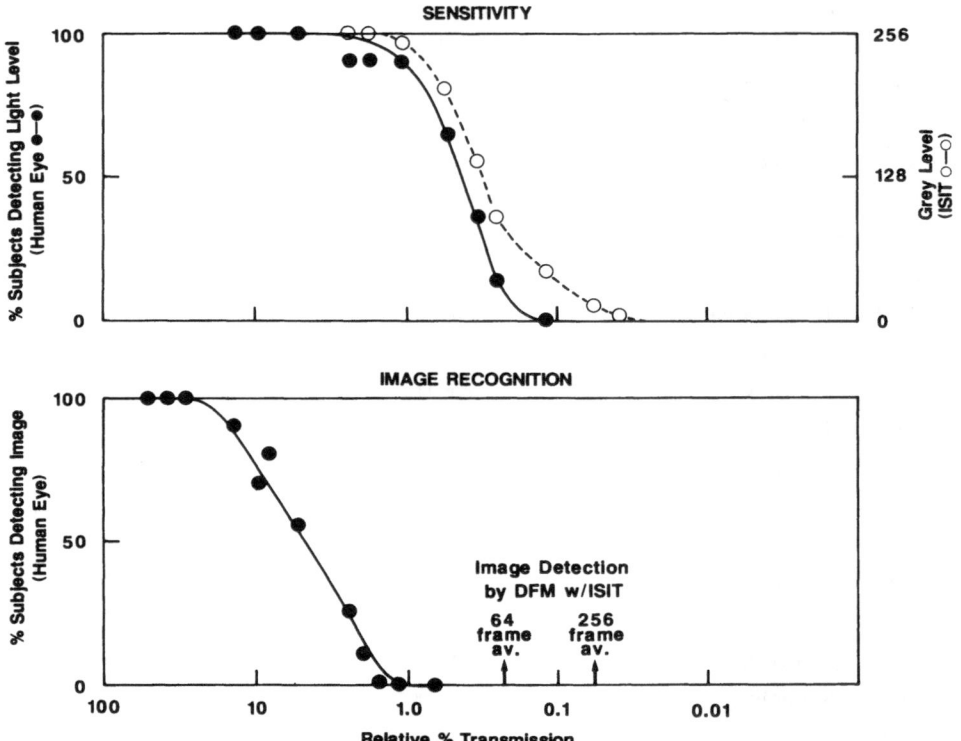

Figure 2. Sensitivity of low light level camera (Dage Intensifier Silicon Intensifier Target (ISIT Tube) employed in conjunction with a video digitizing computer compared to the human eye. Random light levels or test scenes at various illumination levels were viewed either through the microscope or on a TV monitor by ten human subjects dark adapted for about 5 min. The fraction of subjects detecting the level or scene was recorded. Top panel: sensitivity to uniform light level. For camera detection, 100% transmission set to gray level 256 (maximum signal). Bottom panel: detection of test scene, a stage micrometer. Arrows on abscissa indicate at what relative intensity test pattern can be seen using the stated number of frames to compose the averaged image.

particular component of interest. The contamination of the image of each section with out-of-focus emission from other parts of the specimen is removed by sophisticated computation. Finally, the three-dimensional structure of the entire object is reconstructed and viewed from various angles using computer graphic techniques (see Arndt-Jovin et al. 1985, for examples). The spatial disposition of α-actinin plaques in smooth muscle cells has been studied in this way (Fay et al, 1986). These techniques have also been used to study the 3-D structure of the polytene chromosome

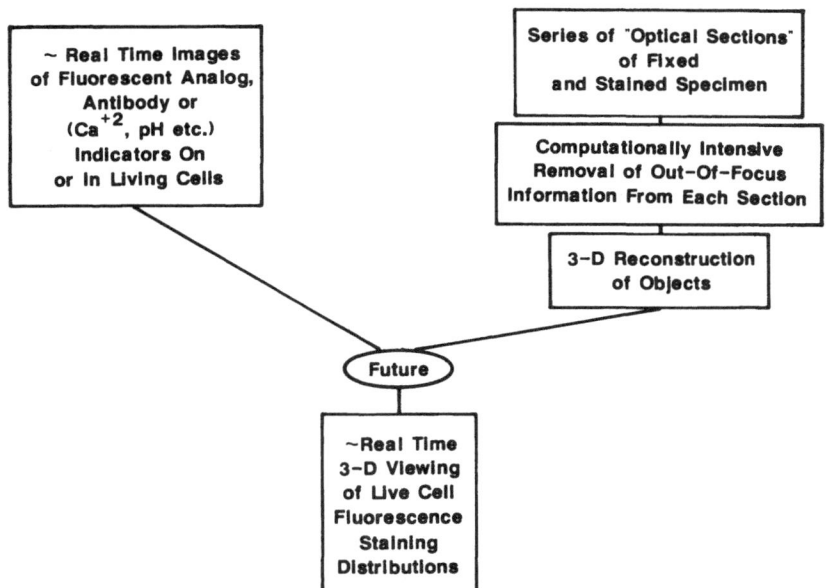

Figure 3.   BRANCHES OF DIGITIZED FLUORESCENCE MICROSCOPY
Evolution of the applications of digitized fluorescence microscopy.

in <u>Drosophila</u> (Mathog et al., 1984).

As indicated in the Introduction, DFM is proving to be invaluable in gathering images of the fluorescence staining patterns of living cells. Cell surface components such as the low density lipoprotein receptor (Webb and Gross, 1986) and a major plasma membrane glycoprotein, GP80 (see below), can be imaged in living cells by employing ligands or antibodies directed toward the particular membrane component. Cytoskeletal components can likewise be imaged in living cells using fluorescent analogs of actin, tubulin, etc. (Taylor and Wang, 1980).

Calcium and pH sensitive fluorescent dyes have been used to monitor the intracellular variation in free calcium or proton concentrations in response to a variety of cellular stimuli including, for example, growth stimulating hormones. Calcium chelating dyes such as FURA-2 (Grynskiewicz, 1985; Waggoner, 1986) are loaded into cells as membrane permeant ester. The ester groups are rapidly cleaved by cytoplasmic esterases liberating charged carboxylate groups to bind Ca^{++} and inhibiting escape of dye back into the medium. The excitation spectra of these dyes is typically altered upon chelation of Ca^{++}. Considering the FURA dyes for example, the excitation spectrum drops at 380 nm while it increases at 350 nm as increasing amounts of Ca^{++} are bound. Thus, a ratiometric measurement at these excitation wavelengths can be used to monitor cytoplasmic free Ca^{++} independent of the amount of indicator dye trapped by the cells and the cell thickness.

Eventually, the two branches indicated in Figure 3 will be combined to yield the three dimensional distributions of various fluorescently labeled cellular constituents in real time.

Translational transport can be quantitated for molecules in membranes and in the cytoplasm cells using the Fluorescence Recovery After Photobleaching (FRAP) techniques (for brief review, see Jacobson et al., 1983). These methods are based on photobleaching discrete regions of specimens bearing fluorescent molecules to destroy their emission and then measuring the recovery of fluorescence due to diffusive or flow influx of unbleached fluorophores from the surrounding reservoir into the irradiated area. The photobleached region can be a single spot, a line, or a more complex pattern. The kinetics of recovery are related to the diffusion coefficient (or flow velocity) of the labeled molecules, and the degree to which the fluorescence regains the prebleach value gives the mobile fraction of the measured population of probe molecules. Imaging methods combined with photobleaching can produce more detailed information on such transport processes. In particular, deviations from purely isotropic diffusion produced by anisotropic diffusion or flow processes can be detected.

In spot photobleaching, fluorescence detection is normally done by a photomultiplier, so that unless the laser beam (Koppel, 1979) or stage is scanned, no spatial information on the recovery process can be obtained. In the "video-FRAP" or "image photobleaching" technique, the photomultiplier is replaced by a low light level image detector which is calibrated to operate as a time-resolved spatial photometer (Kapitza et al., 1985). Spatial information on the fluorescence redistribution following photobleaching process is thus obtained. Images are put in quantitative form by plotting the array of intensity values in the form of isointensity contour maps. For example, in anisotropic diffusion deviations occur from the circular isointensity contour lines which characterize a fluorescence recovery after photobleaching dominated by isotropic diffusion. Recovery via a flow in the membrane or cytoplasm will be characterized by translation of the centroid of the bleached region. Recent advances in the analysis of such data may provide more quantitative information. The spatial Fourier transform of the image after photobleaching can be calcu-

144

lated, and each spatial frequency of this transform can be "aged" by multiplying it by a factor that describes the suspected lateral transport process (e.g. isotropic diffusion, anisotropic diffusion, flow). The effect of aging by a known process can then be compared to the redistribution measured after a specific time, by inverting the "aged" transform to give back the stimulated image at that time. The parameters describing the transport process are then adjusted to give an optimum fit of the simulated and experimental images.

REDISTRIBUTION OF PLASMA MEMBRANE COMPONENTS DURING CELL LOCOMOTION

As pointed out above, DFM is a useful tool for studying changing distributions of cellular components during physiological function. In our laboratory, we have been investigating the lateral redistribution of a membrane glycoprotein during mouse fibroblast locomotion across substrates in tissue culture. Fibroblasts in culture crawl slowly over their substratum; the mechanism and regulation of this form of motility remains largely unknown. Cell locomotion is accompanied by changes in the distribution of certain plasma membrane components. Flow of membrane as the cell

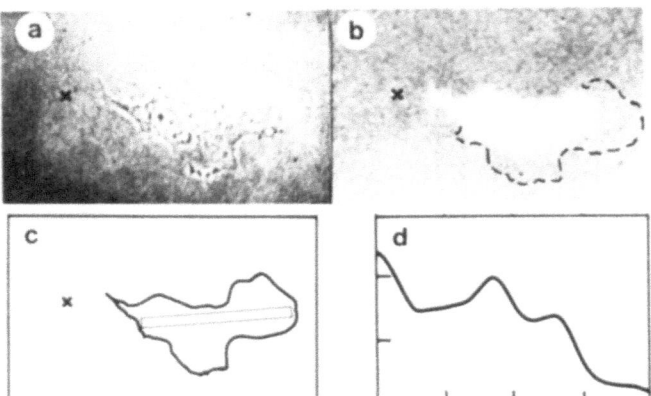

Figure 4. Surface distribution of GP80 in a locomoting fibroblast. The cell was stained with rhodamine-conjugated anti-GP80 monoclonal antibody and observed by DFM. a) Phase contrast image, b) fluorescence image, c) the cell shape and the rectangular area in which the cross-section profile (see (d)) was taken. d) the cross-sectional profile of the fluorescence image. Several profiles parallel to the long axis of the rectangular area were taken and averaged. The abscissa is the relative position along the axis and the ordinate is the relative fluorescence intensity. The scale bar indicates 10 μm.

locomotes has been assumed to explain such redistribution (Bretscher; 1984), however, neither the rate nor characteristics of the flow have been determined. To examine membrane flow during locomotion, we are studying the redistribution of a major membrane protein, GP80, using rhodamine-conjugated monoclonal antibodies as tags for the antigen. When cells are moving, the GP80 distributes in a gradient from the leading edge to the trailing edge of the cell, with the lowest fluorescence intensity in the leading edge. This gradient disappears when cells cease moving. Such gradients in GP80 concentration are also observed in motile macrophages and polymorphonuclear leukocytes. However, other membrane proteins such as the Thy-1 antigen and the transferrin receptor as well as lipid analogs do not display gradated patterns in these motile cells. The photometric nature of the image data allows a quantitative intensity profile of the distribution to be mapped along the direction of cell motion. An example is seen in Figure 4. Figure 4a shows a phase image of fibroblast moving toward the bottom of the photograph; Figure 4b shows the corresponding fluorescence image. In Figure 4c, an outline is shown of the motile fibroblast stained with rhodamine anti-GP80 (4b) with a superimposed rectangle oriented along the direction of locomotion. Ten to fifteen rows of pixels are averaged along the short axis of the rectangle to give the mean intensity profile shown in Figure 4d. Theoretical profiles corresponding to various models for locomotion can be constructed by computer simulation using the membrane flow velocity and lateral diffusion coefficient of the protein as parameters. These models can then be compared to the experimental profiles to assess which of them more accurately fits the data. In this way, we hope to obtain important new information on plasma membrane movements during cell locomotion.

The motion of antibody mediated crosslinked aggregates of GP80 can also be investigated by DFM and compared to the motions of polymeric beads, carbon particles, etc., on motile cells (Harris, 1973; Dembo and Harris, 1981). By adding rat monoclonal anti-GP80 followed by a second layer of polyclonal anti-rat antibodies, patches of GP80 are quickly formed (Figure 5a). The motions of these patches were then followed with time by focusing first on the ventral (substrate attached surface) and then on the dorsal (free) surface of the cell. It is important to realize that such a study would be impossible without DFM for to obtain enough film exposures the sample fluorescence would be heavily bleached. From the successive position coordinates of identified patches, the trajectories of patch movements were reconstructed (see Example, Figure 5b). Several patterns of systematic motion, some of them unexpected, were discovered to occur simultaneously on

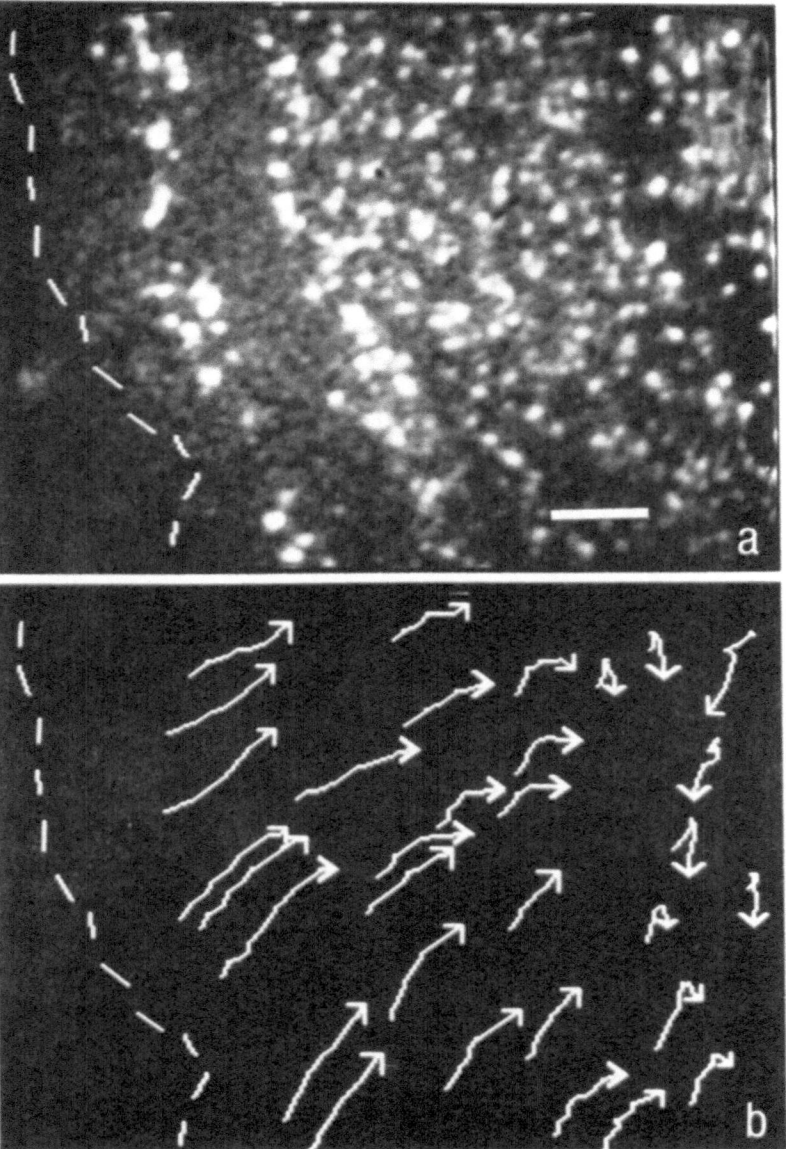

Figure 5. Movement of aggregated membrane proteins. a) DFM image showing patches of crosslinked GP80 near the leading edge of a motile fibroblast. Monoclonal rat anti-GP80 was bound to the cell surface (see Figure 4) and then crosslinked with a fluorescently labeled secondary antibody (rhodamine-anti-rat IgG) to induce formation of patches. This image was taken five minutes after applying secondary antibody and seven such images were acquired during a thirty minute period. The broken line represents the cell edge. b) Trajectories of individual patches undergoing redistribution in the phenomenon called capping. Coordinates of patches were obtained from the series of DFM images and sequential positions of an individual patch were connected to reconstruct the trajectory. Arrowheads at the end of trajectories indicate directions of movement. Note that patches at the left moved continuously away from the cell edge, while during this time some patches at the right turned or reversed direction. The length of a trajectory is proportional to average velocity, the greatest being approximately 0.5 microns/minute. The scale bar indicates 5 microns.

147

live cells. The most frequent movement was directly inward, away from a convex advancing edge. This occurred on both dorsal and ventral lamellar surfaces and was independent of net advance of the motile cell; simultaneously transported polystyrene beads moved with velocities similar to nearby patches. Some of the patches located immediately behind an active cell edge moved approximately parallel to the edge; such patches occasionally turned and moved inward. A third pattern of patch movement was noted on some cells; dorsal surface patches in the central region moved forward toward the cell margin. We believe these complex patterns are reflections of multiple forces exerted at the cell surface during motile phenomena. Thus it will be important to correlate these newly discovered patterns with other structural features of the cell's cytomatrix in order to uncover the origin of these forces.

ACKNOWLEDGEMENTS

This work was supported by grants from the NIH (GM35325), the American Cancer Society (CD-181A), and the American Heart Association (83-1325).

REFERENCES

Arndt-Jovin, Robert-Nicoud, M., Kaufmann, S. and Jovin, T., 1985, Fluorescence Digital Imaging Microscopy in Cell Biology, Science, 230:247.

Bretscher, M. S., 1984, Endocytosis: Relation to Capping and Cell Locomotion, Science, 224:681.

Dembo, M., and Harris, A. K., 1981, Motion of Particles Adhering to the Leading Lamella of Crawling Cells, J. Cell Biol., 91:528.

DiGuiseppi, J., Inman, R., Ishihara, A., Jacobson, K., and Herman, B., 1985, Applications of Digitized Fluorescence Microscopy to Problems in Cell Biology, Biotechniques, 3:394.

Fay, F., Fogarty, K., and Coggins, J., 1986, Analysis of Molecular Distribution in Single Cells Using a Digital Imaging Microscope, in: "Optical Methods in Cell Physiology", P. DeWeer and B. Salszberg, eds., John Wiley and Sons, New York.

Grynkiewicz, G., Poenie, M., Tsien, R., 1985, A New Generation of Ca^{2+} Indicators with Greatly Improved Fluorescence Properties, J. Biol. Chem., 260:3440.

Harris, A. K., 1973, in: "Locomotion of Tissue Cells", CIBA Foundation Symposium 14.

Jacobson, K. E., Elson, D., Koppel, D. and Webb, W., 1983, Summary of an International Workshop on the Application of Fluorescence Photobleaching Techniques to Problems in Cell Biology, Fed. Proc., 42:72.

Kapitza, H.-G., McGregor, G. and Jacobson, K., 1985, Direct Measurement of Lateral Transport in Membranes Using Time-Resolved Spatial Photometry, Proc. Nat. Acad. Sci. USA, 82:4122.

Koppel, D. E., 1979, Fluorescence Redistribution After Photobleaching. A New Multipoint Analysis of Membrane Translation Dynamics, Biophys. J., 28:281.

Mathog, D., Hochstrasser, M., Gruenbaum, Y., Saumweber, H., and Sedat, J., 1984, Characteristic Folding Pattern of Polytene Chromosomes in Drosophila Salivary Gland Nuclei, Nature (London), 308:414.

MacGregor, G., Kapitza, H.-G., and Jacobson, K., 1984, Laser-Based Fluorescence Microscopy of Living Cells, Laser Focus, 20-84.

Taylor, D. L., and Wang, Y-L, 1980, Fluorescently Labelled Molecules as Probes of the Structure and Function of Living Cells, Nature, 284:405.

Waggoner, A., 1986, Fluorescent Probes for Analysis of Cell Structure, Function, and Health by Flow and Imaging Cytometry, in: "Applications of Fluorescence in the Biomedical Sciences", D. L. Taylor, A. Waggoner, R. Murphy, F. Lanni, and R. Birge, eds., Alan R. Liss, Inc., New York.

Webb, W. W., and Gross, D., 1986, Patterns of Individual Molecular Motions Deduced from Fluorescent Image Analysis, in: "Applications of Fluorescence in the Biomedical Sciences", D. L. Taylor, A. Waggoner, R. Murphy, F. Lanni, and R. Birge, eds., Alan R. Liss, Inc., New York.

MEMBRANE FLUIDITY AND DIFFUSIVE TRANSPORT

Josef Eisinger

Department of Physiology and Biophysics
Mount Sinai School of Medicine
New York, NY 10029 USA

INTRODUCTION

The fluid mosaic model (Singer and Nicolson, 1972) provides not only a picture of the basic structure of membranes but also implies certain dynamic characteristics of its components. Thus the lipids of the bilayer are subject to Brownian motion which endows them with rotational and translational mobility. The same is true of those proteins which are not attached to the cytoskeleton, although their mobility is much smaller. In the last few years, considerable effort has gone into the measurement of the rotational and translational mobilities of both lipid analogue probes and of membrane proteins, in model systems, as well as in biological membranes. It has become clear that the lateral mobility of membrane components not only provides an interesting demonstration of the fluidity of a two-dimensional liquid, but that it plays an important role in various functions of the cell membrane. For example, membrane receptors are known to migrate and aggregate in certain small regions of the membrane (coated pits), and the electron transport between cytochromes in the mitochondrial membrane appears to be diffusion-limited (Hackenbrock et al., 1986). These and other cellular functions related to the lateral mobility of membrane components have recently been reviewed by Axelrod (1983).

The most widely used method for studying lateral membrane fluidity has been fluorescence photobleaching recovery (FPR). This technique is based on measuring the time course of a flux of fluorescent particles diffusing in a concentration gradient (Axelrod et al., 1976). Since the concentration gradient is created by bleaching the probes located in a small region of the membrane by means of a focused laser beam, the average diffusion

151

distance in FPR experiments is comparable to the size of the bleached spot - of the order of 1 μm. This distance is large compared to the average separation of membrane proteins, so that the diffusivity determined in an FPR experiment is a measure of the ease with which a test particle, be it a lipid analogue probe or a fluorescently labeled protein, migrates through an archipelago of many stationary or slowly moving obstacles.

In the present paper we present a dynamical model of membranes which allows us to estimate "local" membrane fluidity from the mobility of excimeric membrane probes. The probe considered here is a pyrene-containing fatty acid molecule, which has an excited state lifetime of the order of 10^{-7} seconds and can form an excimer only if it encounters another probe in its ground state before its de-excitation as a monomer. In that time interval the probe migrates only a few nanometers, a distance in which the presence of obstructing proteins makes itself felt to a much lesser extent than in FPR experiments and it will be seen that "local" fluidity is indeed greater than the long-range fluidity measured in FPR experiments. We will also present results of simulations of obstructed diffusion, to illustrate how the modulation of protein density in a membrane affects the diffusivity of a probe and how this could result in heterogeneities in its diffusivity. Theoretical considerations suggest that the presence of diffusivity gradients would speed up the lateral transport of membrane components. Whether or not the cell makes use of such fluidity features, we suggest how fluidity heterogeneities could be made visible by the use of fluorescence microscopy of cell membranes labeled with excimeric or other dynamic fluorescence probes, or, by what might be called "fluidity microscopy".

THE LATERAL DISTRIBUTION OF MEMBRANE PROTEINS

Since the lateral diffusivity of small or large membrane components depends on the density, size and mobility of membrane proteins surrounding them, we wish to provide some rough, but reasonable estimates for these parameters, for use in our simulation models and in the analysis of excimeric probe experiments.

The fractional mass of proteins in biological membranes ranges from 20 to 75 percent, the lower and upper limits being attained in myelin and in mitochondrial membranes, respectively. This range reflects the different functions of membranes, primarily that of an electrical insulator in the

152

first example and that of a matrix for enzymes participating in electron transport, ATP synthesis and solute transport, in the latter (Finean and Michell, 1981). The erythrocyte membrane is an intermediate example and contains 50 percent protein by weight, including integral and peripheral species. In attempting to assess how the presence of these membrane proteins modulates the diffusivity of small or large membrane components it is necessary to provide at least approximate answers to the following two questions: 1) What fraction of the membrane area is occupied by obstacles to diffusion and what is the mobility of these obstacles? 2) What is the size distribution, or at least the average size, of these obstacles?

An answer to the first question may be obtained from experiments similar to the classical Gorter and Grendel (1925) study, which was the first to suggest that the lipid bilayer is the fundamental structure of biological membranes. The logic of this approach is straightforward: One measures the area of a monomolecular film of the lipids extracted from a known number of erythrocytes, after spreading them on a Langmuir trough, and compares it to the total area of the plasma membranes of the erythrocytes. The Gorter and Grendel experiment was recently repeated by Bar et al. (1966), who avoided several errors and assumptions of the original study and measured the ratio of monolayer to membrane area as a function of the film's surface pressure. Since there are no direct measurements of the surface pressure which prevails in membranes of intact red blood cells, an estimate of 31-35 dynes cm^{-1} was obtained from a comparison of the activities of various phospholipases on monomolecular films and on intact erythrocytes (Demel, et al., 1975). For this surface pressure one obtains 1.5 as the ratio of the monomolecular film to the total (bilayer) membrane area, which suggests that 25 percent of the membrane area is occupied by proteins. This estimate assumes of course that membrane proteins are equally distributed between the two bilayer leaflets, while the asymmetrical distribution of proteins, as well as that of lipids, is well established (Steck, 1974). When freeze-fractured membranes are examined by electron microscopy, the outer leaflet was found to contain almost 5 times as many intra-membrane particles (IMP) than the inner one (Weinstein, 1969). While the identity of these particles has not been established, and their size of about 8 nm suggest that they are aggregates and not individual proteins, this asymmetry implies that some 40 percent of their outer and 8 percent of the inner leaflet's area is occupied by proteins.

The answer to the second question, what is the effective average size of the membrane proteins as obstacles to diffusion, is also difficult to answer with certainty. The minimum obstacle of a membrane spanning protein would be of the order of 1 nm in diameter, corresponding to a single γ-helix (e.g. glycophorin), but proteins with several helices (e.g. bacteriorhodopsin) would exclude considerably greater lipid areas. In addition, little is known about the existence and size of lipid solvation shells which may surround the hydrophobic amino-acid residues of intra-membrane proteins. Such shells of immobile, or slowly exchanging, lipids could increase the effective obstacle size appreciably.

If it is assumed, for the sake of discussion, that the average (circular) obstacle diameter is 3 nm, that they are randomly distributed, and that the fractional area covered by obstacles in the outer leaflet is 30 percent, then it is readily seen that the average spacing between obstacles is of the order of 5 nm. Each erythrocyte membrane contains about 7.4 x 10^{10} μmoles of lipids, including cholesterol, (Sweeley and Dawson, 1969) which corresponds to 4.5 x 10^8 molecules, of which some 2.0 x 10^8 are expected to be in the outer leaflet. Since a human erythrocyte has an area of 140 μm^2, this admittedly crude model suggests that each lipid molecule occupies an area of 0.49 nm^2, with an average diameter of 0.8 nm. The important conclusion to be drawn from these considerations is that, in general, there is room for only a few lipid molecules between membrane proteins. This model for the distribution of lipids and proteins should be borne in mind when considering the simulated diffusion of probe molecules, to be discussed in a later section.

THE MOBILITY OF MEMBRANE LIPIDS

The molecules which constitute the lipid bilayer exchange thermal energy with their environment and each other, and the statistical fluctuations of these exchanges are the origin of Brownian motion. The fluctuations in the rotational and positional coordinates of the molecules are responsible for their rotational and translational diffusion.

The rotational diffusion rate of a molecule in an isotropic environment may be estimated from the depolarization of its fluorescence by means of the well-known Perrin equation. The environment of a probe molecule in a bilayer is, however, highly anisotropic. This manifests itself in the observation that its emission anisotropy does not vanish even a long time after excitation (Dale et al., 1977). A detailed analysis of anisotropic rotational diffusion exists (v.d. Meer et al., 1984) but is not required in the present context. It is sufficient to note that for a variety of membrane systems, the rotational relaxation rate of fluorescent membrane probes is at most a factor of 2 greater than that obtained by use of Perrin's equation (v.d. Meer et al., 1986) and is typically of the order of 10^8 s^{-1}.

The translational mobility of membrane lipids is of primary interest here and can be estimated from the diffusivity of suitable probe molecules. It can be analyzed from two distinct points of view:

In macroscopic diffusion theory the lateral diffusion coefficient (D) appears as a proportionality factor between a particle flux (J) and the particle concentration gradient, that is

$$J = -D \text{ grad } c \tag{1}$$

where c is the concentration of particles. As was already pointed out in the Introduction, equation (1), which is known as Fick's equation, is the natural way of analyzing fluorescence photobleaching experiments (Axelrod et al., 1976).

In microscopic diffusion theory, the lipid molecules are considered to execute a random walk in two dimensions and one calculates how far a molecule travels as a function of the number of steps it takes in random directions. If λ is the size of each step and τ is the time between steps, it is readily shown that in time t the molecule travels a mean square distance (Berg, 1983)

$$\langle x^2 \rangle = 4 \text{ DT} \tag{2}$$

with

$$D = \lambda^2/4 = \nu\lambda^2/4 \tag{3}$$

where $\nu = 1/\tau$ is the frequency with which the random steps are taken. The equivalence of the macroscopic and microscopic definitions of D (Equations 1 and 3 respectively) was established by Einstein (1906).

It is clear that for the analysis of experiments in which the rate of a diffusion-dependent, intermolecular interaction is measured, the microscopic, or Markovian point of view is appropriate. An example of this approach is provided by excimeric membrane probes, because they signal the formation of an intermolecular complex by a readily observable change in their fluorescence spectrum. Their kinetic equations may be written as

155

$$A + A^* \xrightarrow{\quad K \quad} (AA)^*$$

$$A^* \xrightarrow{\quad k_M, k_M' \quad} A + h\nu_M \tag{4}$$

$$(AA)^* \xrightarrow{\quad k_E, k_E' \quad} 2A + h\nu_E$$

where K is the rate at which an excited probe, A^*, and one in its ground-state, A, become nearest neighbors and form an excited dimer, or excimer, $(AA)^*$. k and k' are the radiative and non-radiative decay rates of excited monomers and excimers (subscripts M and E, respectively). If, as is the case for pyrene excimers at room temperature, the inverse reaction rate to K is negligible, one obtains the following expressions for the monomer and excimer quantum yields (Eisinger et al., 1986):

$$\Phi_M = \frac{k_M}{k_M + k_M' + K} \tag{5}$$

$$\Phi_E = \left[\frac{k_E}{k_E + k_E'} \right] \left[\frac{K}{k_M + k_M' + K} \right] \tag{6}$$

The excimer formation rate K depends on the mobility of the probes and on their concentration, which is conveniently measured by the molar probe/lipid ratio, x. In the limits of very low and very high probe ratios, respectively, the yields given by equations (5) and (6) approach their intrinsic values

$$\Phi_M^* = \frac{k_M}{k_M + k_M'} \tag{7}$$

$$\Phi_E^* = \frac{k_E}{k_E + k_E'} \tag{8}$$

and from equation (5) - (8) the normalized monomeric and excimeric fluorescence yields and their ratio are

$$\frac{\Phi_M}{\Phi_M^*} = \frac{1}{1 + K\tau_M} \tag{9}$$

$$\frac{\Phi_E}{\Phi_E^*} = \frac{K\tau_M}{1+K\tau_M} \qquad (10)$$

$$\frac{\Phi_E}{\Phi_M} = \frac{\Phi_E^*}{\Phi_M^*} K\tau_M \qquad (11)$$

where we have written $\tau_M = (k_m + k_m')^{-1}$ for the monomer lifetime, which has a value of about 10^{-7}s for pyrene-containing probes.

In order to relate the excimer formation rate to the concentration and diffusivity of the probes, it is noted that if the probes make random steps at a frequency υ, and $n(p_E,x)$ is the average number of steps taken by an excited probe with an infinite lifetime before forming an excimer, then

$$K = \frac{\upsilon}{n(p_E,x)} \qquad (12)$$

$n(p_E,x)$ is a function of the probe ratio x, and of p_E, the probability that a nearest neighbor pair (A*,A) forms an excimer during an interval υ^{-1}.

In order to evaluate $n(p_E,x)$ one may model the dynamics of an assembly of lipids and probes by the so-called milling crowd model with the following characteristics (Eisinger et al., 1986):

(1) The lipid probe molecules form a regular trigonal array with periodic boundary conditions, so that each molecule has six nearest neighbors and each probe particle changes places with a randomly selected nearest neighbor at a frequency υ.

(2) If L is the number of lattice points in the array, one lattice point, selected at random, is initially occupied by an excited probe, A*, and xL-1 others are occupied by probes, A.

(3) When A* and one of A probes become nearest neighbors, an excimer is formed with a probability p_E.

By performing many simulation experiments according to this algorithm for particular values of p_E and x, one obtains a frequency distribution of the number of steps, N(x), taken before an excimer was formed and from this one derives $n(p_E,x)$ the average number of spatial exchanges leading to an excimer. The results of such computer simulations are illustrated in

157

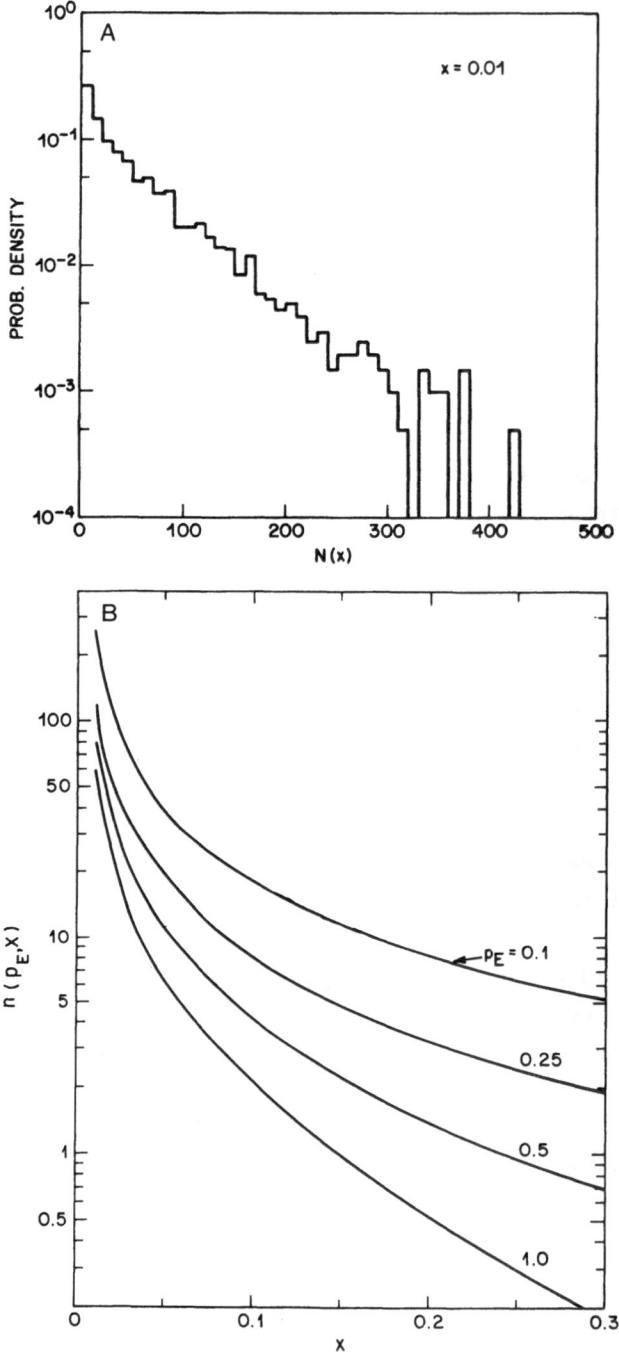

Figure 1 (A). Frequency distribution of the number of spatial exchanges, N(x), required before an excited probe becomes an excimer, for a probe ratio of x=0.01. It is assumed that the monomeric probe has no other decay mechanisms and that an excimer is formed whenever an excited and a ground state probe become nearest neighbors (i.e. p_E=1). The data were obtained by a computer simulation of the two-dimensional milling crowd model (Eisinger et al., 1986). **(B).** $n(p_E,x)$, the average number of spatial exchanges leading to the formation of an excimer, as a function of the probe ratio x and the probability of excimer formation between nearest neighbors (p_E).

Figure 1. Note that the probability distribution for excimer formation as a function of N(x) is also the decay rate of a monomer due to excimer formation, with the time given by $t=N(x)\upsilon^{-1}$.

From equations (9) and (12), the normalized monomeric fluorescence yield is given by

$$\frac{\Phi_M(x)}{\overset{*}{\Phi}_M} = [1 + \frac{\upsilon\tau_M}{n(p_E,x)}]^{-1} \qquad (13)$$

with similar equations for excimeric yield and the excimer/monomer ratio coming from equations (10) and (11) with equation (12). By fitting equation (13) to the measured fluorescence yields as a function of x, one obtains a value of the frequency υ for each assumption for p_E. Once the exchange frequency υ is known, the corresponding diffusion coefficient D is obtained by use of equation (3).

The spatial exchanges of probes and lipids postulated in the milling crowd model are of course not to be taken literally. Rather, they represent a formal description of the rate at which the molecules migrate a distance of one lattice distance, which is identified with the average lipid-lipid spacing, λ.

COMPARISON WITH EXPERIMENT

Figure 2 gives the results of an experiment in which intact human erythrocytes were labelled with different concentrations of the excimeric probe 1'-pyrene-dodecanoic acid (PDA), which is known to reside in the outer leaflet of the plasma membrane. The ordinate is the normalized monomeric fluorescence intensity measured at 378 nm, divided by x. The experimental points are fitted with equation (13), for three assumptions of p_E. The accuracy of the data does not permit the selection of a particular value of p_E with certainty, but the following conclusions may be drawn:

p_E is certainly less than or equal to unity. The value of corresponding to $p_E=1$ therefore represents a minimum value and one may conclude that $\upsilon \gtrsim 2 \times 10^7 s^{-1}$ and, from eq (3), that $D \gtrsim 3 \times 10^{-8}$ cm^{2}s^{-1}.

It is interesting to compare this lower limit for spatial exchange frequency of PDA with the rotational relaxation rate for the same probe.

Table 1. Comparison of Lateral Diffusion Coefficients of Lipid Analogue Probes in Various Bilayer Systems

Bilayer[a]	Probe[b]	$D^{(c)}$ (10^{-9} cm^{2}s^{-1})	Reference
Erythrocyte	PDA	30	Eisinger et al., 1986a
Erythrocyte	diI	8	Bloom and Webb, 1983
Erythrocyte ghost	diI	2	Thompson and Axelrod, 1980
Fluid DMPC	various	45-84	Derzko and Jacobson, 1980
DMPC, P/L=0.005	diO	69	Peters and Cherry, 1982
DMPC, P/L=0.03	diO	7	Peters and Cherry, 1982
IMM	diI	5	Hackenbrock et al., 1986

Notes

(a) PDA; 1'-pyrenedodecanoic acid
DMPC: dimyristoylphosphotidylcholine; IMM: Inner mitochondrial membrane; P/L: molar ratio of protein (bacteriorhodopsin): lipid

(b) diI: dihexadecylindscarbocyanine iodide; diO: dioctadecyloxatricarbocyanine

(c) In the first line of the Table, D was obtained from excimeric probe experiments, the quoted lower limit corresponding to $p_E=1$ (see text). All other measurements employed FPR, i.e. long-range diffusion.

Its steady state anisotropy was measured to be $\langle r \rangle = 0.013 \pm 0.002$ at 26°C at a probe ratio for which the excimer yield is negligible (Eisinger and Ross, unpublished results). With a limiting anisotropy of 0.13 and a lifetime of 75 ns, this corresponds to a rotational relaxation rate of $0.7 \times 10^8 s^{-1}$ according to the isotropic Perrin equation and about twice that rate if the anisotropic environment of the bilayer (cf. earlier in this chapter) is taken into account. The rotational rate is therefore at least an order of magnitude larger than the rate at which the probe translates one lipid-lipid spacing.

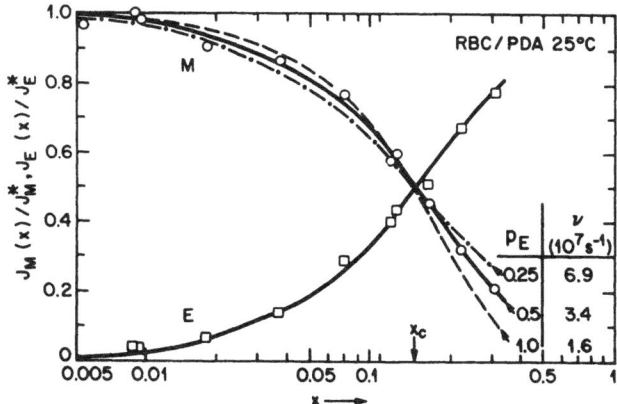

Figure 2. The monomeric (M) and excimeric (E) fluorescence yields of PDA in intact erythrocyte membranes as a function of the probe ratio, the yields are normalized to their low and high probe ratio limits. The experimental points are fitted by equations (9) and (10). The values of υ corresponding to 3 choices for p_E are shown in the lower right hand corner.

The experimental results for lateral diffusion of PDA are given in Table 1, along with typical data for long-range lateral diffusion coefficients of lipid analogue probes in erythrocytes and for vesicles with proteins embedded in the bilayer. The data are consistent with the following conclusions:

(1) In vesicle systems, the presence of proteins slows lipid long-range diffusion drastically.

(2) Long-range lateral mobility of lipid analogues is reduced in erythrocytes compared to pure phospholipid vesicles.

(3) In erythrocytes, local lateral diffusion of lipids is appreciably
greater than long-range diffusion.

While we have confined our attention to the diffusivity of small, or
lipid analogue probes, the diffusivity of proteins is also slowed dramatically when high concentrations of proteins are present in the bilayer.
Thus, the diffusion coefficient of bacteriorhodopsin is reduced by a factor
of 20 when the molar ratio of protein to lipid is raised from 0.005 to 0.3
(Peters and Cherry, 1982). A similar dependence on concentration was
reported for the long-range diffusivity of cytochrome b-c$_1$ in inner mitochondrial membranes (Hackenbrock et al., 1986).

OBSTRUCTED DIFFUSION

The phenomenon of diffusion inhibition in the presence of obstacles is
of course well known to physicists and has given rise to percolation theory.
This and other continuum theories for obstructed diffusion were originally
derived in studies of heat and electrical conductivity in two dimensions
and have recently been reviewed by Saxton (1982). Since it was demonstrated
earlier in this chapter that a typical biological membrane is expected to
accommodate only a small number of lipid molecules between its randomly
distributed proteins, the computer simulation of the diffusive behavior of
lipid analogue probes is a more flexible and appropriate approach than
continuum theory. Such simulations are conveniently performed by employing
a modified form of the milling crowd model which was used to simulate
excimer formation kinetics (Eisinger et al., 1986). The lipids and probes
are again assumed to form a regular trigonal array without vacancies and
they again change places with randomly chosen nearest neighbors at an
exchange frequency, υ. After i exchanges, the distance travelled by a
probe, d(i), is determined and the experiment is repeated a large number of
times (at least 100). In this way, the mean square diffusion distance,
$\langle d^2(i)\rangle$ is obtained as a function of i. When $\langle d^2(i)\rangle$ is plotted against i,
one obtains a straight line, whose slope is according to equation (2)
proportional to the diffusion coefficient:

$$D = \langle d^2(i)\rangle/4i \qquad\qquad\qquad (14)$$

One may now introduce a random distribution of stationary obstacles
into the lipid array and repeat the simulations. This is illustrated in
figure 3a, where hexagonal obstacles whose sides equal one lattice spacing

were used. Lattice points which belong to an obstacle are forbidden sites for the diffusing test particle: if they are selected in the random choice of a partner for exchange, no exchange takes place and the migrating particle "loses its turn".

The obstacle population is conveniently characterized by the obstacle size and the total fractional area of the array which is occupied by them. The values of these parameters depend of course on how one chooses to define the obstacle area, A. Two examples are used here. In the first, A is set equal to the hexagon's area and the fractional area, f, is obtained by dividing the total area of all obstacles by the lattice area. The second method sets the fractional obstacle area, f_{LP}, equal to the ratio of the number of lattice points belonging to an obstacle, to their total number of lattice points in the array. These two definitions

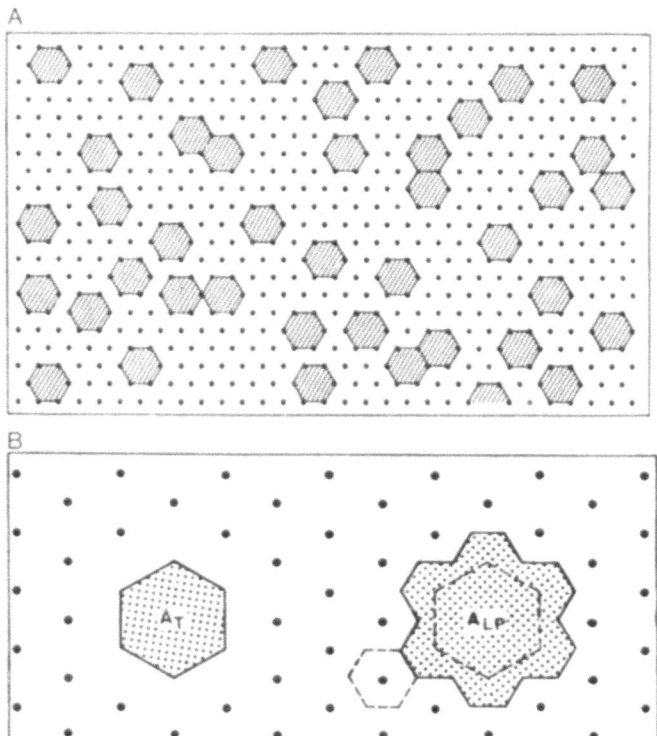

Figure 3 (A). The trigonal array of lipids/probes employed in the milling crowd simulations is shown here with hexagonal obstacles of size j=1. Their fractional coverage is f=0.21.
(B). The figure illustrates two possible definitions for obstacle area, as explained in the text, the fractional obstacle areas, f and f_{LP}, correspond to the definitions illustrated by the left and right diagrams.

are illustrated in Figure 3 which shows that the second (f_{LP}) implies a much more intimate contact between probe and obstacle.

When $\langle d^2(i)\rangle$ is plotted against i for a particle diffusing among obstacles, one obtains a curve whose slope is everywhere smaller than that for free diffusion, even for a small number of spatial exchanges. For

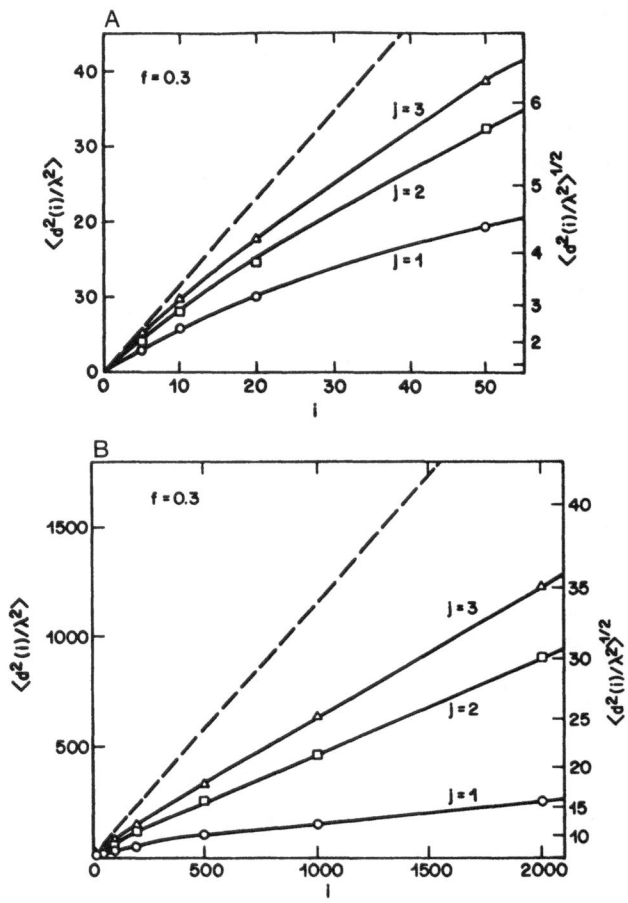

Figures 4 (A), (B). Comparison of free and obstructed diffusion by a lipid analogue probe, simulated according to the milling crowd model. The dashed and solid curves refer to diffusion in the absence and presence of hexagonal obstacles, respectively, j being the side of the hexagon in units of the lipid lattice spacing. The fractional area covered by obstacles is f+0.3 for both (a) and (b). The mean-square-diffusion distance (left) and root-mean-square-distance (right) are shown as functions of the number of random steps (i) taken by the lipid probe. Note that for large i the curves become straight. This shows that the diffusion equation (2) is valid for long-range obstructed diffusion, but that the effective diffusion coefficient is reduced from its unobstructed value by the ratio of the limiting slopes of the curves to the slope of the straight line which characterizes free diffusion.

large diffusion distances (large i), the slope approaches a constant value, which represents, according to equation (12), 4D. (cf. Figure 4). For large values of f and small obstacles, its value may be considerably reduced from the free diffusion value. Equation (14) may therefore be used to define an effective diffusion coefficient, which depends on the distance through which the test particle has diffused for small i but attains its constant, long-range value when i is large. It should be noted that the effective diffusion coefficient for i exchanges is obtained by dividing $<d^2(i)>$ by 4i and not from the derivative of the curve at i, because the test particle has no memory of how long it has been diffusing.

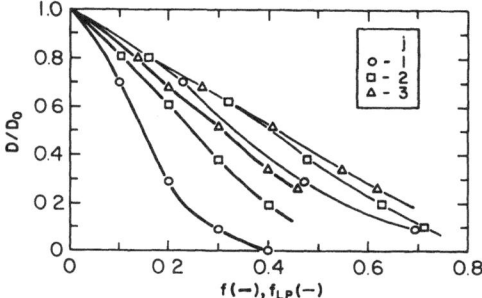

Figure 5. The effective long-range diffusion coefficients for obstructed diffusion (D), normalized to the free diffusion coefficient (D_o), are shown as a function of the fractional area occupied by the obstacles. For the heavy curves the fractional area is given by f, and for the light curves, by f_{LP}. The definitions of f and f_{LP} are given in the text and in figure 3B. Three sizes of hexagonic obstacles were used with j referring to the length of a side of the hexagon in units of a lattice constant, i.e. the average lipid-lipid spacing. Note that the dependence on obstacle size is much greater if the obstacle area is measured in the manner illustrated on the left side of figure 3B.

In figure 5 are shown the asymptotic or long-range diffusion coefficients as functions of the fractional obstacle area, for different sizes of obstacles, calculated by both of the proposed methods for determining obstacle size, f and f_{PL}. Note that the former method leads to a much greater dependence on size. Continuum theories of obstructed diffusion predict no dependence on obstacle size below the percolation limit and by definition, no diffusion above the percolation limit (Saxton, 1982).

DIFFUSIVE TRANSPORT

The simulations of obstructed diffusion and the results of a wide range of experiments (c.f. Table 1) demonstrate clearly that a distribution of obstacles, comparable in density and sizes to that of proteins in membranes, can reduce diffusivity dramatically. Other mechanisms which modulate diffusivity may of course be operative. It has, for instance, been shown that loosening the attachment between band 3 proteins and the cytoskeleton greatly increases the mobility of the protein (Golan and Veatch, 1980). Diffusivity is also modulated by the partitioning of lipids or cholesterol and there exists electron microscopic evidence that coated pits (~0.1-0.2 μm) in macrophage membranes are cholesterol depleted (Montesano et al., 1980).

Mention should finally be made of another intriguing mechanism for the local modulation of lipid fluidity, which was suggested by Hirata and Axelrod (1978). They observed that the β-adrenergic agonist L-isoproterenol, upon binding to its membrane receptor triggers the methylation and translocation of phospholipids from the inner to the outer leaflet of the bilayer and that this results in increased fluidity of the membrane, as measured by the depolarization of diphenylhexatriene (DPH) fluorescence.

Whatever such modulations of fluidity are due to, it is worth inquiring if they play a part in lateral membrane transport. For example, is the aggregation of membrane receptors speeded up by "fluidizing" the area surrounding a coated pit?

To investigate theoretically the effectiveness of such "diffusive transport", Fick's Equation (1) was first rewritten with D not a constant, but a spatially variable parameter and solved for cases of biological interest (Eisinger and Halperin, 1986). Consider, for example the mean-time-to capture, $(\bar{t}_c)$ of particles initially distributed in a circular region of radius b surrounding a central circular trap (radius a). The trap is considered to absorb all particles which come in contact with its circumference. If the particles' diffusion coefficient is constant in the annular region between a and b, and b>>a (Berg and Purcell, 1977),

$$\bar{t}_c = \frac{b^2}{2D} \left(\ln \frac{b}{a} - \frac{3}{4} \right) \tag{15}$$

which shows that the average diffusion time of a particle before it is trapped, varies inversely with D. One may next enquire if $\bar{t}_c$ would be further reduced if in the annular region between a and b the particle experienced a diffusivity gradient; specifically, let $D=D_{max}$ for a<r<c and $D=D_{min}$ for c<r<b, where r is the particle's radial coordinate. Solutions for this discontinuous gradient are shown in Figure 6. Figure 7 shows a hypothetical example: A comparison of cases (b) and (c) shows how the same mean-time-to capture is achieved "more efficiently" in the presence of a diffusivity gradient.

Quite generally, the diffusive transport of particles from a source, (which may be a distributed one, as in the example above) to a sink is speeded up if source and sink are linked by a high diffusivity domain. While there is at present no convincing experimental evidence that membranes make use of this stratagem, the existence of fluidity heterogeneities is worth investigating.

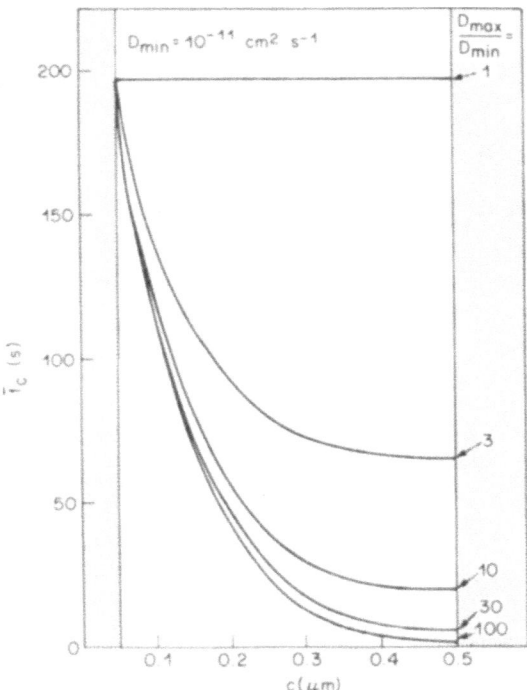

Figure 6. The mean-time-to-capture ($\bar{t}_c$) by a trap with radius a, for particles initially distributed in an annular region a<r<b. The diffusion coefficient is $D_{min}=10^{-11}$ cm²s⁻¹ in the region c<r<b and is D_{max} in the region a<r<c, with an a=0.05 μm and b=0.5 μm. The family of curves for different D_{max}/D_{min} ratios may be used for other diffusivity gradients, since $\bar{t}_c$ varies inversely with D_{min} (c.f. equation 21 in Eisinger and Halperin, 1986).

FLUIDITY IMAGING

The fluidity of a membrane can in principle be mapped by using dynamic fluorescence probes, i.e., probes whose spectral or polarization characteristics depend on the local fluidity.

If excimeric probes are inserted in the membrane and fluorescence microscopic images of the cell are obtained at both the monomeric and excimeric emission wavelengths, they can be combined to form an image whose brightness (or color) is proportional to the excimer/monomer intensity ratio at each position which was seen to be indicative of the local fluidity (c.f. equation. 11). With the stability and sensitivity of the best modern photo-imaging systems it should be possible to map the local lateral fluidity of the membrane with a resolution of about 0.2 μm. Alternatively, one could insert polarization probes into the cell's membrane and excite them with polarized light in the fluorescence microscope. By combining the emitted light images recorded with polarizers parallel and perpendicular to the polarization vector of the exciting light, one may then combine them to create a depolarization image of the cell's membrane.

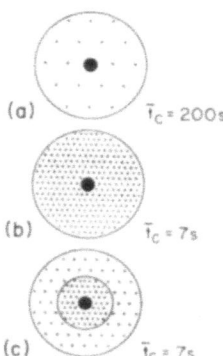

(a) $\bar{t}_c = 200s$

(b) $\bar{t}_c = 7s$

(c) $\bar{t}_c = 7s$

Figure 7. The diagrams (a), (b) and (c) illustrate diffusive transport of particles from an annular region (outer diameter 1 μm) to a circular sink (diameter 0.1 μm) at the center. The mean-time-to-capture, $\bar{t}_c$, is obtained from equation (13) and the curves of figure 6. The values of the diffusion coefficients in the regions surrounding the particle sink are indicated by the density of the stippling, the light, medium and heavy stipplings corresponding to $D = 1 \times 10^{-11}$, 1×10^{-10} and 3×10^{-10} cm^{2}s^{-1}, respectively. Note that in the presence of a diffusivity gradient (case c), $\bar{t}_c$ is the same as in case b, where diffusivity is constant but has the maximal diffusivity in a much larger region surrounding the sink. The diffusivity discontinuity is concentric with the sink and has a diameter of 0.5 μm.

Membranes may of course contain fluidity heterogeneities which are unconnected with molecular transport and a fluidity microscope may be useful in identifying domains which differ, say, in lipid composition, protein density or in the strength of the proteins' attachment to the cytoskeleton. The existence of such domains and of lateral fluidity heterogeneities in general can also be investigated by measuring the time dependence of the excimer production rate which is obtained from a comparison of the monomeric fluorescence decay of excimeric probes at low and high probe ratios. In the same way the anisotropy decay of polarization probes can in principle be used to investigate if domains which differ in the wobble diffusivity of the probe exist.

CONCLUSION

We have reviewed the meaning of membrane fluidity parameters which can be derived from experiments employing dynamic fluorescence probes and have shown that differences between long-range and local diffusibility can be explained as a consequence of obstructed diffusion. We have indicated how the steady state measurements which yield average values for the mobility of such probes can be extended to obtain spatial and temporal information. The possibility that heterogeneities in membrane fluidity play a part in lateral membrane transport was considered in the light of theoretical findings and a strategy was outlined for observing them experimentally. The technical requirements for obtaining fluidity microscopic images of single cells with optical resolution are available, as is a wide selection of suitable dynamic fluorescent probes. A fluorescence microscope and imaging system for testing these ideas is now being designed in our laboratory.

REFERENCES

Axelrod, D., Koppel, D. E., Schlessinger, J., Elson, E., and Watt, W. W., 1976, Mobility Measurement by Analysis of Fluorescence Photobleaching Recovery Kinetics, Biophys. J., 16:1055.

Axelrod, D., 1983, Lateral Motion of Membrane Proteins and Biological Function, J. Membr. Biol., 75:1.

Bar, S. B., Deamer, D. W., and Cornwell, D. G., 1966, Surface Area of Human Erythrocyte Lipids: Reinvestigation of Experiments on Plasma Membranes, Science, 1010.

Berg, H. C., and Purcell, E. M., 1977, Physics of Chemoreception, <u>Biophys. J.</u>, 20:193.

Bloom, J. A., and Webb, W. W., 1983, Lipid Diffusibility in the Intact Erythrocyte Membrane, <u>Biophys. J.</u>, 42:295.

Dale, R. E., Chen, L. A., and Brand, L., 1977, Rotational Relaxation of the 'Microviscosity' Probe Diphenylhexatriene in Paraffin Oil and Egg Lecithin Vesicles, <u>J. Biol. Chem.</u>, 252:7500.

Demel, R. A., Kessel, W. S. M., Zwaal, R. F. A., Roelofson, B., Comfurius, P., and V. Deenen, L. L. M., 1975, Relationship Between Various Phospholipase Actions on Human Red Cell Membranes and the Interfacial Phospholipid Pressure in Monolayers, <u>Biochim. Biophys. Acta</u>, 406:97.

Derzko, Z., and Jacobson, K., 1980, Comparative Lateral Diffusion of Fluorescent Lipid Analogues in Phospholipid Multibilayers, <u>Biochem.</u>, 19-6050.

Eisinger, J., Flores, J., and Petersen, W. P., 1986, A Milling Crowd Model for Local and Long-Range Lateral Diffusion. Mobility of Excimeric Probes in the Membrane of Intact Erythrocytes, <u>Biophys. J.</u>, 49:987.

Eisinger, J., and Halperin, B. I., 1986, Effects of Spatial Variation in Membrane Diffusibility and Solubility on the Lateral Transport of Membrane Components, <u>Biophys. J.</u>, 50:513.

Einstein, A., 1906, Zur Theorie der Brownschen Bewegung, <u>Ann. d. Physik</u>, 19:371.

Finean, J. B., and Michell, R. H., 1981, Isolation, Composition and General Structure of Membranes, <u>in</u>: "Membrane Structure", Finean and Michell, ed., Elsevier/North Holland Biomedical Press.

Golan, D. E., and Veatch, W., 1980, Lateral Mobility of Band 3 in the Human Erythrocyte Membrane Studied by Fluorescence Photobleaching Recovery: Evidence for Control of Cytoskeletal Interactions, <u>Proc. Nat. Acad. Sci. USA</u>, 77:2537.

Gorter, E., and Grendel, F., 1925, On Bimolecular Layers of Lipoids of the Chromocytes of Blood, <u>J. Exp. Med.</u>, 41:439.

Hackenbrock, C. R., Chazotte, B., and Gupte, S. S., 1986, The Random Collision Model and a Critical Assessment of Diffusion and Collision in Mitochondrial Electron Transport, <u>J. Bioenergetics Biomemb.</u>, 18:327.

Hirata, F., and Axelrod, J., 1978, Enzymatic Methylation of Phosphatidyl-ethanolamine Increases Erythrocyte Membrane Fluidity, <u>Nature</u>, 275:219.

v. d. Meer, W., Pottel, H., Herreman, W., Ameloot, M., Hendrickx, H. and Schroder, H., 1984, Effect of Rotational Order on the Decay of the Fluorescence Anisotropy in Membrane Suspensions. A New Approximate Solution of the Rotational Diffusion Equation, <u>Biophys. J.</u>, 46:515.

v. d. Meer, W., v. Hoeven, R. P., and v. Blitterswijk, W. J., 1986,

Steady-State Fluorescence Polarization Parameters by an Extended Perrin Equation for Restricted Rotation of Fluorophores, _Biochim. Biophys. Acta_, 854:38.

Montesano, R., Vassalli, P., and Orci, L., 1981, Structural Heterogeneity of Endocytic Membranes in Macrophages as Revealed by the Cholesterol Probe, Filipin, _J. Cell Sci._, 51:95.

Peters, R., and Cherry, R. J., 1982, Lateral Diffusion in an Archipelago. Effects of Impermeable Patches on Diffusion in a Cell Membrane, _Proc. Nat. Acad. Sci. USA_, 79:4317.

Singer, S. J., and Nicolson, G. L., 1972, The Fluid Mosaic Model of the Structure of Cell Membranes, _Science_, 175:720.

Saxton, M. J., 1982, Lateral Diffusion in an Archipelago. Effects of Impermeable Patches on Diffusion in a Cell Membrane, _Biophys. J._, 39:165.

Steck, T. L., 1974, The Organization of Proteins in the Human Red Blood Cell Membrane, _J. Cell Biol._, 62:1.

Sweeley, C. C., and Dawson, G., 1969, Lipids of the Erythrocyte, _in_: "Red Cell Membrane: Structure and Function:, Jamieson and Greenwalt, ed., Lippincott, Philadelphia.

Thompson, N. L., and Axelrod, D., 1980, Reduced Lateral Mobility of a Fluorescent Lipid Probe in Cholesterol-Depleted Erythrocyte Membranes, _Biochim. Biophys. Acta_, 597:155.

Weinstein, R. S., 1969, Electron Microscopy of Surfaces of Red Cell Membranes, _in_: "Red Cell Membrane: Structure and Function", Jamieson and Greenwalt, ed., Lippincott, Philadelphia.

CONFORMATIONAL FLEXIBILITY OF THE POLYPEPTIDE HORMONE BOMBESIN: TIME RESOLVED AND STATIC FLUORESCENCE STUDIES

Lanfranco Masotti, Paolo Cavatorta*, Arthur G. Szabo[o], Giorgio Farruggia, and Giovanna Sartor

Institute of Biological Chemistry and
*Department of Biophysics-GNCB,
University Parma, 43100 Parma, Italy
[o]Division of Biological Sciences, National Research Council of Canada, Ottawa, Canada, K1A OR6

INTRODUCTION

The three groups of peptides, the bombesins, sauvagine and the dermorphins were first isolated from amphibian skin (Erspamer and Melchiorri, 1980a; Erspamer and Melchiorri, 1980b; Erspamer et al., 1980; Montecucchi et al., 1981; Montecucchi, 1981) and preceded the discovery of analogous peptides in mammalian tissues. Skin peptides have revealed interesting and unexpected actions both on the CNS and the endocrine system showing a wide range of activities. Their use in contributing to the understanding of the mechanisms underlying the central regulation of several visceral functions, the peripheral regulation of gut secretion and mobility, and the control of the release of hypophysary hormones is the basis for the ever growing interest in this group of peptides.

Bombesin (BBS) in particular is a tetradecapeptide whose primary sequence is: Pyr-Gln-Arg-Leu-Gly-Asn-Gln-Trp-Ala-Val-Gly-His-Leu-Met-NH$_2$ The last seven residues of the C-terminal portion are common to all the members of the bombesin family; eight have been shown to be fundamental for their pharmacological potency (Broccardo et al., 1975). His and Trp at position 3 and 7, respectively, from the C-terminus are considered vital for their physiological activity (Erspamer and Melchiorri, 1983). Bombesin-like peptides and bombesin binding sites occur in the CNS and may have a direct effect on its neurons (Moody et al., 1978). High levels of intracellular bombesin-like peptides have been found in human lung small cell carcinomas (Cuttitta et al., 1985). Bombesin shows a central control of visceral function by affecting thermoregulation (Brown et al., 1981), acts on gastric acid secretion (Brown et al., 1978), on pancreatic secretion (Dubrasquet et al., 1982), on glucoregulation (Brown and Vale, 1981) and modulating food intake (Martin and Gibbs, 1980) and drinking behaviour (De Caro et al., 1980); it also stimulates somatostatin release (Abe et al., 1981; Brown and Vale, 1979) and affects the secretion of adenohypophyseal hormones (Rivier et al., 1978; Westendorf and Schombrunn, 1982).

Polypeptide hormones are synthesized and stored in specialized endocrine cells and then released and circulated to target tissues. Often the hormone message is conveyed to the target cell through a specific receptor located on the surface of the membrane, and then degraded rapidly, in many cases via receptor binding and internalization. If we are to understand the various processes at the molecular level during this complex life cycle of the hormone, we clearly need to define the conformations that the polypeptide assumes during such processes.

This paper reports studies on the conformation of bombesin in aqueous solutions and when interacting with different lipid structures as studied by fluorescence and circular dichroism. It is recognized that polypeptide hormones of less than 30 amino acids have conformational flexibility in water and in apolar solvents that enables them to carry on their physiological activities. We chose here two different lipid systems, lysolecithin and DMPS, because the two lipids organize in micelles and vesicles respectively and can therefore give information on the influence of different lipid environments on the conformation of the peptide.

MATERIALS AND METHODS

Crystalline bombesin was purchased from Sigma (St. Louis, MO) and Serva (Heidelberg, FRG) and its purity was assayed by HPLC. The peptide concentration was determined spectrophotometrically assuming an ε = 5600 lit.mol^{-1}.cm^{-1} at 280 nm and a molecular weight of 1620. Lysolecithin and dimyristoylphosphatidylserine (DMPS) were purchased from Avanti Polar Lipids (Birmingham, AL) and used without further purification. All other reagents used were of analytical grade purity. The buffer used was 0.01 M sodium cacodylate at the desired pH. Lysolecithin was dissolved in buffer at a concentration higher than 180 µM, the critical micellar concentration. Unilamellar DMPS vesicles were prepared as described previously (Cevc et al., 1981). The interaction of bombesin with the micelles or vesicles was studied by addition of small aliquots of the lipids to a known volume of the peptide.

Fluorescence measurements were carried out with a Perkin Elmer MPF-44A spectrophotofluorimeter equipped with a DCSU2 correction unit. The excitation wavelength used was 280 nm. The excitation and emission bandpass was 3 nm. Fluorescence quantum yields were evaluated using N-acetyltryptophanamide (NATA) as a standard (Szabo and Rayner, 1980) with a value of 0.14. Circular dichroism spectra were obtained using a Jasco 500 A spectropolarimeter equipped with a computer for spectral smoothing. Normally eight spectra were stored and averaged. 0.2 mm cells were used and placed close to the photomultiplier in order to minimize scattering distortions. The molar ellipticity was calculated using a mean residue weight of 115. Absorption measurements were carried out using a Perkin Elmer 576 or Cary 219 spectrophotometer. In all experiments the temperature was kept constant at 20°C using thermostated cell holders.

Time resolved fluorescence measurements were performed using the time correlated single photon counting technique with instrumentation described previously (Zuker et al., 1985). A Spectra Physics sync-pumped argon ion dye laser and cavity dumper system served as the excitation source operating at 825 KHz and with a pulse width of 15 ps. The excitation was vertically polarized at a wavelength of 295 nm. The emission was detected after passing through a polarizer set at 55° and a monochromator, with a 4 nm bandpass, on a Hamamatsu 1564 U microchannel plate photomultiplier. No correction for emission wavelength dependency of the instrument response function was required with this photomultiplier. The channel width typically was 21.6 ps/channel and fluorescence decays were collected in 1024 channels of a multichannel analyzer. Usually the data was collected until 25000 counts were accumulated in the maximum channel of the sample decay curve. The stop/start ratio was 100:1. A blank was also measured for each sample for the same accumulation time which was controlled by the clock in the MCA. The instrument response function was determined by measuring the scattered light from a glycogen sample having an optical density of 0.3 at the excitation wavelength. The blank signal was subtracted from the sample decay curve giving appropriate consideration to the weighting of the resultant curve. The data were analyzed using the non-linear least squares iterative convolution method based on the Marquardt algorithm. Adequacy of the exponential decay fitting was judged by the inspection of the plots of weighted residuals and other statistical parameters (MacKinnon et al., 1977).

RESULTS

Circular Dichroism

The far UV circular dichroism spectra of an aqueous solution of
bombesin at pH 7 was recorded and is presented in figure 1a and 1 b. This
spectrum was characterized by a maximum negative signal centered near 200
nm with an ellipticity $[\theta] = -10^4$ deg cm^2 dmol^{-1}. The addition of DMPS
vesicles (Figure 1a) to a lipid/peptide ratio (L/P) = 5 resulted in a small
positive maximum at 189 nm with a shift of the negative maximum to 203 nm
and an increase of ellipticity in the region around 220 nm. At a L/P = 50
for DMPS the spectrum was significantly different from that of bombesin in
aqueous solution having a clear maximum at 190 nm ($[\theta] = 7.5 \times 10^3$ deg cm^2
dmol^{-1}) and a minimum at 206 nm ($[\theta] = -6.5 \times 10^3$ deg cm^2 dmol^{-1}) with a
shoulder centered near 220 nm ($[\theta] = -3.5 \times 10^3$ deg cm^2 dmol^{-1}). Further
addition of DMPS vesicles caused no further spectral nor ellipticity
changes. In the case of the addition of lysolecithin micelles (Figure 1b),
a spectrum comparable to that of a L/P = 5 for DMPS was not achieved until
a L/P = 80. Addition of lysolecithin up to a L/P = 200 resulted in a
spectrum which appeared to be comparable to that for DMPS at a L/P = 50
except that the ellipticity at 190 nm was only 5×10^3 deg cm^2 dmol^{-1}. The

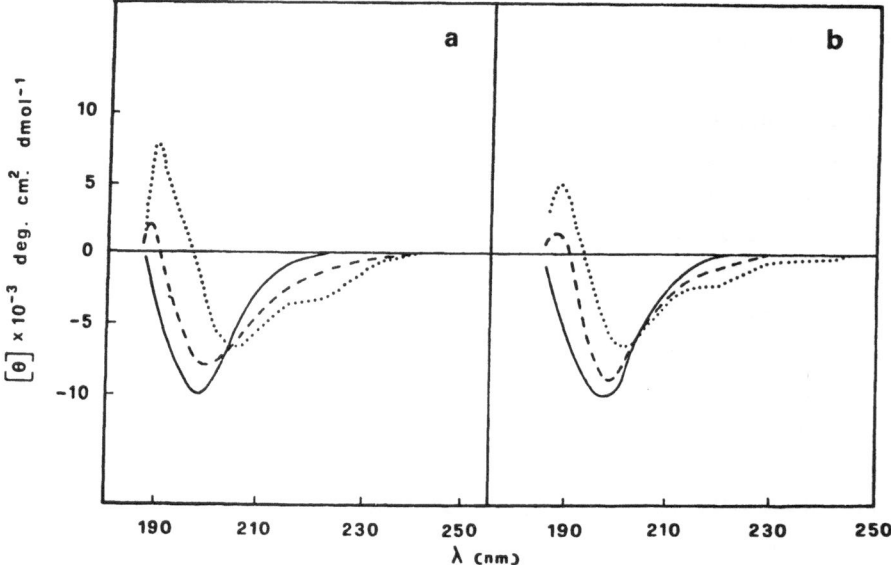

Figure 1. Circular dichroism spectra of bombesin (solid line) and
a) bombesin-DMPS complexes, pH 7: L/P = 5 (dashed line), L/P = 100 (dotted
line); **b)** bombesin-lysolecithin complexes, pH 7: L/P = 80 (dashed line);
L/P = 200 (dotted line).

minimum was found at 203 nm ($[\theta]$ = -6.4 x 10^3 deg cm² dmol⁻¹) with a weaker shoulder centered near 220 nm whose ellipticity was less than that for DMPS at a L/P = 50.

Static Fluorescence

The corrected fluorescence spectrum of bombesin (10 mM sodium cacodylate buffer, pH 7, 20°C) had a maximum at 351 nm, with a half band width of 57 nm (Figure 2). The fluorescence quantum yield was found to be 0.09. Addition of 6 M GdHCl caused an increase in fluorescence intensity of 30% but no change in spectral shape or maximum. It must be noted that the absorption spectrum of the solution also changed, shifting slightly to the red. Figure 3 shows the quantum yield behavior with pH. Again, no significant spectral shift was observed. The pH titration curve was sigmoidal in character and had a mid-point near pH = 6.5.

When DMPS vesicles were progressively added to bombesin (1.25 x 10⁻⁵M)

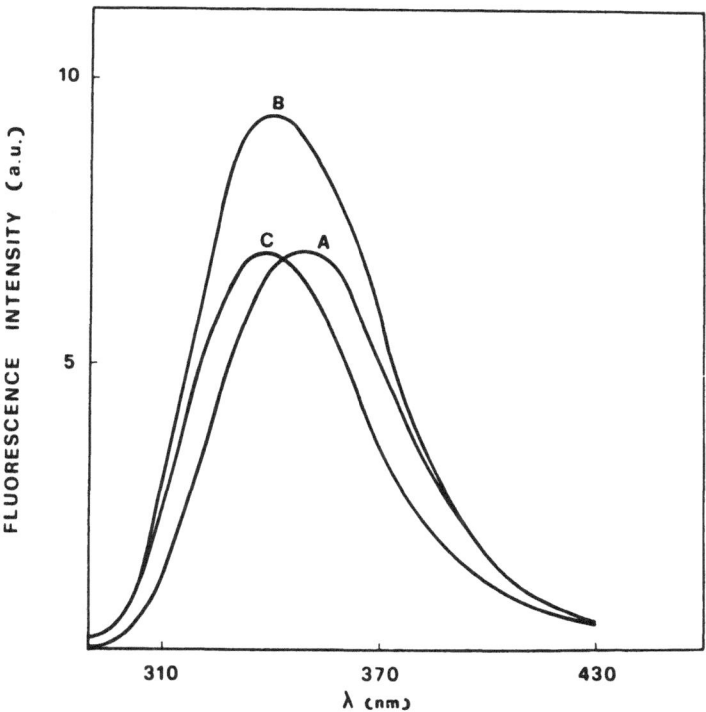

Figure 2. Fluorescence spectra of bombesin, 1.25 x 10⁻⁵ M; **A)** pH 7, 10 mM cacodylate buffer, 20°C; **B)** bombesin-lysolecithin, pH 7, L/P = 200, 20°C; **C)** bombesin-DMPS, pH 7, L/P = 100, 20°C.

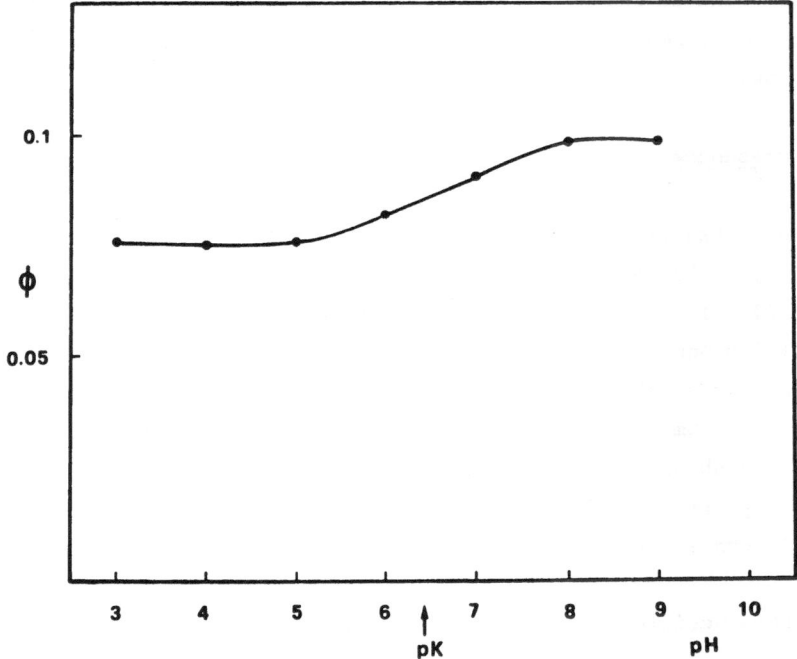

Figure 3. Effect of pH on the fluorescence quantum yield of bombesin.

at pH 7, the spectral maximum shifted to shorter wavelength until at a L/P = 70. At higher ratios, no further change was observed and the maximum was found at 339 nm (Figure 4a). The fluorescence intensity of the maximum appeared to remain relatively constant during the addition. At 339 nm the fluorescence intensity increased by 11% at a saturating DMPS concentration.

In the case of the addition of lysolecithin micelles, again a shift to higher energy was observed, but a L/P = 200 was required before no further changes were observed. The spectral maximum was found near 341 nm (Figure 2), a 2 nm higher wavelength than that found for the DMPS-bombesin complex, along with an increase in fluorescence of 25%.

At pH 8.5 the titration behavior of bombesin with DMPS vesicles (Figure 5) was different from that found at pH 7. The position of their spectral maxima for the DMPS-bombesin complex (338 nm) and bombesin (350 nm) were virtually identical with their spectra at pH 7. However, there was a significant increase in fluorescence yield on addition of DMPS vesicles to bombesin at pH 8.5 compared to that at pH 7. At 338 nm the

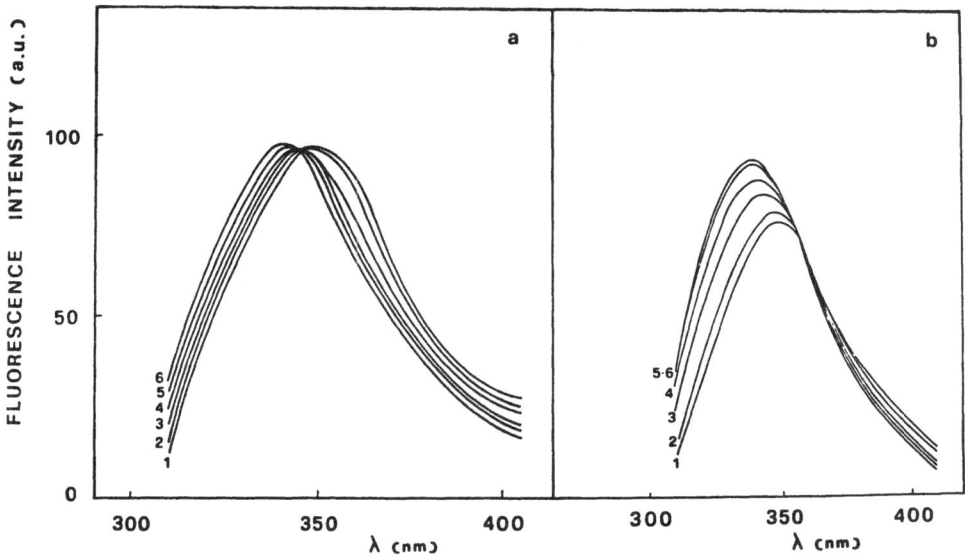

Figure 4. **a)** Fluorescence titration of bombesin, 1.25 x 10⁻⁵ M with DMPS vesicles at pH 7; 1) L/P=0, 2) L/P=10, 3) L/P=24, 4) L/P=37, 5) L/P=53, 6) L/P=70. **b)** Fluorescence titration of bombesin with DMPS vesicles at pH 8.5; 1) L/P=0, 2) L/P=10, 3) L/P=24, 4) L/P=37, 5) L/P=53, 6) L/P=70.

fluorescence intensity increased by 35%, and the quantum yield increased to a value of 0.12. At pH 8.5 the results for the lysolecithin titration of bombesin was the same as that seen at pH 7.

When bombesin was titrated with DMPS vesicles at pH 5, only a small shift of the spectral maximum to 345 nm was observed and there was no change in fluorescence intensity.

Table 1. Fluorescence Anisotropy Values of Bombesin and Bombesin-Lipid Complexes at pH 7.

Sample	Anisotropy
BBS (H₂O)	0.024
BBS-lysolecithin L/P = 200	0.062
BBS-DMPS L/P = 100	0.098

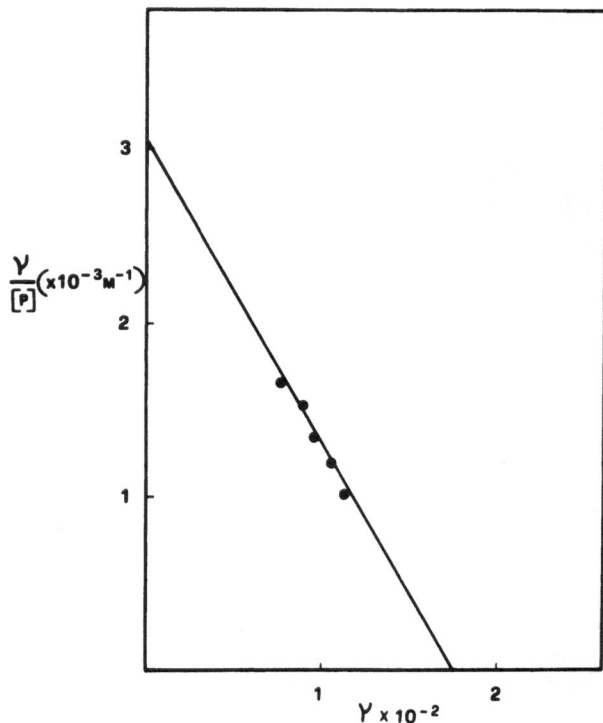

Figure 5. Scatchard plot of the interaction of bombesin with DMPS vesicles at pH 8.5.

The Scatchard plot analysis of the fluorescence titration data for DMPS at pH 8.5 (Figure 5) gave an apparent binding constant K_b = 1.75 x 10^6 M^{-1}, with a stoichiometry of 57 DMPS molecules/peptide. From this binding data a ΔG^o = -6.8 kcal mole^{-1} could be calculated.

When KCl was added to a DMPS-bombesin complex (L/P = 100) at pH 7, the fluorescence spectrum shifted to longer wavelength until at 0.3 M KCl and the maximum was observed at 346 nm (Figure 6). Further addition of KCl resulted in precipitation of the solutes and no satisfactory spectra could be recorded.

The steady state fluorescence anisotropy of bombesin in aqueous solution and of the complexes at pH 7 was measured. The results are summarized in Table 1. Further addition of either DMPS vesicles or lyso-lecithin micelles caused no further change in the anisotropy values. Similar values were found at pH 8.5.

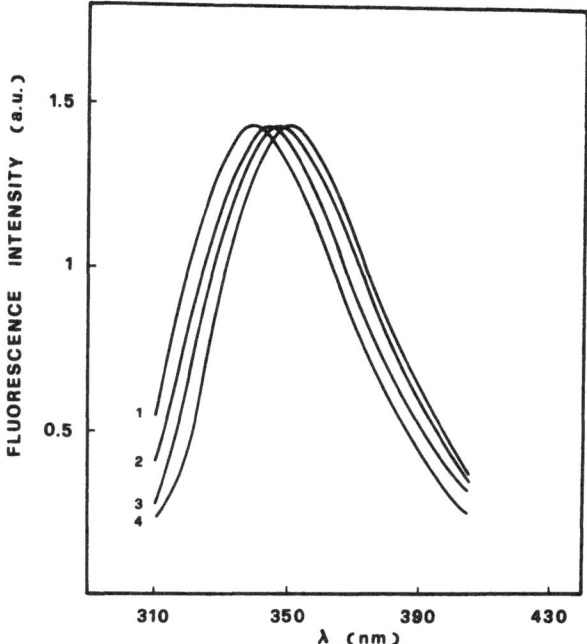

Figure 6. Effect of KCl on the fluorescence spectrum of bombesin-DMPS complexes, L/P=100; 1) 0 KCl; 2) 0.15 M KCl; 3) 0.3 M KCl; 4) bombesin alone.

Time Resolved Fluorescence

The fluorescence decay of bombesin in aqueous solution, bombesin-DMPS complex (L/P = 100), and bombesin-lysolecithin complex (L/P = 100), all at pH 7, were first measured at 325 nm and 380 nm. Figure 7 shows a log plot of the fluorescence intensity decay, measured at 380 nm, for a bombesin-DMPS complex (L/P = 100) and that of the instrument response function measured at the excitation wavelength (295 nm). The sample fluorescence decay behavior was clearly multi-exponential. The plots of weighted residuals for the "best fit" to sums of one, two, and three exponentials to the data are shown in figure 8a. Only when the data was fit to a sum of three exponentials with τ_1 = 4.97 ns, τ_2 = 1.71 ns, and τ_3 = 0.31 ns with a_1 = 0.23, a_2 = 0.45, and a_3 = 0.33, was the plot of weighted residuals randomly distributed. Other statistical criteria such as χ^2, and SVR (MacKinnon et al., 1977) also indicated a satisfactory fit had been achieved. An important criterion for deciding whether the data might be fit to a distribution function rather than a sum of discrete exponential components was the examination of the correlation matrix of the decay parameters (Szabo, Zuker, Ridgeway and Krajcarski, unpublished results). This matrix showed

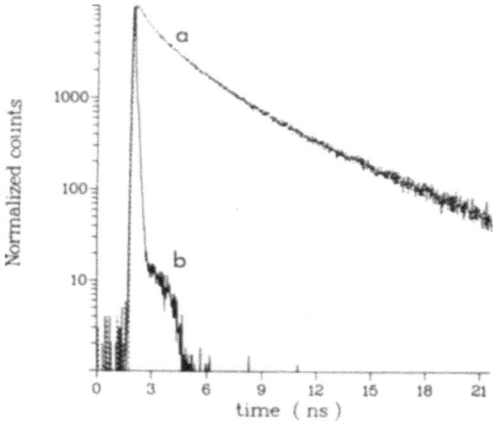

Figure 7. Log plot of a) fluorescence decay profile of bombesin-DMPS, L/P=100, pH 7, 20°C; channel width = 21.6 ps, 380 nm; b) instrument response function measured at 295 nm with a glycogen scatter.

that the decay parameters were not highly correlated. Similar data analysis was performed on several different samples at 325 nm and 380 nm. In every case only a sum of three exponential decay components provided a satisfactory fit to the data. These results are summarized in Table 2.

Clearly it can be seen that the decay times of the three components changed when bombesin interacted with the lipids. Most striking, however, was the observation of a marked difference between the fractional fluorescence values at 325 nm and 380 nm for the bombesin-DMPS complex. For bombesin in solution and lysolecithin complexes there was almost no difference in the fractional fluorescence values either at the two wavelengths or between these two samples.

The similarity of the lifetime values at these two well separated wavelengths suggested that the fluorescence spectrum might be resolvable into three distinct spectral components. Hence the fluorescence decay profiles were measured over the entire spectral range, 310 nm - 400 nm for each sample and the data were analyzed using a global analysis (Knutson et al., 1982) procedure. In the case of bombesin in aqueous buffer it was not possible to obtain a satisfactory fit when the data at all wavelengths were simultaneously analyzed. Excellent fits to three exponential decay components were found for the spectral region 310 nm - 335 nm, and 340 nm - 400 nm but each set had different decay times (Table 3). In the case of bombesin in lipid complexes, however, excellent global analyses for three

182

Table 2. Fluorescence Decay Parameters for Bombesin and Bombesin Lipid Complexes at pH 7.

Sample	λ(nm)	τ_1(ns)^o	τ_2(ns)^o	τ_3(ns)^o	F_1*	F_2*	F_3*
BBS	325	3.00	1.14	0.14	0.66	0.30	0.04
	380	3.23	1.58	0.22	0.64	0.34	0.02
BBS-DMPS	325	4.14	1.29	0.28	0.39	0.45	0.16
L/P = 100	380	4.97	1.71	0.31	0.56	0.38	0.05
BBS-lysolecithin	325	5.55	2.00	0.49	0.63	0.32	0.04
L/P = 200	380	5.70	2.08	0.36	0.60	0.39	0.02

^{o} Standard errors were, τ_1) ± 0.02 ns, τ_2) ± 0.02 ns, τ_3) ± 0.01 ns
* F_1, F_2, F_3, are the fractional fluorescence contributions of each component to the total fluorescence where, $F_i = a_i \cdot \tau_i / \Sigma a_i \cdot \tau_i$.

exponential decay components were obtained over the entire spectral range. All statistical parameters consistently showed that the data were adequately represented by a sum of three exponential decay components. Figure 8b presents the plots of weighted residuals for the global analyses of the bombesin-DMPS complex at 325 nm, 350 nm, and 380 nm. The weighted residuals were clearly randomly distributed in each case. The decay time data are summarized in Table 3.

Satisfactory results could not be obtained at 310 nm for bombesin-DMPS complexes owing to an inappropriate estimation of a scattered light component at that wavelength originating from the vesicle preparation. It may be that, even for bombesin in aqueous solution, the data were less than satisfactory at 310 nm and hence the necessity of splitting the decay analysis of this sample into two spectral regions. Every precaution had been taken to eliminate parasitic light. Time resolved measurements on the

Table 3. Fluorescence Decay Times Obtained After Global Analysis Over the Fluorescence Spectra of Bombesin and Bombesin-Lipid Complexes, pH 7*

Sample		Spectral Range (nm)	τ_1(ns)	τ_2(ns)	τ_3(ns)
BBS		310 - 335	3.00	1.15	0.14
		340 - 400	3.19	1.52	0.21
BBS-lysolecithin	L/P=200	310 - 400	5.67	2.08	0.37
BBS-DMPS	L/P=100	320 - 390	4.60	1.52	0.30

*Standard errors were, τ_1) ± 0.01 ns, τ_2) ± 0.01 ns, τ_3) ± 0.01 ns

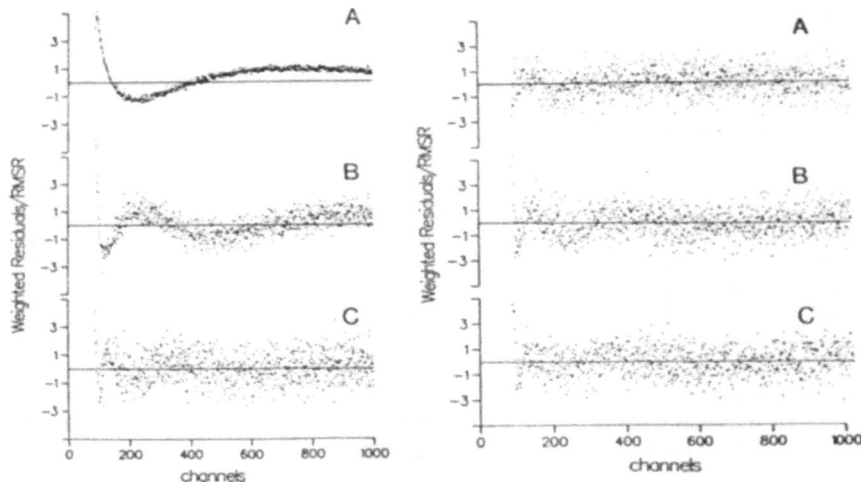

Figure 8. **Left:** Plots of weighted residuals after iterative convolution of the data in figure 8 for the 'best fits' for a) single exponential decay function; b) double exponential decay function; c) triple exponential decay function. **Right:** Plots of weighted residuals of 'best fits' after global analysis for three exponential decay components for the bombesin-DMPS complex, L/P = 100 pH 7; for a) 325 nm; b) 350 nm; c) 380 nm.

bombesin-DMPS complex at pH 8.5 have not been performed but are of obvious interest.

In the usual way that steady state fluorescence spectra and spectrally resolved fluorescence decay measurements (Szabo and Rayner, 1980) have been treated, we obtained the fluorescence spectrum of each decay component. The fractional fluorescence values of the three decay components at each wavelength of observation for each sample are summarized in Figure 9. It was evident that the values of the respective components were very similar for both bombesin in aqueous solution and bombesin-lysolecithin complexes (pH 7) over the entire spectral range. In the case of the bombesin-DMPS complex the fractional fluorescence values were significantly different from the other two samples and varied over the measured spectral range. The three decay components for bombesin in aqueous solution had a similar spectral shape and maxima near 350 nm (Figure 10). In the case of the bombesin-lysolecithin complex, the 5.67 ns component had a spectral maximum near 340 nm, the 2.08 ns component appeared to have a maximum at a slightly higher wavelength, 345 nm, while the 0.37 ns component had a spectral maximum centered near 330 nm. The 5.67 ns component dominated the spectrum (Figure 10). The time resolved spectrum for the bombesin-DMPS complex (Figure 11) was markedly different from the others presented above. The 4.60 ns component had a maximum near 340 nm, while the spectrum of the 1.52

184

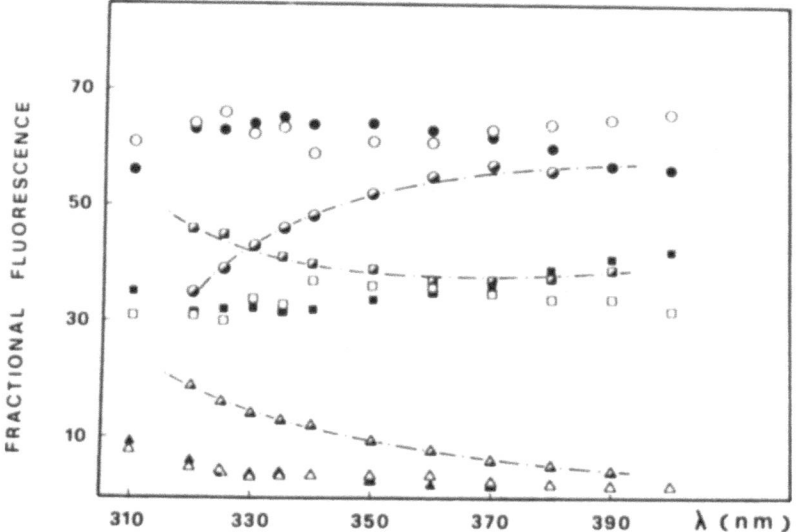

Figure 9. Variation of fractional fluorescence values for each bombesin sample and for each decay component for the respective samples. Bombesin; 3.10 ns (O); 1.35 ns (□); 0.17 ns (Δ).
Bombesin-lysolecithin, L/P = 200; 5.67 ns (●); 2.08 ns (■); 0.37 ns (Δ).
Bombesin-DMPS, L/P = 100; 4.60 ns (◐); 1.52 ns (◪); 0.30 ns (▲).
In the figure only the curves relative to Trp families which interconvert, are evidenced.

ns component, whose intensity was comparable to the longer decay time component, had a maximum close to 335 nm. The 0.30 ns component also had an increased intensity when compared to the corresponding component for bombesin. Its spectral maximum appeared to be close to 325 nm.

DISCUSSION

The fluorescence titration of bombesin with lipids demonstrated that when DMPS was added to a L/P = 70 or, for lysolecithin, L/P = 200 the fluorescence signal reached a plateau value. Additionally, after further addition of the lipids, there was no further change in the steady state fluorescence anisotropy values nor in the CD spectra. These observations indicated that at those L/P ratios there was negligible free bombesin remaining in the sample. The Scatchard plot analysis for bombesin-DMPS (pH 8.5) showed that a unique lipid-peptide complex had been obtained. Hence the data collected at those L/P ratios or higher may be considered to be representative of a single type of complex.

The CD spectrum of bombesin in aqueous solution was typical of that obtained for a protein or peptide in an extended random coil conformation. The appearance of a positive ellipticity at 190 nm, a decrease in negative ellipticity at 200 nm and a negative ellipticity maximum at 208 nm in bombesin-DMPS complexes, can be attributed to the formation of a conformation of the peptide which has a significant degree of α-helical structure on interaction with the lipids. In the lysolecithin complexes qualitatively similar changes in the CD spectra again suggest the formation of an α-helical conformation in these complexes. However, the ellipticity values were quantitatively different from those of the DMPS complex, which implied that the structure of the bombesin in the two complexes was not identical.

The fluorescence behavior of bombesin was typical for that of a tryptophan residue which was exposed to an aqueous environment (Burstein et al., 1973). Initially it might be considered that the increase in fluorescence by 30% on the addition of 6M GdHCl might have suggested the loss of some secondary structure which was not indicated by the CD spectra. We noted that the fluorescence of the standard tryptophan derivative NATA in GdHCl also increased by a similar fraction. However, the fluorescence decay behavior of NATA in aqueous solution and in 6M GdHCl were identical, obeying single exponential decay kinetics with a lifetime of 2.95 ns. Furthermore a spectral shift of the absorption spectra resulting in a change in the optical density at the excitation wavelength could account for this fluorescence intensity change (Cavatorta, Szabo, Krajcarski and Masotti, unpublished results). The effect of pH on the tryptophan fluorescence, showing sigmoidal behavior with an apparent $pK_a = 6.5$, indicated that there was some interaction between an ionizable group such as a histidine residue and the tryptophan. The single histidine residue (His 12) in bombesin is located four residues from the tryptophan (Trp 8) and the decrease in fluorescence at pH 5 suggests a proton transfer quenching mechanism of the tryptophan excited singlet state at that pH (Szabo and Rayner, 1980). This interacion between the histidine and tryptophan residues will be discussed further below, but it is interesting to point out that both residues have been shown to be essential for bombesin's biological activity (Erspamer and Melchiorri, 1983). In solution, therefore, bombesin appears to exist in an extended, flexible conformation with very little if any ordered secondary structure. In the peptide-lipid complexes some α-helical structure was formed with more helical character being present in the bombesin-DMPS complex. The short length of the bombesin chain makes it difficult to predict at what level the helix formation occurs. But, because of the helicogenic character of all of the residues except the first three at the N-terminus, it is reasonable to

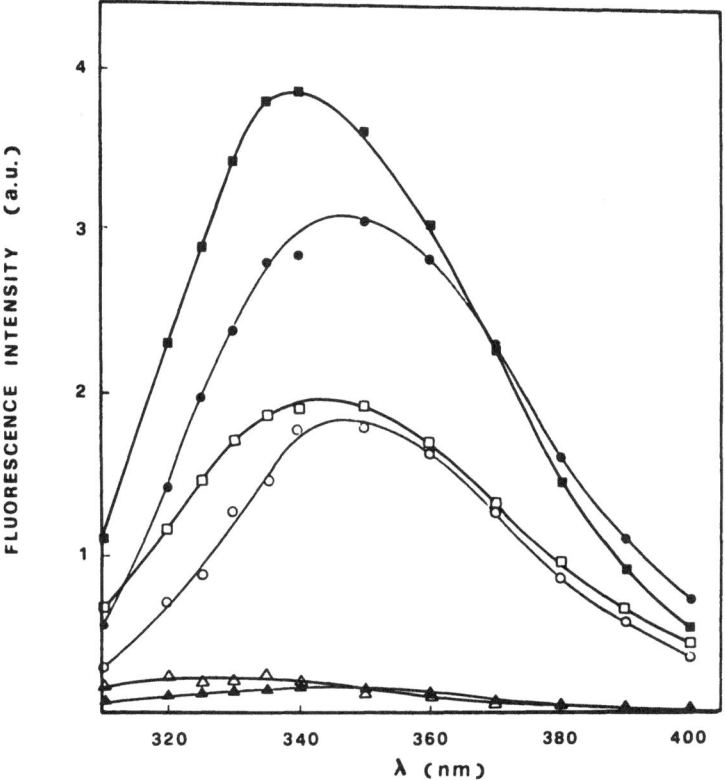

Figure 10. Time resolved fluorescence spectra of bombesin, pH 7; 3.1 ns (■); 1.35 ns (□); 0.17 ns (Δ); bombesin-lysolecithin complex, L/P = 200; 5.67 ns (●); 2.08 ns (O); 0.37 ns (▲).

suggest that the peptide would be able to assume considerable α-helical structure.

The amphipatic character of bombesin, evaluated using the methods described by Kyte and Doolittle (Kyte and Doolittle, 1982), indicated that the N-terminal portion of bombesin was hydrophilic while the C-terminal residues had a high degree of hydrophobic character. It is therefore reasonable to assume that, at least in the DMPS complexes, the arginine residue (Arg 3) near the N-terminus would bind to the negatively charged head group of the DMPS vesicles and, in the case of the electrically neutral lysolecithin micelles, the hydrophilic end would be exposed to the solvent. Then the C-terminal octapeptide would penetrate deepest into the lipid bilayer (DMPS) or lipid interior and could be organized in a helical structure, and in such a structure with approximately four residues per

187

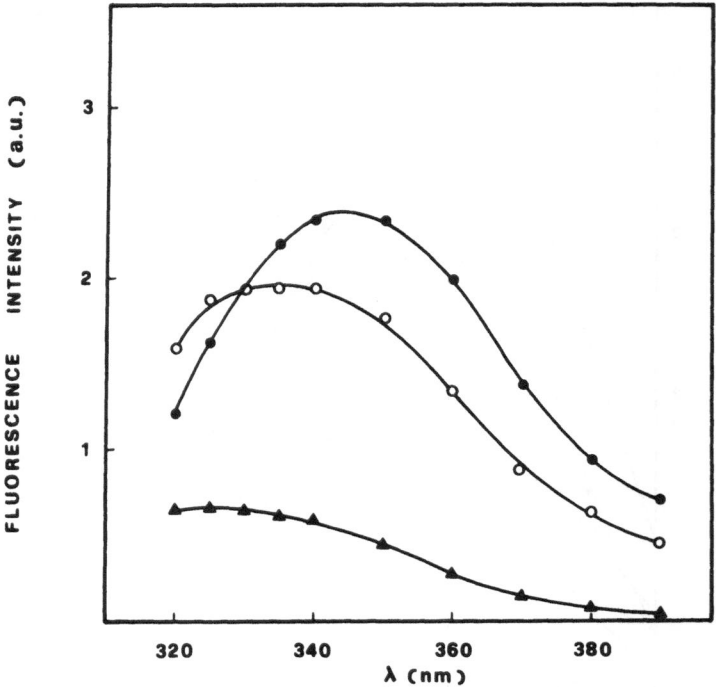

Figure 11. Time resolved fluorescence spectra of bombesin-DMPS complex, L/P = 100, pH 7; 4.60 ns (●); 1.52 ns (O); 0.30 ns (▲).

turn the histidine and tryptophan residues would come in close contact with one another. The free energy change, calculated for the bombesin-DMPS interaction, was very similar to the values found for the amphipathic helices formed in the lipid complexes of apolipoproteins or glucagon (Jahnig, 1983). These proteins are thought to associate with membranes forming α-helices, whose hydrophobic structure penetrated into the lipid interior with their hydrophilic portion remaining exposed to the aqueous environment.

The changes in fluorescence spectra, occurring on addition of lipid, demonstrate that bombesin interacts with both neutral and charged lipids. The shift of bombesin fluorescence maxima from 351 nm in aqueous solutions to 341 nm and 339 nm on the addition of lysolecithin and DMPS (pH 7 and 8.5), respectively, may be attributed to the tryptophan experiencing a non-polar, hydrophobic environment such as would be found in the acyl chain environment in the interior of lipid systems. The effect of KCl addition to the bombesin-DMPS complex implies that an electrostatic interaction plays a role in the formation of the complex. This may be due either to a competition of the K⁺ cations with the positively charged arginine residue of bombesin (also the positively charged histidine at pH values less than 7) for the negatively charged head group of DMPS. Also it is possible that the salt may later the structure of the vesicles preventing penetration of the peptide into the bilayer interior. Such an effect has been observed for DMPS vesicles on interaction with the small model peptide lysine-tryptophan-lysine (Szabo, Yamashita and Krajcarski, unpublished results). Similarly, the lack of a significant spectral shift on binding at pH 5 may be the result of a change in the DMPS vesicle structure since it is known that the phase transition of these vesicles is found at 41°C at pH 5 while at pH 7 it is observed at 29°C (Cevc et al., 1981). However, we consider it more likely that at pH 5 the histidine residue, which would also be protonated, would bind to the surface of the DMPS vesicles along with the arginine residue, hence preventing penetration of the tryptophan and C-terminal octapeptide into the bilayer interior. Therefore only a small shift of 6 nm in the fluorescence spectrum was observed at this pH. In the case of lysolecithin the lack of any KCl effect suggested that in these micelles the interaction with bombesin was entirely hydrophobic. To test the alternative effects of lipid structure or peptide protonation on the interaction of bombesin requires additional measurements.

The above suggests that the protonation of the histidine residue may play an important role in determining the nature of the complex. At pH 7 only a fraction of the peptide would have a protonated histidine residue; therefore, an important fraction of the bombesin may interact with DMPS by penetrating into the bilayer interior and the spectral shift to 339 nm may reflect this type of structure. The increase in fluorescence intensity at pH 8.5, compared to that at pH 7 for DMPS, is likely due to the fact that the histidine is no longer protonated and a majority of the bombesin could enter into the bilayer interior. Additionally, any quenching of the tryptophan fluorescence caused by the protonated histidine residue would not be observed.

The increase in fluorescence anisotropy from 0.024 to 0.098 in DMPS complexes indicated that bombesin assumed a less flexible and perhaps more ordered structure in the lipids. We therefore conclude that, in the case of DMPS at pH 7 and 8.5, the positively charged arginine residue bound to the negatively charged head groups. The C-terminal octapeptide sequence may form an α-helical structure, inserting into the acyl chain environment of the bilayer. At pH 5 the peptide bound solely to the surface of the bilayer with little penetration into the interior. In lysolecithin, by contrast, bombesin was able to penetrate into the acyl chain region of the micelles and assumed a conformation with a lesser amount of α-helical structure.

The time resolved fluorescence results provided additional insights into the structure and conformational heterogeneity of bombesin, both free in solution and in the lipid complexes. The observation of three fluorescence decay components indicated that the tryptophan, or at least that segment of the peptide containing the tryptophan, existed in at least three average conformational states each with a characteristic decay time. This would result if, because of interactions with different residues of the peptide, the tryptophan would have non-radiative deactivation pathways. In free solution the spectral components had similar maxima and there was very little of the short lifetime component. In the bombesin-lysolecithin complex the increased value of each lifetime component was consistent with the entry of the tryptophan residue into the acyl chain environment with concomitant reduced importance of non-radiative deactivation by the solvent. The time resolved emission spectrum (Figure 10) showed that each component underwent a spectral shift to higher energy as expected in the hydrophobic interior of the lysolecithin micelle. In the case of DMPS complexes, the spectra of each component clearly showed that the tryptophan was located in the acyl chain interior of the bilayer. But now the proportion of each of the components changed considerably from the other two cases presented. The shorter lifetimes observed in DMPS complexes, compared to those in lysolecithin, may be the result of increased quenching by the protonated histidine. Alternatively, since, as we suggested above, if the histidine is protonated the C-terminal segment cannot enter the bilayer interior, it is more likely that the formation of a tighter α-helical structure in the DMPS complexes results in a charge transfer interaction between the excited singlet state of the tryptophan and the histidine residue reducing the excited state lifetime. The more structured DMPS vesicle bilayer would require a higher degree of helical structure in order that the peptide can penetrate into its interior matrix. These time resolved fluorescence studies clearly showed that the structure of the bombesin-lipid complexes were quite different in DMPS and in lysolecithin, which had been suggested earlier by the static spectroscopic measurements.

In this work we have shown how the combination of CD spectra, steady state and time resolved fluorescence can provide important information on the structure and conformational heterogeneity of peptides and on their interaction with different types of lipid environments. Clearly more studies are required and are currently in progress: the results reported in this paper, however, give more evidence in support of the view that biologically active peptides have the ability to assume specific conformations which are required to exert their biological activity.

ACKNOWLEDGEMENTS

The authors wish to give special acknowledgement to Professor V. Erspamer for interesting and stimulating discussions. This work was performed with the financial assistance of a NATO travel grant; a grant from the National Science and Engineering Research Council of Canada (NSERC) and a grant from MPI (40%) Rome. The expert technical assistance of D. T. Krajcarski is also acknowledged as well as the development of the computer programming by M. Zuker and J. Ridgeway of the National Research Council Biomathematics section.

REFERENCES

Abe, H., Chihara, K., Minamitani, N., Iwasaki, J., Chiba, T., Matsukura, S., and Fujita, T., 1981, Stimulation by Bombesin of Immunoreactive Somatostatin Release into Rat Hypophyseal Portal Blood, Endocrinology, 109:229.

Broccardo, M., Falconieri Erspamer, G., Melchiorri, P., Negri, L., and de Castiglione, R., 1975, Relative Potency of Bombesin-Like Peptides, Br. J. Pharmac., 55:221.

Brown, M., and Vale, W., 1979, Somatostatin: Central Nervous System Actions on Glucoregulation, Endocrinology, 104:1709.

Brown, M., and Vale, W., 1981, in: Diabetes Mellitus: Diagnosis and Treatment, H. Rifin, and P. Raskin, eds., R. J. Brady Co., Bowie, MD, pp. 55-62.

Brown, M., Rivier, J., Kobayashi, R., and Vale, W., 1978, in: Gut Hormones, S. R. Bloom, ed., Churchill Livingstone, Edinburgh, pp. 550-558.

Brown, M., Taché, Y., Rivier, J., and Pittman, Q., 1981, Peptides and Regulation of Body Temperature, Adv. Biochem. Psychopharmacol., 94:132613c.

Burstein, E. A., Vedenkina, N. S., and Ivkova, M. N., 1973, Fluorescence and the Location of Tryptophan Residues in Protein Molecules, Photochem. Photobiol., 18:263.

Cevc, G., Watts, A., and Marsh, D., 1981, Titration of the Phase Transition of Phosphatidylserine Bilayer Membranes. Effects of pH, Surface Electrostatics, Ion Binding and Head-Group Hydration, Biochemistry, 20:4955.

Cuttitta, F., Carney, D. N., Mulshine, J., Moody, T. W., Fedorko, J., Fischler, A., and Minna, J. D., 1985, Bombesin-like Peptides can Function as Autocrine Growth Factors in Human Small-cell Lung Cancer, Nature, 316:823.

De Caro, G., Massi, M., and Micossi, L. G., 1980, Modifications of Drinking Behaviour and of Arterial Blood Pressure Induced by Tachykinesis in Rats and Pigeons, Psychopharmacology, 68:243.

Dubrasquet, M., Rozé, C., Ling, N., and Florencio, H., 1982, Inhibition of Gastric and Pancreatic Secretions by Cerebroventricular Injections of Gastrin-Releasing Peptide and Bombesin in Rats, Regul. Peptides, 3:105.

Erspamer, V., and Melchiorri, P., 1980a, Amphibian Skin Peptides and Mammalian Neuropeptides, in: Growth Hormone and Other Biologically Active Peptides, A. Pecile, and E. E. Muller, eds., Excerpta Medica, Amsterdam, pp 185-200.

Erspamer, V., and Melchiorri, P., 1980b, Active Polypeptides from Amphibian Skin to Gastrointestinal Tract and Brain of Mammals, Trends Pharmacol. Sci., 1:391.

Erspamer, V. and Melchiorri, P., 1983, Actions of Amphibian Skin Peptides on the Central Nervous System and the Anterior Pituitary, in: Neuroendocrine Perspectives, E. Muller, and R. M. MacLeod, eds., Vol. 2. Elsevier, New York, pp. 37-106.

Erspamer, V., Falconieri Erspamer, G., Improta, G., Negri, L., and de Castiglione, R., 1980, Sauvagine, A New Polypeptide from Phyllomedusa sauvagei Skin, Naunyn-Schmiedeberg's Arch. Pharmacol., 312:265.

Jahnig, F., 1983, Thermodynamics and Kinetics of Protein Incorporation into Membranes, Proc. Natl. Acad. Sci. USA, 80:3691.

Knutson, J. R., Wallbridge, D. G., and Brand, L., 1982, Decay-Associated Fluorescence Spectra and the Heterogeneous Emission of Alcohol Dehydrogenase, Biochemistry, 21:4671.

Kyte, J. and Doolittle, R. F., 1982, A Simple Method for Displaying the Hydropathic Character of α Protein, J. Mol. Biol., 157:105.

MacKinnon, A. E., Szabo, A. G., and Miller, D. R., 1977, The Deconvolution of Photoluminescence Data, J. Phys. Chem., 81:1564.

Martin, C. F., and Gibbs, J., 1980, Bombesin Elicits Satiety in Sham Feeding Rat, Peptides, 1:131.

Montecucchi, P. C., de Castiglione, R., Piani, S., Gozzini, L., and

Erspamer, V., 1981, Identification of Dermorphin and Hyp[6]-Dermorphin in Skin Extracts of the Brazilian Frog Phyllomedusa Rhodei, Int. J. Peptide Protein Res., 17:316.

Montecucchi, P. C., de Castiglione, R., Piani, S., Gozzini, L., and Erspamer, V., 1981, Amino Acid Composition and Sequence of Dermorphin, A Novel Opiate-Like Peptide from the Skin of Phyllomedusa Sauvagei, Int. J. Peptide Protein Res., 17:275.

Moody, T. W., Pert, C. B., Rivier, J., and Brown, M. R., 1978, Bombesin: Specific Binding to Rat Brain Membranes, Proc. Natl. Acad. Sci. USA, 75:5372.

Rivier, C., Rivier, J., and Vale, W., 1978, The Effect of Bombesin and Related Peptides on Prolactin and Growth Hormone Secretion in the Rat, Endocrinology, 102:519.

Szabo, A. G., and Rayner, D. M., Fluorescence Decay of Tryptophan Conformers in Aqueous Solution,1980, J. Am. Chem. Soc., 102:554.

Westendorf, J. M., and Schombrunn, A., 1982, Bombesin Stimulates Prolactin and Growth Hormone Release by Pituitary Cells in Culture, Endocrinology, 110:352.

Zuker, M., Szabo, A. G., Bramall, L., Krajcarski, D. T., and Selinger, B., 1985, Delta Function Convolution Method (DFCM) for Fluorescence Decay Experiments, Rev. Sci. Instrum., 56:14.

USEFULNESS OF THE LINKAGE CONCEPT TO UNDERSTANDING THE REGULATION OF RAT LIVER PHOSPHOFRUCTOKINASE

Gregory D. Reinhart

Department of Chemistry
University of Oklahoma
Norman, OK 73019 USA

INTRODUCTION

Although this tribute to Gregorio Weber has quite appropriately
focused on the use of fluorescence methodologies in the study of macromolecules, most of which he pioneered, it is perhaps also fitting to
discuss other contributions to our current thinking regarding protein
biophysics, that do not necessarily involve the exploitation of the fluorescence phenomenon, to which the Professor has made seminal contributions
as well. I refer specifically to several papers (Weber, 1971, 1972, 1975),
written in the early 1970's, pertaining to protein ligand binding and the
ramifications of the principles of linkage that must be satisfied when
evaluating binding behavior. After a brief review of the points made in
these papers, I would like to then discuss how the concept of linkage, as
elaborated by Dr. Weber, has proven to be invaluable in providing a greater
insight into the mechanisms governing the allosteric regulation of an
enzyme of considerable interest to those studying the metabolic control of
hepatic carbohydrate metabolism; namely, rat liver phosphofructokinase
(PFK).

Jeffries Wyman was actually the first to recognize the principle of
linkage (and coin the name) as it applies, in particular, to the binding of
small molecular weight ligands to proteins (Wyman, 1948, 1964, 1967).
Linkage refers to the fact that the chemical potential of a bound ligand
will be altered by the subsequent binding of another ligand to the same
protein molecule by exactly the same amount as would the chemical potential
of the second ligand if the order of ligand binding were reversed. Thus
the binding potentials of the two ligands are "linked" in that each is

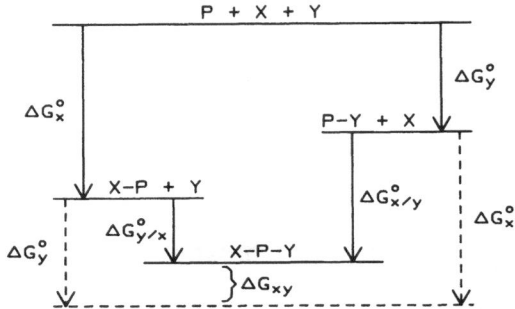

Figure 1. Free energy diagram of the two ways in which two ligands, X and Y, can bind to a protein, P, yielding doubly ligated protein, X-P-Y, as originally depicted by Weber (1971, 1972, 1975). ΔG°_{x} and ΔG°_{y} refer to the standard free energy of binding X and Y respectively to P with the length of the arrow being proportional to magnitude. $\Delta G^{\circ}_{x/y}$ and $\Delta G^{\circ}_{y/x}$ represent the standard free energy of binding X and Y, respectively, to singly-ligated forms of P, X-P and P-Y, respectively. The horizontal dashed line corresponds to the free energy realized if the binding of the two ligands are independent. The "free energy of interaction" or "coupling free energy" between X and Y, ΔG_{xy}, is given by the difference between the actual free energy of the doubly-ligated form and the anticipated free energy if the binding of X and Y were independent.

influenced by that of the other ligand to the same degree.

Weber made many contributions to our understanding of the ramifications that linkage has on protein function, not the least of which was recasting the argument in terms of binding free energies. In addition to avoiding the somewhat arcane terminology employed by Wyman, free energies could more clearly be related to the types and natures of the forces that are ultimately responsible for the interactions within a protein molecule and between the protein and ligand.

To visualize the free energy relationships occurring when two ligands, X and Y, bind to separate binding sites on a single protein molecule, P, Weber employed standard free energy "arrow" diagrams similar to the one presented in Figure 1. When a ligand binds to a protein, the standard free energy of the system will decrease by an amount equal to the standard free energy of binding. Different ligands of course bind with differing affinities, hence the change in free energy depicted in the diagram for the binding of X and the binding of Y, each to unligated protein, are different. When a protein with one ligand bound in turn binds the other ligand, e.g. X-P + Y ———> X-P-Y, the free energy change may be different than it was when Y bound to free protein. In other words, the affinity of the protein for Y may be altered by the binding of X. However, the end product of doubly-ligated protein, X-P-Y, is the same regardless of the order of ligand binding. Consequently:

$$\Delta G^{\circ}_x + \Delta G^{\circ}_{y/x} = \Delta G^{\circ}_y + \Delta G^{\circ}_{x/y} \tag{1}$$

where ΔG°_x and ΔG°_y equal the standard free energies of binding X and Y, respectively, to P. $\Delta G^{\circ}_{x/y}$ and $\Delta G^{\circ}_{y/x}$ equal the standard free energies of binding X and Y, respectively, to P-Y and X-P respectively.

If the binding of either ligand has no effect on the binding of the other, i.e. $\Delta G^{\circ}_x = \Delta G^{\circ}_{x/y}$ and $\Delta G^{\circ}_y = \Delta G^{\circ}_{y/x}$, then the binding sites are properly termed "independent". If the total free energy change that results from achieving X-P-Y is _less_ in absolute magnitude than that to be realized from independent binding, then the prior binding of X antagonizes the binding of Y, and vice versa, to an extent equal to the difference between the actual free energy change and what would have been realized in the case of independent binding. This difference was defined by Weber to be equal to the "free energy of interaction" or equivalently the "coupling free energy" between X and Y, ΔG_{xy}. Combining this definition with equation 1 yields:

$$\Delta G_{xy} = (\Delta G^{\circ}_x + \Delta G^{\circ}_{y/x}) - (\Delta G^{\circ}_x + \Delta G^{\circ}_y) = (\Delta G^{\circ}_y + \Delta G^{\circ}_{x/y}) - (\Delta G^{\circ}_x + \Delta G^{\circ}_y)$$

$$\text{or:} \quad \Delta G_{xy} = \Delta G^{\circ}_{y/x} - \Delta G^{\circ}_y = \Delta G^{\circ}_{x/y} - \Delta G^{\circ}_x \tag{2}$$

Consequently the magnitude of the influence that X has on the binding of Y must be the same as the magnitude of the influence that Y has on the binding of X. For ligands that are mutually antagonistic, ΔG_{xy} will have a positive sign by the above definition. All of the arguments are analogous for the situation in which ligands facilitate one another's binding, in which case a negative coupling free energy will result.

In my opinion perhaps the most important contribution Weber made to the discussion of linkage was his consideration of the likely magnitude of coupling free energies. Reasoning that typical binding energies in the range of 4 to 8 kcal/mol arise from a multitude of energetically small individual contributions, the modification of a total binding interaction achieved through a conformational alteration induced by the binding of a small ligand at some distant site on the protein should be expected to be only a fraction of the binding energy itself. A survey of the literature corroborated his reasoning. In the few cases where coupling free energies could be estimated, all were less than ±2 kcal/mol. Although ±2 kcal/mol

represents a substantial change (a change in binding constant of approximately 1½ orders of magnitude), such a realization predicted that in most
cases not only should coupling free energies exist <u>theoretically</u>, but also
that they should be measurable <u>experimentally</u>. Since coupling free energies are a direct measure of the consequences of conformational alteration
induced by ligand binding, their measurement should provide valuable
information pertaining to the biophysical causes of these perturbations.

These ideas are actually quite straightforward in the abstract; essentially just a re-statement of the definition of State Functions that is
covered in most undergraduate freshman-level chemistry courses. However,
Weber quite correctly sensed the need to reiterate these principles within
the specific context of protein-ligand binding because they are almost
universally corrupted; sometimes with intent, sometimes out of naiveté; in
efforts to "simplify" our interpretation of protein-ligand binding behavior.
In the discussion that follows, I will attempt to illustrate how remaining
mindful of the fundamental principles of linkage has <u>enlightened</u> rather
than obscured our understanding of the allosteric behavior of rat liver
PFK.

LINKAGE AND AGGREGATION OF LIVER PFK

Our first use of the linkage concept to understand facets of rat liver
PFK regulation occurred in studies of the kinetic ramifications of the high
molecular self-association reactions to which the enzyme is prone. The
smallest active form of rat liver PFK is a tetramer of identical subunits,
each with a molecular weight of 82,000 (Reinhart and Lardy, 1980a). The
enzyme is capable of further self-association reactions, however, ultimately
forming enzyme aggregates with sedimentation coefficients as large as 50S;
implying molecular weights of several million (Reinhart and Lardy, 1980b).
From gel filtration experiments of crude rat liver cytosol preparations and
the concentration dependence of the aggregation phenomenon in the presence
of physiological concentrations of various ligands, we had become convinced
that the high molecular weight aggregate forms of liver PFK could exist,
and were probably predominant, <u>in vivo</u> (Reinhart and Lardy, 1980c). The
question before us, then, was how did this high molecular weight aggregation affect the activity of the enzyme under physiological conditions?

Virtually all mechanisms for activation or inhibition of rat liver
phosphofructokinase involve an alteration of the apparent affinity of the

enzyme for the substrate fructose 6-phosphate (Fru-6-P) rather than altering maximal velocity (Reinhart and Lardy, 1980a). Since the concentration
of Fru-6-P in vivo is always substantially less than that needed for
saturation, and the saturation profile is highly cooperative, such mechanisms are able to provide considerable control on turnover. Moreover, it
had been previously shown that the maximal specific activity of high
molecular weight aggregates of rabbit muscle PFK differed little from that
of the tetramer (Aaronson and Frieden, 1972; Lad, et al., 1973; Pavelich
and Hammes, 1973). The issue of physiological relevance of the high
molecular weight aggregation to the activity of rat liver PFK could therefore be restated as: to what extent does aggregation alter the concentration of Fru-6-P required to produce half-maximal velocity, K_a?

This question is not as easy to answer for phosphofructokinase as it
might at first seem. At the concentrations of PFK normally employed for
assay of steady-state kinetic behavior, 0.1 µg/ml or less, the enzyme
exists almost exclusively in the tetrameric form (Reinhart and Lardy,
1980b). The most straightforward way to ascertain the kinetic ramification
of further enzyme aggregation is to measure specific activity at low
concentrations of Fru-6-P as a function of enzyme concentration over a
range of PFK concentrations in which appreciable aggregation takes place
(in the case of rat liver PFK, 1 to 2 orders of magnitude greater than the
normal assay concentration). However, PFK is strongly activated allosteri-
cally by both of the products of the reaction it catalyzes, namely ADP and
fructose 1,6-biphosphate (Fru-1,6-BP). Although one can design assays in
which both of these products are coupled to other enzymatic reactions, it
is very difficult to maintain the concentration of these effectors at a
negligible, or at least constant, level when the overall rate of turnover
of the PFK reaction is varied by the necessary one to two orders of magnitude. Indeed, in previous reports in the literature purporting to describe
aggregation induced activation of rabbit muscle PFK (Hofer, 1971; Hulme and
Tipton, 1971), the observed activation is almost surely attributable to an
increase in the steady-state concentration of these allosteric activators.

Largely because of these technical factors that complicated a direct
measurement of the K_a as a function of enzyme concentration, we decided to
approach this problem by exploiting the most fundamental aspect of the
linkage concept, that being the fact that the influence that the binding of
ligand X has on the subsequent binding of ligand Y must be the same as the
influence that the initial binding of Y has on the subsequent binding of
ligand X. Weber (1975) has discussed the fact that formally no distinction

need be made, when considering these concepts, between a small molecular
weight ligand and a macromolecular subunit binding to the protomeric form
of a protein. The relative affinity for a particular ligand displayed by
two enzyme forms with different subunit stoichiometries can be assessed by
determining the relative tendency of that ligand to promote the protein
self-association reaction. In the case of a protein subject to a monomer -
dimer equilibrium, this principle of linkage states that the extent that
dimerization changes the average affinity of ligand X must be mirrored in
the ability of X to promote dimerization. That is to say, if the dimer
form has a higher affinity for X than does the monomer, then in the presence of X the protein dissociation constant must decrease. Similar arguments can be made for protein aggregation equilibria of more complex
stoichiometries such as is the case for rat liver PFK. Since a direct
measure of the relative affinity of the aggregate forms for Fru-6-P was
difficult to measure, we chose instead to examine the influence that
Fru-6-P has on PFK self-association.

We have directly evaluated the self-association of rat liver PFK by
measuring the fluorescence polarization of pyrenebutyric acid covalently
attached to PFK. Pyrenebutyric acid is a long-lifetime fluorescent probe
that was first used by Weber and coworkers to study the hydrodynamics of
several large proteins (Knopp and Weber, 1969; Rawitch, et al., 1969) and
is the most suitable probe for studying the high molecular weight aggregates formed by rat liver PFK. The covalent attachment of pyrenebutyric
acid to a protein can be accomplished readily by first converting pyrenebutyric acid to the sulfuric anhydride by reaction with SO_3 as described by
Knopp and Weber (1969). The anhydride spontaneously reacts with free amino
groups on the enzyme to form stable amide linkages.

The fluorescence polarization of pyrenebutyryl-phosphofructokinase
(PB-PFK), at 7 µg/ml (1 X 10^{-7} M subunits) in an isotonic, pH 7 buffer,
varies considerably depending upon whether Fru-6-P, MgATP, or no specific
ligand is present as is evident from the Perrin plots shown in Figure 2.
Moreover, the relatively invariant Y-axis intercepts of these plots indicates that the polarization differences are not due to variations in local
motion of the pyrene moiety as originally explained by Weber in his landmark 1952 papers describing the utilization of fluorescence polarization
for studying protein hydrodynamics (Weber, 1952a,b). The protein titration
experiments shown in Figure 3, although unfortunately incomplete due
primarily to insufficient availability of rat liver PFK, clearly indicate
that at this concentration, which is comparable to the concentration found

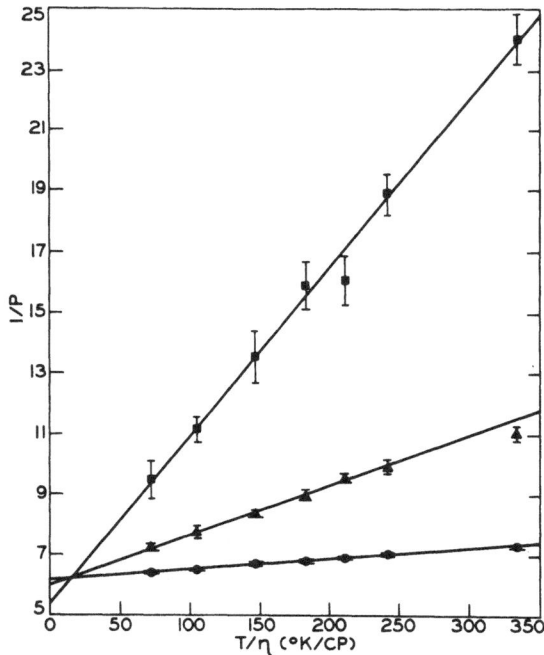

Figure 2. Perrin plot of pyrenebutyryl-label rat liver phosphofructokinase (PB-PFK). The quantity temperature/viscosity (T/η) was varied isothermally by the addition of glycerol. PB-PFK was incubated for 3 hours at 7 μg/ml prior to the experiment. Values represent averages of at least 12 measurements with the standard deviation indicated. Isotonic pH 7 buffer contained either 5 mM Fru-6-P (•), 3 MgATP (▲), or no additional ligands (■). Reprinted with permission from: Reinhart and Lardy, 1980a. Copyright by the American Chemical Society.

in the cell (Reinhart and Lardy, 1980; Donofrio, et al., 1984), the enzyme exists largely as the tetramer when saturated with MgATP. Fru-6-P, on the other hand, causes the tetramers to further associate to a considerable extent at 10 μg/ml. With neither ligand present, the dissociation constant is much greater and the tetramers dissociate almost completely, and catalytic activity is lost (data not shown). Even in the presence of the nonhydrolyzable ATP analog adenylylimidodiphosphate (AMPPNP), which also stabilizes the tetramer, increasing Fru-6-P concentration increases the average size of the enzyme indicating the formation of higher molecular weight aggregates (see Figure 4).

We have interpreted these and other results (Reinhart, 1980b,c) as implying that the overall association constant of rat liver PFK subunits is greater when Fru-6-P is bound than when MgATP is bound but that the association constant in the presence of either ligand is very much greater than the association constant when both ligands are absent. Moreover, since

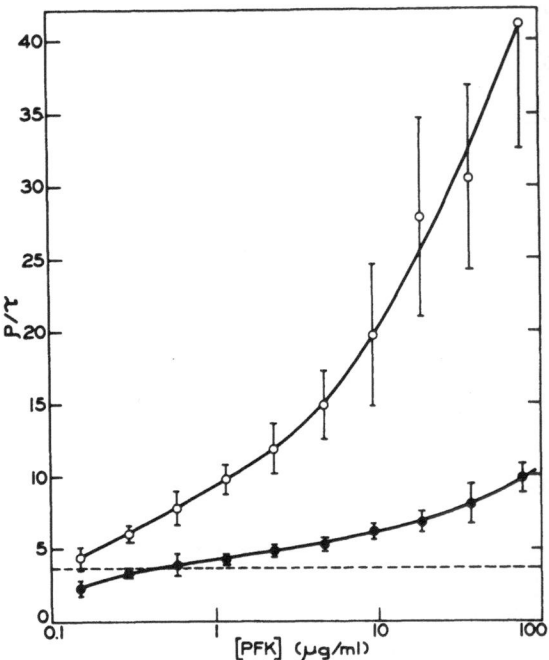

Figure 3. Dependence of the ratio of rotational correlation time to
fluorescent lifetime (ρ/τ) on the concentration of PB-PFK in the presence
of either 5 mM Fru-6-P (o) or 3 mM MgATP (•). The large standard deviations at large ρ/τ are a consequence of the close proximity of polarization
values to P_0. The dashed line corresponds to the theoretical ρ/τ of the
tetramer of PFK based upon a Stokes' radius of 67Å (Pavelich and Hammes,
1973) and a fluorescent lifetime of 230 ns (Reinhart and Lardy, 1980a).
Reprinted with permission from Reinhart and Lardy, 1980a. Copyright 1980
by the American Chemical Society.

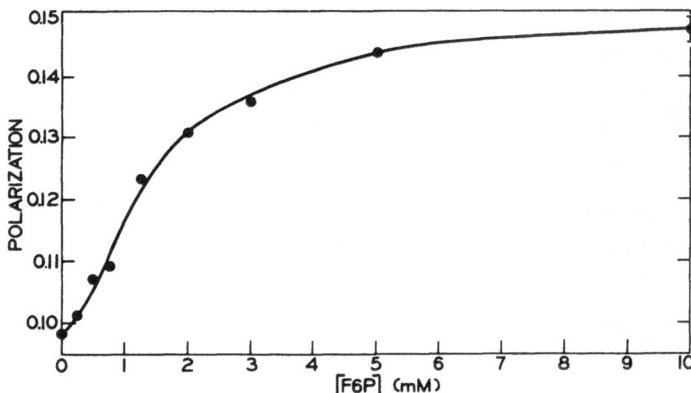

Figure 4. Polarization of 8 µg of PB-PFK per ml in the presence of 3 mM
MgAMPPNP and varying amounts of Fru-6-P (indicated as F6P). The representative standard deviation is indicated for 10 mM Fru-6-P for each curve.
Under these conditions, the fluorescence polarization of the PB-PFK tetramer is approximately 0.095. Reprinted with permission from: Reinhart
and Lardy, 1980b. Copyright 1980 by the American Chemical Society.

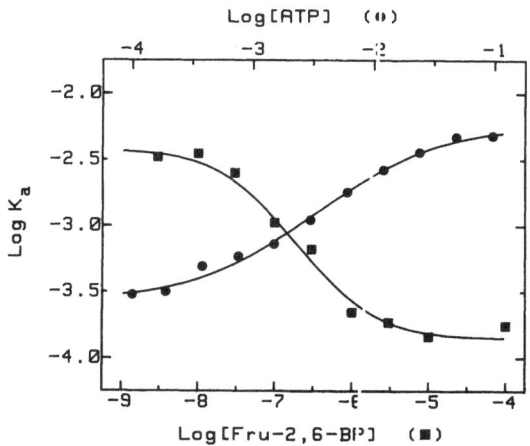

Figure 5. Ability of a linked-function model to describe allosteric inhibition by MgATP (•) and activation by Fru-2,6-BP (■) of rat liver PFK. K_a represents the concentration of Fru-6-P required to produce half-maximal velocity and hence is the effective average <u>dissociation</u> constant of Fru-6-P. Fru-2,6-BP activation data were obtained with [MgATP] = 24 mM. In each case, the curve drawn represents the best fit of the data to equation 4 described in the text.

Fru-6-P is able to further associate PFK already stabilized in its tetramer form, the specific association constant of the tetramers forming high molecular weight aggregates is greater when Fru-6-P is bound than it is when Fru-6-P is not bound. Linkage allows us to therefore conclude that the affinity of the high molecular weight aggregates for Fru-6-P must be greater than the affinity of the tetramers for Fru-6-P. Since the affinity for Fru-6-P is the factor limiting activity of rat liver PFK at the low concentrations of Fru-6-P that occur <u>in vivo</u>, we conclude that the high molecular weight aggregates are more active than tetramers <u>in vivo</u>. In other experiments we concluded that the high molecular weight aggregates do form <u>in vivo</u> (Reinhart and Lardy, 1980c). Consequently we feel that modulation of the tetramer - high molecular weight equilibrium is a regulatory motif that is potentially very important physiologically.

LINKAGE BETWEEN MULTIPLE ALLOSTERIC LIGANDS

Our next utilization of the concept of linkage in the study of rat liver phosphofructokinase had the objective of developing a systematic description of the allosteric behavior induced by the several important regulatory ligands of the enzyme acting in concert. This objective has necessitated the extension of linkage theory to systems involving three different types of ligands. In so doing we have managed to not only provide a good model upon which to describe rat liver PFK's behavior, but more importantly we have begun to clarify several important mechanistic insights that would not have been possible without using this approach.

PFK catalyzes the reaction:

$$\text{Fru-6-P} + \text{MgATP} \longrightarrow \text{Fru-1,6-BP} + \text{MgADP} \tag{3}$$

Note that MgATP plays the dual role of both substrate and allosteric inhibitor. Virtually all of the allosteric ligands of rat liver PFK effect their regulation almost exclusively by altering the apparent affinity displayed by the enzyme for the substrate Fru-6-P (Reinhart and Lardy, 1980a). Most ligands have little if any effect on either maximal velocity or the binding of MgATP to the active site. Consequently an allosteric inhibitor, such as MgATP, increases the concentration of Fru-6-P required to produce half-maximal velocity, K_a, (i.e. the apparent affinity for Fru-6-P <u>decreases</u>) whereas an allosteric activator, such as fructose 2,6-biphosphate (Fru-2,6-BP), acts by decreasing K_a thereby <u>increasing</u> the affinity of the enzyme for Fru-6-P. Consequently the allosteric regulation of rat liver phosphofructokinase is well suited to a linked-function analysis.

The specific analysis that we have used to analyze the effect of a single allosteric ligand, usually MgATP, on the apparent dissociation constant for Fru-6-P from PFK is derived directly from the analysis set forth by Weber in describing the interactions between two ligands, X and Y, binding to discrete sites on a single protein (Reinhart, 1983), using somewhat altered notation:

$$K_a = K_a^o \left[\frac{K_{ix}^o + [X]}{K_{ix}^o + Q_{ax}[X]} \right] \tag{4}$$

In this notation, X = the allosteric ligand, K_a = the concentration of Fru-6-P (A) producing half-saturation, $K_a^o = K_a$ when $[X] = 0$, K_{ix}^o = the dissociation constant for the allosteric ligand when $[\text{Fru-6-P}] = 0$, and Q_{ax} = the coupling constant between Fru-6-P and X. Q_{ax} is related to the coupling free energy between Fru-6-P and X, ΔG_{ax}, according to:

$$\Delta G_{ax} = -RT\ln(Q_{ax}) \tag{5}$$

where R = the gas constant and T = temperature in °K.

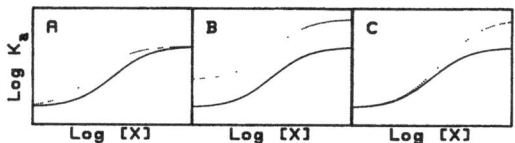

Figure 6. Various ways in which a second allosteric ligand, Y, can influence the relationship between the apparent dissociation constant for substrate A, K_a, and the first allosteric ligand, X. The solid curve in each panel was calculated from equation 4 using arbitrary values of the parameters K_a^o, K_{ix}^o, and Q_{ax}. If Y decreases the dissociation constant for X in the absence of A, K_{ix}^o, then the curve will shift to the left as indicated by the dashed curve in panel A. If Y has an antagonistic effect on the dissociation constant of A that is independent of X, then the curve will be displaced vertically as indicated by the dashed curve in panel B. If Y augments the antagonism between A and X, then the dashed curve in panel C will result. In all three cases, Y will increase the apparent K_a. Reprinted by permission from: Reinhart, 1985. Copyright 1985 by the American Chemical Society.

Despite the obvious complexities associated with the kinetic behavior of rat liver PFK, most notably the very high degree of positive cooperativity exhibited in the Fru-6-P binding profile, equation 4 does a surprisingly good job describing the allosteric effects of both inhibitory allosteric ligands, such as MgATP, and activating allosteric ligands, such as Fru-2,6-BP, as shown in Figure 5. Much of this success is due to the fact that we have essentially factored [Fru-6-P] out of the analysis by considering K_a to be the dependent variable. Since, to a first approximation, sigmoidal binding profiles are nearly symmetrical about the half-saturation point, this approach minimizes the impact of cooperativity on the analysis.

The real interest in understanding PFK's allosteric regulation lies in the effects of ligands when acting in combination. Most of the many effector ligands of PFK exist at some relevant steady-state concentration, and rarely, if ever, does a changing metabolic situation affect the concentration of only one. In addition, the study of PFK regulation has a practical constraint that arises from the fact that MgATP is also a substrate of the enzyme. Even though the allosteric properties of the enzyme are not thought to affect MgATP's interaction with the active site, and hence its role as a substrate can be ignored, in every kinetics experiment two ligands of the enzyme, i.e. the substrate Fru-6-P and MgATP as an allosteric ligand, must always be present. What is needed, therefore, is a more "global" linkage analysis that can account for the actions and interactions of several ligands potentially binding to PFK simultaneously. At the same time, one can expect to develop an appreciation of the intricacies of regulatory responses that can have implications regarding their mechanism of action. We have begun this task by extending the linkage arguments to systems of three ligands; i.e. the substrate Fru-6-P (A) and two allosteric ligands X and Y.

The equation analogous to equation 4 which describes K_a as a function of both [X] and [Y] has been derived previously (Reinhart, 1985), and the derivation need not be reiterated here. However, one can intuitively anticipate that this equation contains intrinsic dissociation constants, K_{ix}^O and K_{iy}^O, for X and Y respectively as well as a unique coupling constant for each of the four possible multi-ligated forms the enzyme that can be generated. Therefore, three two ligand coupling terms exist: Q_{ax}, Q_{ay}, and Q_{xy}; and one three ligand coupling term, Q_{axy}, (where the subscripts indicate the combination of ligands involved in the particular enzyme form) should appear in the equation. The actual relationship between K_a and [X] and [Y] is given by (Reinhart, 1985):

$$K_a = K_a^O \left[\frac{K_{ix}^O K_{iy}^O + K_{iy}^O [X] + K_{ix}^O [Y] + Q_{xy}[X][Y]}{K_{ix}^O K_{iy}^O + K_{iy}^O Q_{ax}[X] + K_{ix}^O Q_{ay}[Y] + Q_{axy}[X][Y]} \right] \qquad (6)$$

The usual procedure for evaluating the effects of two allosteric ligands is to vary the concentration of one while leaving the concentration of the other fixed and then repeating the experiment at several different concentrations of the "fixed" ligand. Equation 6 predicts that with either [X] or [Y] fixed, the dependence of K_a on the variable ligand will conform to the form of equation 4. For the sake of consistency of notation, if we consider [Y] to be fixed, then fitting the data obtained when X is varied to equation 4 will result in apparent values for the three parameters $K_a^{O'}$, $K_{ix}^{O'}$, and Q_{ax}', where the "prime" signifies that these values are actually functions of [Y] rather than absolute.

The actions of a second allosteric ligand can be considered from the perspective of how if affects these attributes of the first allosteric ligand. The second ligand, Y, may change the affinity of the enzyme for X, and hence change $K_{ix}^{O'}$, or it may alter the **effectiveness** of X after X has bound, in which case the coupling constant, Q_{ax}', will be altered. In addition Y may not affect the actions of X at all but rather have its own, independent influence on the affinity of the substrate, in which case Y will affect $K_a^{O'}$. These effects can be discerned qualitatively by examining the shifts induced by varying Y in the curves describing the dependence of K_a on the concentration of X as shown in Figure 6.

It can be shown (Reinhart, 1985) that these intuitive expectations regarding the possible effects of Y on $K_a^{o\prime}$, $K_{ix}^{o\prime}$, and $Q_{ax}^\prime$, as described in Figure 6, can be quantitatively described by the following relationships:

$$K_a^{o\prime} = K_a^o \left[\frac{K_{iy}^o + [Y]}{K_{iy}^o + Q_{ay}[Y]} \right] \tag{7}$$

$$K_{ix}^{o\prime} = K_{ix}^o \left[\frac{K_{iy}^o + [Y]}{K_{iy}^o + Q_{xy}[Y]} \right] \tag{8}$$

$$Q_{ax}^\prime = \frac{(K_{iy}^o + [Y])(K_{iy}^o Q_{ax} + Q_{axy}[Y])}{(K_{iy}^o + Q_{ay}[Y])(K_{iy}^o + Q_{xy}[Y])} \tag{9}$$

We have utilized this approach to evaluate the combined actions of MgATP and H⁺, and MgATP and Fru-2,6-BP on the K_a for Fru-6-P exhibited by rat liver PFK.

The allosteric regulation of PFK is very sensitive to pH. Indeed, virtually all of the allosteric properties of PFK from rabbit skeletal muscle disappear at or above pH 8. Rat liver PFK continues to exhibit allosteric characteristics such as sigmoidal Fru-6-P saturation profiles and MgATP inhibition even at high pH although the K_a for Fru-6-P increases dramatically as pH is decreased. Since both MgATP and H⁺ must always be present in any _in vitro_ assay of PFK activity, we felt it necessary to evaluate the combined actions of these two effector ligands first.

The dependence of K_a on [MgATP] was measured at several pH values between 5.5 and 9.0, and the data are presented in Figure 7. At each pH, the data were fit to equation 4 (X = MgATP) to obtain apparent values for $K_a^{o\prime}$, $K_{ix}^{o\prime}$, and $Q_{ax}^\prime$. These values are presented in Figure 8 as a function of pH. In addition, all of the data presented in Figure 7 were fit to equation 6 (X = MgATP, Y = H⁺), and the resulting parameter values were used to describe the trends in $K_a^{o\prime}$, $K_{ix}^{o\prime}$, and $Q_{ax}^\prime$ that are predicted equations 7-9 as shown in Figure 8. The data appear to be reasonably well described by these equations. Not only are we now in a better position to predict K_a at any [MgATP] and pH, but certain mechanistic relationships between the ligands are clarified. For example, the data in Figure 8 suggest that pH primarily affects the affinity of PFK for Fru-6-P directly rather than indirectly by altering the binding affinity of MgATP. This conclusion stands in direct contrast to the conclusion that lowering pH increases the inhibition of PFK by increasing the affinity of the enzyme for MgATP which has been proposed to explain the effects of pH on the rabbit muscle isozyme (Pettigrew and Frieden, 1979a,b).

207

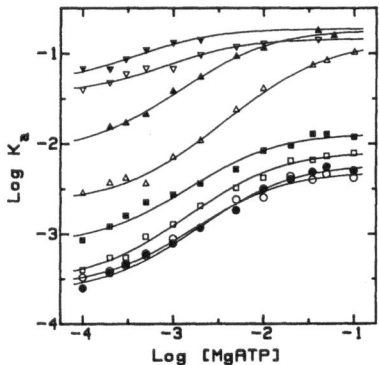

Figure 7. Influence of pH on the relationship between MgATP and the apparent K_a for Fru-6-P. Curves represent the best fit of the data for each pH to equation 4 given in the text. pH values are 9.0 (o), 8.5 (●), 8.0 (◻), 7.5 (◼), 7.0 (△), 6.5 (▲), 6.0 (▽), and 5.5 (▼). Compare the pattern of these curves to the alternatives illustrated in Figure 6 and note the similarity to panel B. Reprinted by permission from: Reinhart, 1985. Copyright 1985 by the American Chemical Society.

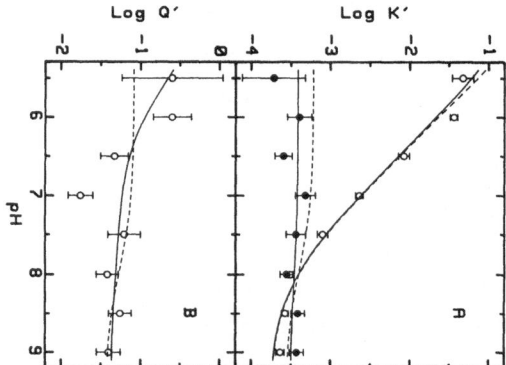

Figure 8. Influence of pH on the parameters describing the inhibition of rat liver phosphofructokinase by MgATP. Panel A: Logarithm of the apparent dissociation constant for Fru-6-P in the absence of MgATP, $K_a^{o'}$, (o); logarithm of the apparent dissociation constant of MgATP in the absence of Fru-6-P, $K_{ix}^{o'}$ (●). Panel B: Logarithm of the apparent coupling term between MgATP and Fru-6-P, Q_{ax}' (o). Values (in molar units) were obtained by fitting the data for each pH presented in Figure 7 to equation 4. Error bars indicate plus or minus standard error for each determination. Solid curves were plotted according to equations 7-9 using parameters determined by fitting all of the data shown in Figure 7 to equation 6 described in the text. The dashed curves were calculated in a similar manner except that Q_{ay} and Q_{axy} were set equal to 0 in which case equations 6-9 reflect the circumstance in which the binding of Fru-6-P and H⁺ are mutually exclusive. Reprinted by permission from: Reinhart, 1985. Copyright 1985 by the American Chemical Society.

208

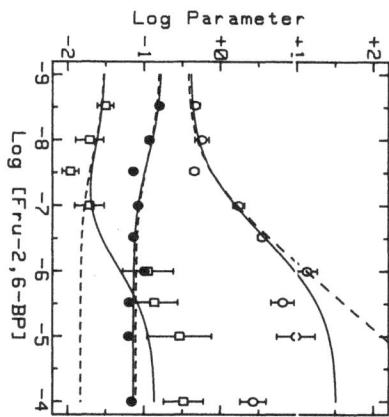

Figure 9. Influence of Fru-2,6-BP on the parameters describing allosteric inhibition by MgATP at pH 9. Procedures followed were as described in Figures 7 and 8 except that the dashed curves result from setting Q_{xy} and Q_{axy} equal to 0 reflecting a model in which the binding of Fru-2,6-BP and MgATP are mutually exclusive. The parameters $K_a^{o'}$ (●) and $K_{ix}^{o'}$ (o) are expressed in millimolar units whereas Q'_{ax} (□) is unitless. [Fru-2,6-BP] is expressed in molar units. Reprinted by permission from: Reinhart, and Hartleip, 1986. Copyright 1986 by the American Chemical Society.

One of the advantages of analyzing the actions of multiple ligands in this manner is that certain mechanistic alternatives can be directly compared. It is quite common, particularly in the case of PFK, for investigators to invoke mechanistic schemes in which the binding of certain ligand combinations are impossible, the binding of these ligands being mutually exclusive. For example, it has been proposed that for rabbit muscle PFK, MgATP and H$^+$ bind to an "inhibited" form and Fru-6-P binds to an "active" form of the enzyme (Pettigrew and Frieden, 1979a,b). Such a model can be directly evaluated because it implies that the coupling free energy between Fru-6-P and MgATP __and__ between Fru-6-P and H$^+$ are infinite, or in the notation of equation 6: $Q_{ax} = Q_{ay} = 0$. [$Q_{axy} = 0$ also because certainly all three ligands cannot bind simultaneously in this model.] Although an exclusive relationship between H$^+$ and Fru-6-P cannot be ruled out by the present data, the value of the coupling free energy between ATP and Fru-6-P is convincingly finite and thus not compatible with this proposed mechanism. Of course our data pertain only to the liver isozyme, which clearly differs from the muscle isozyme with respect to pH effects, so we cannot refute the mechanism as it pertains to rabbit muscle PFK.

The recent discovery that Fru-2,6-BP is a very potent activator of rat liver PFK (Furuya and Uyeda, 1980a,b; Uyeda, et al., 1981a,b; Pilkis et al., 1981; Van Schaftingen et al., 1981; Hers and Van Schaftingen, 1982) whose formation is subject to hormonal control has led to much interest in the mode of action of this allosteric ligand in particular. We have recently performed a similar analysis of the combined effects of MgATP and Fru-2,6-BP and the data are summarized in Figure 9. Once again the simple monomeric, three ligand linked-function model does a good job of describing the effect of Fru-2,6-BP on the MgATP inhibition characteristics and vice versa. In addition we were able to evaluate the proposal that Fru-2,6-BP and Fru-6-P bind to an "active" form of PFK whereas MgATP binds to an "inhibited" form. In fact, we examined virtually all possible exclusive binding relationships and found that all could be statistically dismissed as being inferior to models in which all three ligands can bind and interact with finite coupling free energies of actually quite moderate magnitudes. Despite its almost ubiquitous utilization, we have found no evidence, therefore, in either of these two systems to support the concept that PFK exists fundamentally in two functional states.

LINKAGE AND DYNAMICS OF RAT LIVER PFK

The third way in which we have found the linkage concept to be useful in the study of phosphofructokinase once again relates to its ability to illuminate possible underlying molecular mechanisms that give rise to allosteric behavior. In this regard, linkage analysis of kinetic or ligand binding behavior cannot supplant direct physical measurements, but it can serve to help focus such efforts.

Weber was the first to point out the fact that the coupling free energy between two ligands could be viewed as the standard free energy of a disproportionation reaction involving the various ligated forms of an enzyme or protein (Weber, 1975). Let us consider the dissociation of ligand A from an enzyme, E, with no other ligands bound, the standard free energy for which is the negative of the standard free energy of association ΔG_a°, as defined earlier:

$$EA \rightleftharpoons E + A \qquad \Delta G^\circ = -\Delta G_a^\circ \qquad (10)$$

Next we can consider the association of A with enzyme to which an allosteric ligand, X, has already bound, the standard free energy for which is equal to $\Delta G_{a/x}^\circ$ as previously defined:

$$A + XE \rightleftharpoons XEA \qquad \Delta G^\circ = \Delta G_{a/x}^\circ \qquad (11)$$

Summing reactions 10 and 11 yields the following reaction for which the standard free energy is the sum of the standard free energies of the individual reactions:

$$EA + XE \rightleftharpoons E + XEA \qquad \Delta G^\circ = \Delta G^\circ_{a/x} - \Delta G^\circ_a = \Delta G_{ax} \qquad (12)$$

Equation 12 is the disproportionation reaction concerning the distribution of bound A and X ligands for which the standard free energy is equal to the coupling free energy between A and X. The coupling constant Q_{ax} is therefore the equilibrium constant of equation 12 as seen by equation 5. If an antagonistic relationship exists between A and X, then $Q_{ax} < 1$ and the equilibrium in 12 favors the binding of A and X to separate protein molecules. If a facilitating relationship exists between the two ligands, then $Q_{ax} > 1$ and X and A will tend to bind to the same molecule. Note that equation 12 is not a ligand binding reaction per se, in that it involves the same number of unoccupied binding sites and no free ligand on either side of the reaction.

We have studied the temperature dependence of Q_{ax} in an effort to estimate the relative magnitudes of the ΔH and ΔS components of the coupling free energy. Since Q_{ax} is in fact an equilibrium constant, the following relationship should hold provided that ΔH does not vary significantly with temperature:

$$2.3 \cdot \log(Q_{ax}) = -(\Delta H/R)(1/T) + (\Delta S/R) \qquad (13)$$

A van't Hoff plot of $\log(Q_{ax})$ vs. $1/T$ should yield ΔH and ΔS from the slope and intercept, respectively, if the data conform to a straight line. Curvature in such a plot would indicate a change in the heat capacity between the reactants and products of equation 12, causing ΔH to vary with temperature.

Figure 10 shows a van't Hoff plot of preliminary results that we have obtained by measuring the coupling constant between Fru-6-P and MgATP when bound to rat liver phosphofructokinase at high pH between 5° and 35°C. The data are well described by a straight line with positive slope, indicating that ΔH is fairly constant in this temperature range and equal to -7.7 kcal/mol. In addition, ΔS, obtained from the intercept of the best-fit line, is equal to +30 entropy units. These data indicate, therefore, that at 25°C, the coupling free energy between Fru-6-P and MgATP is equal to +1.1 kcal/mol and that this coupling is derived from an entropic contribution of +8.8 kcal/mol and the enthalpic contribution of -7.7 kcal/mol. Although these data are preliminary, and the exact magnitudes of ΔS and ΔH vary somewhat

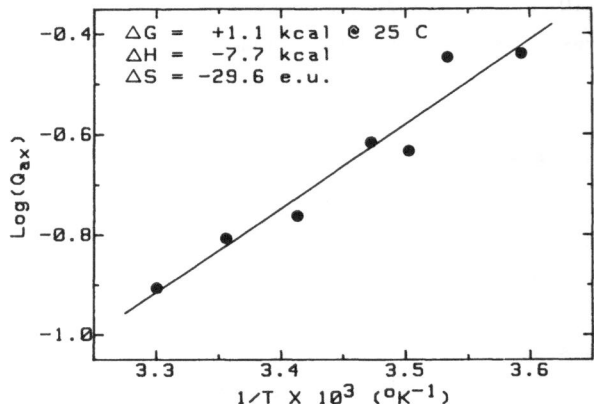

Figure 10. Van't Hoff plot of the logarithm of the coupling constant between Fru-6-P and MgATP on rat liver PFK, Q_{ax}, versus 1/Temperature. ΔH and ΔS values indicated were determined from the slope and intercept of the best line fit to the data as described by equation 13 in the text. ΔG is equal to the coupling free energy at 25°C.

with the preparation of PFK being evaluated, the fundamental observation is consistent; namely that both ΔH and ΔS are negative indicating that it is ΔS that produces the inhibitory relationship between Fru-6-P and MgATP.

This observation is quite remarkable when compared to the usually perceived mechanism of allosteric action. In the traditional view, an inhibitory allosteric ligand effects a diminished substrate affinity by distorting, through a conformational change, the substrate's binding site. This mechanism would seem to suggest an antagonistic coupling that is largely enthalpic in character because the strength of the dipolar and charge-charge interactions of ligand binding is most strongly affected by precise positioning. In contrast, however, we see that our data suggest that <u>an adverse entropy is the cause of the inhibitory influence of MgATP</u> and that if enthalpy were the sole contributor to the coupling interaction then MgATP would actually be an activator.

What then is the source of the adverse entropy associated with reaction 12 as written? At this point one can only speculate, but currently we favor the possibility that entropy is lost via constraint of the dynamical features of protein structure upon ligand binding, and that when both ligands bind, they have a synergistic effect so that the constraints imposed by both ligands, relative to free enzyme, are greater than the sum of the constraints imposed by each ligand individually. Ligand binding typically constrains protein flexibility by effectively anchoring the several points of the binding domain with which the ligand interacts upon binding. Consequently

212

both enzyme species appearing on the left side of reaction 12 would be expected to have less entropy than the free enzyme appearing on the right. In order for the overall entropy to be negative, therefore, the doubly-ligated form on the right must be severely constrained.

Supporting this notion is the fact that the van't Hoff plot shown in Figure 10 is straight, indicating little change in heat capacity over the solvation range examined. When entropy effects arise from the differential solvation of reactants and products, for example as is true for protein-ligand binding reactions, typically they are accompanied by substantial changes in heat capacity that produce noticeable curvature in van't Hoff plots (Sturtevant, 1977). It is reasonable to expect the solvation changes to indeed be small since free ligand is absent from reaction 12 and the number of occupied and unoccupied binding sites on both sides are balanced. If entropy is not lost from the solvent, the protein species involved in 12 are all that remain as possibilities.

Equally interesting is the negative enthalpy suggested by Figure 10, which indicates that in the absence of the entropic effect, whatever its origin, MgATP would actually activate rat liver phosphofructokinase by increasing the apparent affinity of the enzyme for Fru-6-P. Since the magnitude of the entropy contribution to the coupling free energy between Fru-6-P and MgATP will vary with temperature, so will the efficacy of MgATP as an inhibitor of the enzyme, and one can predict that as temperature is lowered, a cross-over point should be encountered, beyond which MgATP will actually act as an activator. If the linear dependence depicted in Figure 10 were to hold beyond the temperature range investigated, such a cross-over point would unfortunately fall below the freezing point of water, at approximately -13°C, and consequently such cross-over behavior has proven impossible to demonstrate in this system. Hopefully, entropy-driven coupling antagonism exists in other allosteric enzymes that may be more suitable for demonstrating a temperature-dependent alteration in the nature as well as the magnitude of the allosteric effect.

ACKNOWLEDGEMENTS

Much of the work described herein has been supported by grant GM 33216 from the National Institutes of Health.

REFERENCES

Aaronson, R. P., and Frieden, C., 1971, Rabbit Muscle Phosphofructokinase: Studies on the Polymerization, J. Biol. Chem., 247:7502.

Donofrio, J. C., Thompson, R. S., Reinhart, G. D., and Veneziale, C. M., 1984, Quantification of Liver and Kidney Phosphofructokinase by Radio-immunoassay in Fed., Starved, and Alloxan-Diabetic Rats, Biochem. J., 224:541.

Furuya, E., and Uyeda, K., 1980a, An Activation Factor of Liver Phospho-fructokinase, Proc. Natl. Acad. Sci. U.S.A., 77:5861.

Furuya, E., and Uyeda, K., 1980b, Regulation of Phosphofructokinase by a New Mechanism, J. Biol. Chem., 255:11656.

Hers, H. G., and Van Schaftingen, E., 1982, Fructose 2,6-Bisphosphate 2 Years After Its Discovery, Biochem. J., 206:1.

Hofer, H. W., 1971, Influence of Enzyme Concentration on the Kinetic Behavior of Rabbit Muscle Phosphofructokinase, Hoppe-Seylers Z. Physiol. Chem., 352:997.

Hulme, E. C., and Tipton, K. F., 1971, The Dependence of Phosphofructo-kinase Kinetics upon Protein Concentration, FEBS Lett., 12:197.

Knopp, J. A., and Weber, G., 1969, Fluorescence Polarization of Pyrene-butyric - Bovine Serum Albumin and Pyrenebutyric - Human Macroglobulin Conjugates, J. Biol. Chem., 244:6309.

Lad, P. M., Hill, D. E., and Hammes, G. G., 1973, Influence of Allosteric Ligands on the Activity and Aggregation of Rabbit Muscle Phosphofructo-kinase, Biochemistry, 12:4303.

Pavelich, M. J., and Hammes, G. G., 1973, Aggregation of Rabbit Muscle Phosphofructokinase, Biochemistry, 12:1408.

Pettigrew, D. W., and Frieden, C., 1979a, Binding of Regulatory Ligands to Rabbit Muscle Phosphofructokinase. A Model for Nucleotide Binding as a Function of Temperature and pH, J. Biol. Chem., 254:1887.

Pettigrew, D. W., and Frieden, C., 1979b, Rabbit Muscle Phosphofructo-kinase. A Model for Regulatory Kinetic Behavior, J. Biol. Chem., 254:1896.

Pilkis, S. J., El-Maghrabi, M. R., Pilkis, J., Claus, T. H., and Cumming, D. A., 1981, Fructose 2,6-Bisphosphate. A New Activator of Phospho-fructokinase, J. Biol. Chem., 256:3171.

Rawitch, A. B., Hudson, E., and Weber, G., 1969, The Rotational Diffusion of Thyroglobulin, J. Biol. Chem., 244:6543.

Reinhart, G. D., 1983, The Determination of Thermodynamic Allosteric Parameters of an Enzyme Undergoing Steady-State Turnover, Arch. Biochem. Biophys., 224:389.

Reinhart, G. D., 1985, Influence of pH on the Regulatory Kinetics of Rat Liver Phosphofructokinase: A Thermodynamic Linked-Function Analysis, Biochemistry, 24:7166.

Reinhart, G. D., and Hartleip, S. B., 1986, Relationship between Fructose 2,6-Bisphosphate Activation and MgATP Inhibition of Rat Liver Phosphofructokinase at High pH. Kinetic Evidence for Individual Binding Sites Linked by Finite Couplings, Biochemistry, 25:7308.

Reinhart, G. D., and Lardy, H. A., 1980a, Rat Liver Phosphofructokinase: Kinetic Activity under Near-Physiological Conditions, Biochemistry, 19:1477.

Reinhart, G. D., and Lardy, H. A., 1980b, Rat Liver Phosphofructokinase: Use of Fluorescence Polarization to Study Aggregation at Low Protein Concentration, Biochemistry, 19:1484.

Reinhart, G. D., and Lardy, H. A., 1980c, Rat Liver Phosphofructokinase: Kinetic and Physiological Ramifications of the Aggregation Behavior, Biochemistry, 25:1491.

Sturtevant, J. M., 1977, Heat Capacity and Entropy Changes in Processes Involving Proteins, Proc. Natl. Acad. Sci. U.S.A., 74:2236.

Uyeda, K., Furuya, E., and Luby, L. J., 1981a, The Effect of Natural and Synthetic D-Fructose 2,6-Bisphosphate on the Regulatory Kinetic Properties of Liver Phosphofructokinase, J. Biol. Chem., 256:8394.

Uyeda, K., Furuya, E., and Sherry, A. D., 1981b, The Structure of "Activation Factor" for Phosphofructokinase, J. Biol. Chem., 256:8679.

Van Schaftingen, E., Jett, M. F., Hue, L., and Hers, H. G., 1981, Control of Liver Phosphofructokinase by Fructose 2,6-Bisphosphate and Other Effectors, Proc. Natl. Acad. Sci. U.S.A., 78:3483.

Weber, G., 1952a, Polarization of the Fluorescence of Macromolecules. I. Theory and Experimental Method, Biochem. J., 51:145.

Weber, G., 1952b, Polarization of the Fluorescence of Macromolecules. II. Fluorescent Conjugates of Ovalbumin and Bovine Serum Albumin, Biochem. J., 51:155.

Weber, G., 1971, The Binding of Small Ligands by Proteins, in: "Horm. Steroids: Proc. 3rd Internatl. Congr.", Excerpta Medica, Amsterdam, p. 58.

Weber, G., 1972, Ligand Binding and Internal Equilibria in Proteins, Biochemistry, 11:864.

Weber, G., 1975, Energetics of Ligand Binding to Proteins, Adv. Prot. Chem. 29:1.

Wyman, J., 1948, Heme Proteins, <u>Adv. Prot. Chem.</u>, 4:407.

Wyman, J., 1964, Linked Functions and Reciprocal Effects in Hemoglobin: A Second Look, <u>Adv. Prot. Chem.</u>, 19:223.

Wyman, J., 1967, Allosteric Linkage, <u>J. Amer. Chem. Soc.</u>, 89:2202.

FLUORESCENCE STUDIES OF THE CALCIUM-DEPENDENT FUNCTIONS OF CALMODULIN

Sonia R. Anderson and Dean A. Malencik

Department of Biochemistry and Biophysics
Oregon State University
Corvallis, OR 97331 USA

BINDING OF SMALL PEPTIDES BY CALMODULIN

Introduction

Calmodulin is a major intracellular Ca^{2+} receptor in eucaryotic cells. The binding of Ca^{2+} stabilizes one or more conformations of the calmodulin molecule recognized by calmodulin-dependent enzymes such as cyclic nucleotide phosphodiesterase (Cheung, 1967); adenylate cyclase (Cheung et al., 1975 ; Brostrom et al., 1975); phosphorylase kinase ([Grand et al., 1981; cf review by Malencik & Fischer, 1983); and myosin light chain kinase (cf reviews by Stull, 1980; Small & Sobieszek, 1980; Perry et al., 1984). Large increases in catalytic activity typically result from the association of these enzymes with the calcium-calmodulin complex. A variety of small non-protein ligands also undergo calcium-dependent interactions with calmodulin. The best known of these are the phenothiazine drugs and other pharmacological agents first described by Levin and Weiss (1977). Since evidence suggests that these molecules recognize parts of calmodulin which are the same as (or closely related to) the association sites for calmodulin-dependent enzymes, they have been embraced as models for the characterization of the enzyme-calmodulin interface. However, the information obtainable from them is necessarily incomplete.

The recent determination of the three-dimensional structure of the calcium-calmodulin complex [$CaM(Ca^{2+})_4$] (Babu et al., 1985) has led to renewed interest in models for enzyme binding. Small peptides have proved instructive in the characterization of various types of protein-protein interaction. For example, synthetic peptide substrates helped to establish a class of recognition sequences surrounding the serine and threonine residues which can be phosphorylated by the cAMP-dependent protein kinase (cf. review by Carlson et al., 1979). In vitro inhibition of the calmodulin-dependent activity of cyclic nucleotide phosphodiesterase by adenocorticotropin (ACTH) and β-endorphin provided the first clue that calmodulin also interacts with small peptides (Weiss et al., 1980). Although these particular peptide-calmodulin associations do not necessarily occur in vivo, they captured our attention as potentially useful models providing information both on calmodulin and on specific cell surface receptors.

Development of this subject involved two major advancements. The first was the introduction of direct fluorometric assays for the determination of peptide binding by calmodulin. The original cyclic nucleotide phospho-diesterase assay is tedious and the results require correction for calmodulin binding by the enzyme. The second was the discovery of small peptides whose affinities for calmodulin rival those of calmodulin-dependent enzymes. The enzyme-calmodulin complexes have dissociation constants in the range of 1 nM to 10 nM while the complexes with ACTH and β-endorphin only have dissociation constants in the micromolar range.

Fluorescence Binding Methods

Intrinsic Fluorescence of Tryptophan-Containing Peptides. Teale (1960) and Weber (1961) showed that excitation at 295 nm largely excludes the fluorescence of tyrosine from the intrinsic emission spectra of proteins. Since calmodulin contains tyrosine but no tryptophan, we have applied this principle to binding measurements on tryptophan-containing peptides--beginning with ACTH and porcine glucagon (Malencik & Anderson, 1982). These determinations give information on the stoichiometry of complex formation, on equilibrium constants, and on the microscopic environment of the bound chromophore. However, the range of excitation wavelengths used--290 to 295 nm, depending on the sensitivity and resolution needed in individual cases--usually limits the measurements to solutions containing at least 1 μM peptide. Titrations of tryptophan containing peptides with calmodulin are thus suitable for the accurate determination of dissociation constants greater than ~0.2 μM. This restriction follows the advice of Weber (1965), who deduced that the protein and ligand concentrations used in this type of binding measurement should be of the same magnitude as the dissociation constant.

The intrinsic fluorescence of Polistes mastoparan, a toxic peptide from the social wasp, proved especially responsive in these studies.

Val-Asp-Trp-<u>Lys</u>-<u>Lys</u>-Ile-Gly-Gln-His-Ile-Leu-Ser-Val-Leu-NH$_2$

<u>Polistes</u> mastoparan belongs to a group of high affinity calmodulin binding peptides whose discovery is described in the next section. Its fluorescence spectrum undergoes a large shift on complex formation with calmodulin (Figure 1), demonstrating that the local environment of the tryptophan residue in the bound peptide is comparable to the protected environments found in many protein molecules (Malencik & Anderson, 1983a). This conclusion was supported by the results of acrylamide quenching studies (McDowell et al., 1985). Determination of the fluorescence anisotropy $[(I_{\shortparallel}-I_{\perp})/(I_{\shortparallel}+2I_{\perp})]$ is an independent method for the detection of complex formation which is based on molecular weight changes and is responsive to binding even when the fluorescence spectrum and quantum yield do not change. Figure 2 shows the stoichiometric increase in anisotropy occurring when <u>Polistes</u> mastoparan is titrated with calmodulin in the presence of 1 mM CaCl$_2$. Saturation is reached on the addition of one mol calmodulin per mol peptide, indicating that calmodulin has a single binding site for the peptide of substantially higher affinity than any other sites. At the concentrations used, little interaction between calmodulin and <u>Polistes</u> mastoparan occurs in the absence of calcium.

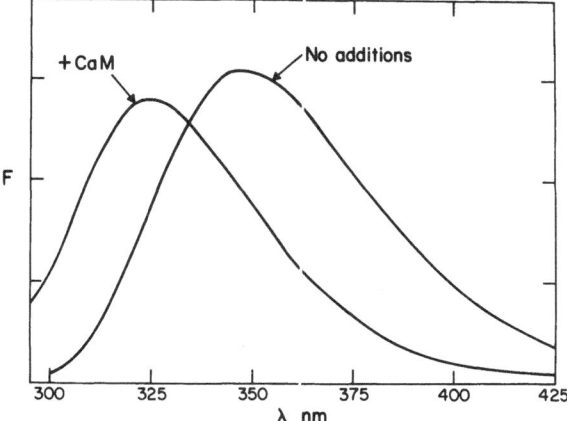

Figure 1. Effect of calmodulin binding on the fluorescence spectrum of <u>Polistes</u> mastoparan. Calmodulin (10.0 μM) was added to <u>Polistes</u> mastoparan (8.0 μM) in a solution containing 0.20 N KCl, 5.0 mM Mops, 1.0 mM CaCl$_2$, pH 7.3 (25.0°). Excitation: 290 nm. The band widths of excitation and emission were 3 and 5 nm, respectively. Reprinted with permission from Malencik, D. A. & Anderson, S. R., <u>Biochem. Biophys. Res. Comm.</u>, <u>114</u>, 50-56. Copyright (1983) Academic Press.

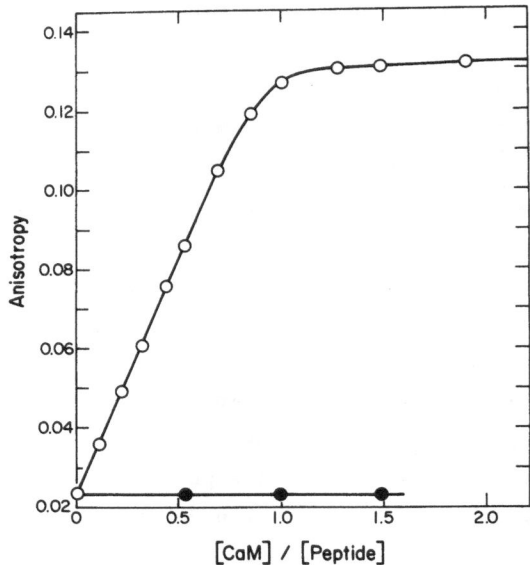

Figure 2. Stoichiometric titration of <u>Polistes</u> mastoparan with calmodulin. The fluorescence anisotropies were measured using an excitation wavelength of 294 nm, with a bandpass of 2 nm. The emitted light passed through Corning glass CS 0-54 filters. Conditions: 5.0 µM <u>Polistes</u> mastoparan in 0.20 N KCl, 5.0 mM Mops, pH 7.3 (25.0°). (o) 0.10 mM $CaCl_2$ added (•) 0.1 mM EDTA added. Reprinted with permission from Malencik, D. A. & Anderson, S. R., <u>Biochem. Biophys. Res. Comm.</u> <u>114</u>, 50-56. Copyright (1983) Academic Press.

Three distinctive wavelengths in the emission spectrum of <u>Polistes</u> mastoparan--320, 340, and 360 nm--were selected for a detailed examination of the calcium dependence. Stepwise additions of calcium were made to an initially calcium-free solution containing 5.0 µM each of calmodulin and the peptide. The accompanying changes in fluorescence intensity suggest a spectroscopically distinct intermediate present at maximum concentration when half of the calcium binding sites of calmodulin are saturated (Figure 3A; Malencik & Anderson, 1986a). The addition of the first two equivalents of calcium results in a shift of the fluorescence spectrum to lower wavelengths while the addition of the third and fourth leads to a decrease in quantum yield. The measurements at 360 nm are about equally affected by these changes, showing a uniform decline in fluorescence. The possibility that calmodulin has a second peptide binding site which is occupied when the calcium-calmodulin complex is limiting is ruled out by the linear stoichiometric titrations obtained when varying concentrations of calmodulin are added to a solution containing 5.0 µM peptide and 1 mM $CaCl_2$ (Figure 3B).

That Ca^{2+} binding by calmodulin takes place in at least two stages has

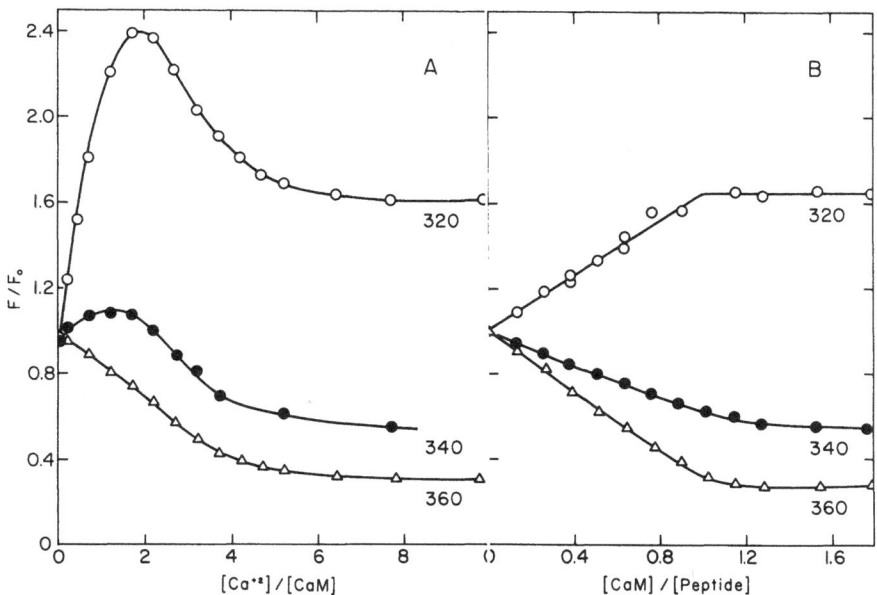

Figure 3. Fluorescence intensity titrations performed at 320, 340, and 360 nm. Panel A shows the effects of varied concentrations of <u>calcium</u> on a solution containing 5.0 μM each of <u>Polistes</u> mastoparan and calmodulin. Panel B shows the effects of varied concentrations of <u>calmodulin</u> on a solution containing 5.0 μM <u>Polistes</u> mastoparan plus 1 mM $CaCl_2$. F is the observed fluorescence intensity; F_o is the fluorescence intensity of the peptide in the absence of calmodulin. Conditions: 5.0 μM <u>Polistes</u> mastoparan, 5.0 μM calmodulin, 0.23 M KCl, 9.7 mM Mops, pH 7.2 (25°). Excitation: 290 nm. The band widths of excitation and emission were 2.5 and 4 nm respectively. Reprinted with permission from Malencik, D. A. & Anderson, S. R., <u>Biochem. Biophys. Res. Comm.</u> <u>135</u>, 1050-1057. Copyright (1986) Academic Press.

been demonstrated in nuclear magnetic resonance (Seamon, 1980; Ikura et al., 1983; cf. review by Forsén et al., 1986), fluorescence (Kilhoffer et al., 1981), and circular dichroism spectra (Crouch & Klee, 1980; Klee, 1977). The association of calmodulin with the first two Ca^{2+} ions-- considered to occur at the C-terminal binding sites, III and IV--is strongly cooperative under diverse conditions (Forsén et al., 1986). Our observations on the binding of calcium by the <u>Polistes</u> mastoparan-calmodulin complex are unique in that a property of the bound ligand, rather than of calmodulin, was monitored. The intermediate detected in Figure 3A corresponds to either the $CaM(Ca^{2+})_2$ complex in the cooperative model or to a mixture of $CaM(Ca^{2+})_1$ and $CaM(Ca^{2+})_2$ in the random binding model suggested by Burger et al. (1984). The contrasting effects of the partially saturated species and $CaM(Ca^{2+})_4$ on the bound peptide may relate to the binding and activation of calmodulin-dependent enzymes. Kincaid & Vaughan (1986) have evidence that the occupation of the first two Ca^{2+}

sites facilitates the association of calcineurin with calmodulin and that further Ca^{2+} binding allows enzyme activation to occur.

The detection of intermediates in calcium binding by calmodulin depends on the use of suitable probes. Both <u>Polistes</u> mastoparan X are sensitive to the appearance of intermediates while porcine glucagon and melittin are not. Due to the avid binding of <u>Polistes</u> mastoparan, the dissociation constant of its complex with calmodulin could not be determined from these measurements. However, we have used the intrinsic peptide fluorescence to determine equilibrium constants for other calmodulin binding peptides: ACTH, porcine and catfish glucagons, dynorphin$_{1-17}$, and gastric inhibitory peptide (cf. review by Anderson & Malencik, 1986).

Dansyl Calmodulin. Dansyl calmodulin is an exceptionally responsive covalent conjugate prepared by labeling calmodulin with 5-(dimethylamino)-1-naphthalenesulfonyl chloride (Malencik & Anderson, 1982). Its fluorescence emission maximum and quantum yield change dramatically on the binding of both calcium and proteins or peptides (Figure 4). These effects are the basis of a highly sensitive method for the determination of both calcium-

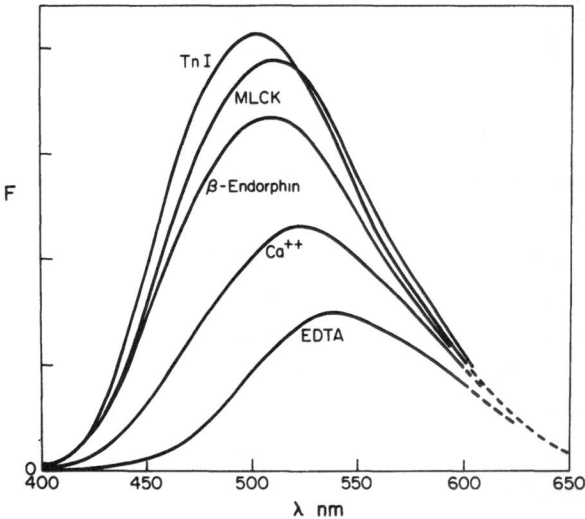

Figure 4. Corrected fluorescence emission spectra of conjugate of porcine brain calmodulin with 5-(dimethylamino)-1-naphthalenesulfonyl chloride. The spectra of 1.0 μM solutions of dansyl calmodulin were recorded in the presence of 1.2 mM EDTA, 0.85 mM $CaCl_2$, and 1.2 μM troponin I, 18 μM β-endorphin, or 2.0 μM myosin light chain kinase plus 0.85 mM $CaCl_2$. Conditions: 0.20 N KCl and 50 mM MOPS, pH 7.3, 25.0°. Excitation: 340 nm. Reprinted with permission from Malencik, D. A. and Anderson, S. R., <u>Biochemistry</u> <u>21</u>, 3480-3486. Copyright (1982) American Chemical Society.

dependent and calcium-independent peptide binding. Varying amounts of peptide are added to a cuvette solution which contains a fixed concentration of dansyl calmodulin ranging upwards from 0.1 µM. The data are fitted to the simple equilibrium

$$\text{CaM·peptide} \xrightleftharpoons{K} \text{CaM + peptide}$$

for which the average degree of saturation of dansyl calmodulin with peptide (Φ) is related to the fluorescence enhancement.

$$\Phi = ((F_{obs}/F_o)-1)/((F_\infty/F_o)-1)$$

F_∞ is the fluorescence of the complex and F_o, that of dansyl calmodulin in the absence of peptide. This method is especially useful for peptides which contain no tryptophan (Kincaid et al. , 1982) independently applied dansyl calmodulin to binding studies of cyclic nucleotide phosphodiesterase and calcineurin.

Competitive Displacement of Smooth Muscle Myosin Light Chain Kinase from Calmodulin. We developed a competitive displacement technique based on the calmodulin-dependent binding of 9-anthroylcholine, a fluorescent dye, by turkey gizzard myosin light chain kinase (Malencik et al., 1982a). The details of this procedure are reviewed by Anderson & Malencik (1986). In essence, the fractional dissociation of (ΔF) of the enzyme-calmodulin complex obtained in the presence of a competing calmodulin binding peptide is determined from changes in the fluorescence of 9-anthroylcholine.

The distribution of calmodulin between the two complexes and the ratio of the dissociation constants for the competing equilibria can be calculated from the values of ΔF, even when the individual constants are beyond the range of direct determination.

$$\text{MLCK·CaM} \xrightleftharpoons{K_{MLCK·CaM}} \text{MLCK + CaM}$$

$$\text{P·CaM} \xrightleftharpoons{K_{P·CaM}} \text{P + CaM}$$

$$\frac{K_{MLCK·CaM}}{K_{P·CaM}} = \frac{\Delta F^2 [MLCK]_o}{(1-\Delta F)([P]_o - \Delta F[MLCK]_o)}$$

Table 1. Sequence Comparisons of Calmodulin Binding Peptides

Substance P (complete)			Arg	Pro	Lys	Pro	Gln	Gln	Phe	Phe	Gly	Leu	Met-NH$_2$	
Glucagon (14-27)	- Leu	Asp	Ser	Arg	Arg	Ala	Gln	Asp	Phe	Val	Gln	Trp	Leu	Met -
ACTH (14-27)	- Gly	Lys	Lys	Arg	Arg	Pro	Val	Lys	Val	Tyr	Pro	Asn	Gly	Ala -
β-Endorphin (reversed, 31-18)	Glu	Gly	Lys	Lys	Tyr	Ala	Asn	Lys	Ile	Ile	Ala	Asn	Lys	Phe -

Alignments are in accord with the model suggested by Malencik and Anderson (1982).

[MLCK]$_o$ is the total concentration of enzyme, which is fixed at a value (usually 0.5 to 1.0 μM) equal to the total concentration of calmodulin. [P]$_o$ is the total concentration of the competing peptide. The above equation is simplified whenever [P]$_o$ >> ΔF[MLCK]$_o$. Under the conditions generally used, $K_{MLCK \cdot CaM}$ = 1.8 ± 0.4 nM (Malencik & Anderson, 1986b). This method is especially effective with high-affinity peptides.

Survey of Calmodulin Binding Peptides

This article emphasizes peptides known to possess intermediate to high affinities for calmodulin. The results were gleaned from observations on more than 60 different small peptides (cf review by Anderson & Malencik, 1986). In 1982 we made all possible forward and reverse sequence comparisons of the peptides then known to undergo efficient (micromolar range) calcium-dependent binding by calmodulin: ACTH, β-endorphin (Weiss et al., 1980; Malencik & Anderson, 1982), porcine glucagon, and substance P (Malencik & Anderson, 1982). Table 1 shows sequence alignments emphasizing similarities among the four peptides. We proposed a rudimentary recognition sequence for calmodulin consisting of a strongly basic tripeptide sequence, with at least two residues which are either Arg or Lys, three positions away from a pair of hydrophobic residues. Experiments with related peptides lacking the specified features generally supported this interpretation. We hypothesized that sequences similar to those found in the peptides occur in accessible positions on the surfaces of enzymes and other proteins which recognize calmodulin and that they are directly involved in the interaction. Improved binding efficiencies were obtained in later experiments with three other members of the glucagon family. Vasoactive intestinal peptide (VIP), gastric inhibitory peptide (GIP), and secretin give calcium-dependent complexes with calmodulin corresponding to dissociation constants of ~50 nM, 90 nM, and 140 nM, respectively (Malencik & Anderson, 1983b). The occurrence of a second cluster of basic amino acid residues is the major distinguishing feature of the latter peptides.

Convergence to high affinity binding came about unexpectedly in 1983, when we sought a peptide conforming to the paradigm, but lacking aromatic residues, for use in proton nuclear magnetic resonance experiments (Malencik & Anderson, 1983a; Muchmore et al., 1986). A survey of commercially available peptides revealed a likely candidate--mastoparan.

Ile-Asn-Leu-Lys-Ala-Leu-Ala-Ala-Leu-Ala-Lys-Lys-Ile-Leu-NH$_2$

This tetradecapeptide from the venom of the Vespid wasp has an affinity for calmodulin surpassing that of smooth muscle myosin light chain kinase. Competitive displacement experiments (refer to preceding section) suggested that the dissociation constant of the calmodulin-mastoparan complex is ~0.3 nM. A larger proportion of hydrophobic residues, a strong potential for α-helix formation, and a palindromic sequence spanning residues 4-11 distinguish mastoparan from the preceding peptides. Polistes mastoparan and mastoparan X also form highly stable complexes with calmodulin (Table 2) (Malencik & Anderson, 1983a). Coincidentally, Barnett et al. (1983) discovered that mastoparan is a potent inhibitor of the calmodulin-dependent activity of cyclic nucleotide phosphodiesterase.

We used the intrinsic fluorescence of mastoparan X in experiments to help localize the peptide binding site of calmodulin. Thrombic fragment 1-106 and half-calmodulin fragment 72-148 form complexes with mastoparan X corresponding to dissociation constants of 0.9 μM and ~0.15 μM, respectively. No interaction is obtained with fragment 107-148 (Malencik & Anderson, 1984). The region of sequence overlap of the two fragments (residues 72-106) may contain a major portion of the peptide binding site. X-ray crystallographic studies subsequently revealed an exposed α-helix in calmodulin, containing amino acid residues 65-92, which is under scrutiny as a possible interaction site for enzymes and pharmacological agents (Babu et al., 1985).

Melittin, a 26-residue peptide from honey bee venom (Table 2), is also a highly effective inhibitor of cyclic nucleotide phosphodiesterase (Barnette et al., 1983; Comte et al., 1983). X-ray studies on crystalline melittin had demonstrated a tetrameric assembly in which each monomer has the overall shape of a bent rod with two α-helical segments (Terwilliger & Eisenberg, 1982). The circular dichroism spectrum of the melittin-calmodulin complex is consistent with α-helix formation, presumably within the peptide (Maulet & Cox, 1983). The α-helical structure of the bound melittin molecule was recently confirmed by the nuclear magnetic resonance spectrum of melittin with completely deuterated calmodulin (Seeholzer et al., 1986).

Of the known naturally occurring calmodulin-binding peptides, melittin displays the highest affinity for calmodulin. In fact, our competition experiments show that the dissociation constant for the melittin-calmodulin complex is considerably smaller than the value (3 nM) estimated by Comte et al. (1983). Gel filtration experiments, fluorescence polarization measurements using dansyl calmodulin, competitive displacement experiments monitored

226

Table 2. Dissociation Constants[a] for Complexes of Calmodulin with Venom Peptides

Phe Leu Pro Leu Ile Leu <u>Arg</u> <u>Lys</u> Ile Val Thr Ala Leu-NH$_2$
      Crabrolin (K = 6.9 nM)

Ile Asn Leu <u>Lys</u> Ala Leu Ala Ala Leu Ala <u>Lys</u> <u>Lys</u> Ile Leu-NH$_2$
      Mastoparan (K ~ 0.3 nM)

Ile Asn Trp <u>Lys</u> Gly Ile Ala Ala Met Ala <u>Lys</u> <u>Lys</u> Leu Leu-NH$_2$
      Mastoparan X (K ~ 0.9 nM)

Val Asp Trp <u>Lys</u> <u>Lys</u> Ile Gly Gln His Ile Leu Ser Val Leu-NH$_2$
      <u>Polistes</u> Mastoparan (K ~ 2.3 nM)

Gly Ile Gly Ala Val Leu <u>Lys</u> Val Leu Thr Thr Gly Leu Pro -
Ala Leu Ile Ser Trp Ile <u>Lys</u> <u>Arg</u> <u>Lys</u> <u>Arg</u> Gln Gln-NH$_2$
      Melittin (K < 0.1 nM)

His Ser Asp Ala Ile Phe Thr Gln Gln Tyr Ser <u>Lys</u> Leu Leu Ala <u>Lys</u> Leu Ala -
Gln <u>Lys</u> Tyr Leu Ala Ser Ile Leu Gly Ser <u>Arg</u> Thr Ser Pro Pro Pro-NH$_2$
      Helodermin (K = 1.8 nM)

[a]Determined by competitive displacement of myosin light chain kinase in solutions containing 0.20 M KCl, 1.0 mM CaCl$_2$, 50 mM Mops, pH 7.3 (25°).

with 9-anthroylcholine, and assays for the hemolytic activity of melittin demonstrate that in solutions containing equal molar amounts of calmodulin, melittin, and smooth muscle myosin light chain kinase, 85-90% of the calmodulin is bound to melittin. This result is independent of the order of mixing (Malencik & Anderson, 1984, 1985; Anderson & Malencik, 1986). The dissociation constants estimated from our studies are consistently in the range of 0.1 nM or less. The complex of calmodulin with melittin apparently resembles that of the <u>Escherichia</u> <u>coli</u> lac repressor-operator complex in stability and responsiveness (Berg et al., 1981).

Crabrolin and helodermin--peptides of unknown function found in the venoms of the European hornet and the Gila monster, respectively--are also tightly bound by calmodulin (Anderson & Malencik, 1986; Malencik & Anderson, 1986c). Overall, the peptides listed in Table 2 share several common features. The first is their occurrence in animal venoms. The strong calcium dependence and stabilities of their complexes with calmodulin are consistent with a possible role in cytotoxicity. However, the hemolytic activity of melittin and the degranulation of mast cells by the mastoparans are traceable to cell surface interactions (cf Banks & Shipolini, 1986; Nakajima, 1986). Second, the close association of basic and hydrophobic sequences seen in the peptides generally follows the model originally envisioned by us. The separation between positively charged residues in helodermin may explain why it binds calmodulin less effectively than melittin

Table 3. Secondary Structures of Calmodulin Binding Peptides

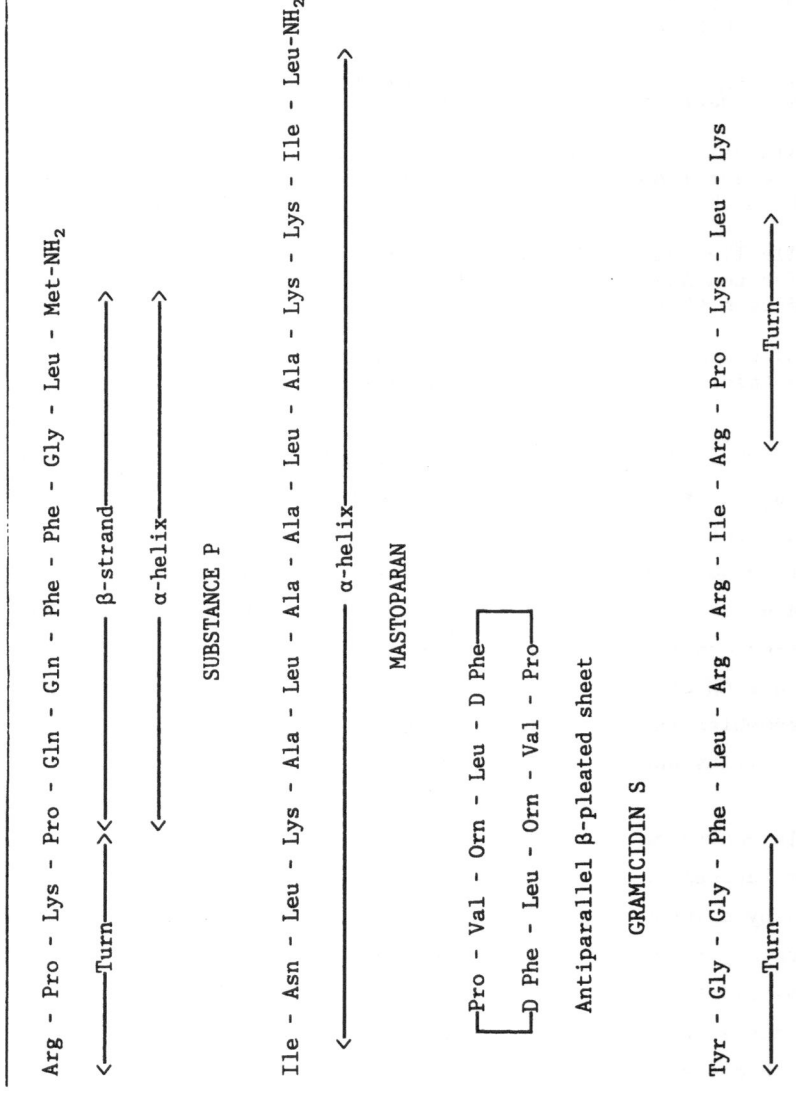

Arg - Pro - Lys - Pro - Gln - Gln - Phe - Phe - Gly - Leu - Met-NH$_2$

<———Turn———><——————— β-strand ———————>

<——————— α-helix ———————>

SUBSTANCE P

Ile - Asn - Leu - Lys - Ala - Leu - Ala - Ala - Leu - Ala - Lys - Lys - Ile - Leu-NH$_2$

<——————————————— α-helix ———————————————>

MASTOPARAN

Pro - Val - Orn - Leu - D Phe

D Phe - Leu - Orn - Val - Pro

Antiparallel β-pleated sheet

GRAMICIDIN S

Tyr - Gly - Gly - Phe - Leu - Arg - Arg - Ile - Arg - Pro - Lys - Leu - Lys

<——————— Turn ———————>

<———Turn———>

DYNORPHIN$_{1-13}$

Structures of substance P, mastoparan, and dynorphin$_{1-13}$ were predicted by the rules of Chou & Fasman (1978). Structure of gramicidin S is from Krauss & Chan (1982).

does. Observations on other peptides belonging to common gene families have indicated that both the total number of basic amino acid residues and their spacing are important in calmodulin binding. Glutamic acid, which appears to be destabilizing, is absent in this group of peptides (cf. review by Anderson & Malencik, 1986). The third common characteristic is a marked tendency towards α-helix formation.

The results obtained with the mastoparans and melittin prompted the synthesis of still other amphipathic, α-helical peptides for use as models in calmodulin binding studies (Cox et al., 1985; DeGrado et al., 1985). The sub-nanomolar range dissociation constants often obtained with these peptides suggest fortuitous optimization of interactions, surpassing that attained by myosin light chain kinase. However, even though the α-helix may be a pre-ferred conformation, it is <u>not</u> an absolute requirement for calmodulin bind-ing.

The inability of calmodulin to associate with an amphipathic β-sheet structure (cf Krauss & Chan, 1982) is exemplified by gramicidin S (Table 3). The calcium-dependent binding of gramicidin S by dansyl calmodulin (K_d = 1.9 μM) and its competition with porcine glucagon were discovered by Yechiel Shalitin during a sabbatical visit to our laboratory. We had sug-gested the β-strand as a conformation involved in peptide binding in conjunc-tion with studies on the glucagon-secretin family (Malencik & Anderson, 1983b).

Finally, a large number of calmodulin-binding peptides correspond to either non-repetitive ("random" coil) structures or to mixed structures that include sequences with similar probabilities of occurrence in the α-helix and β-strand conformations. The dynorphins represent the former while substance P plus members of the glucagon-secretin group typify the latter possibilities (Table 3) (Malencik & Anderson, 1983b). VIP, a 28-residue peptide related to glucagon, forms a complex with calmodulin corresponding to $K_d \sim 50$ nM. The circular dichroism spectrum of the complex is different from that obtained with melittin and calmodulin, indicating little α-helix formation in the bound VIP molecule.

Secondary folding of polypeptides is only one way to generate amphi-pathic structure. Hydrophobic photolabeling demonstrated that association of dynorphin$_{1-13}$ with lipid vesicles is due to its <u>segmental</u> organization into a hydrophobic message (Tyr-Gly-Gly-Phe-) and hydrophilic address (-Leu-Arg-Arg-Ile-Arg-Pro-Lys-Leu-Lys) (Gysin & Schwyzer, 1983). A related pattern was shown for $ACTH_{1-24}$. Since $dynorphin_{1-13}$ and $ACTH_{1-24}$ also bind calmodulin, segmental organization may be involved in some of the peptide-calmodulin interactions.

Notwithstanding the variations in secondary structure, competition experiments--monitored through measurements of intrinsic tryptophan fluorescence--show that the various peptides recognize identical or closely overlapping sites on calmodulin (Anderson & Malencik, 1986; Malencik & Anderson, 1982, 1983b, 1984). There may be more than one way to present the cationic and hydrophobic moieties of peptides in suitable orientations for calmodulin binding. Additionally, some of the local interactions may vary. Prozialeck and Weiss (1982) and Weiss et al. (1982) proposed a model for the finding of phenothiazines and other pharmacological agents displaying some of the features seen in the peptides. They noted that the most potent calmodulin antagonists contain two hydrophobic moieties separated from a positively charged amino nitrogen by at least three carbon atoms.

Calmodulin-Binding Domains of Enzymes: Targets of Protein Phosphorylation?

Cyanogen bromide cleavage of rabbit skeletal muscle (Blumenthal et al., 1985; Edelman et al., 1985), chicken gizzard (Lukas et al., 1986), and turkey gizzard myosin light chain kinases (Malencik & Anderson, 1986, unpublished work) generates low molecular weight calmodulin binding fragments. Inhibition experiments suggest that their affinities for calmodulin are high. The sequences of the fragments (Table 4) are consistent with predictions based on the preceding peptides, showing that we have a good grasp of the stabilizing factors involved in complex formation. The process of enzyme activation may involve these or other weaker interactions. The fragments obtained from skeletal and smooth muscle myosin light chain kinases also differ from each other to about the same extent that the other calmodulin-binding peptides differ.

The fragment of chicken gizzard myosin light chain kinase was prepared from _enzyme_ that had been phosphorylated _in_ _vitro_ by the cAMP-dependent protein kinase. One of the two phosphate moieties introduced by the reaction appeared in the pair of serine residues at the C-terminal end of the peptide (Lukas et al., 1986). _In_ _vitro_ phosphorylation of the calmodulin binding _fragment_ which we have obtained from turkey gizzard myosin light chain kinase (Table 4) results in the incorporation of one mol ^{32}P/mol peptide when the reaction is catalyzed by the cAMP-dependent protein kinase. In contrast to the results obtained with native myosin light chain kinase, where phosphoryla-

Table 4. Calmodulin Binding Domains of Enzymes

Skeletal Muscle Myosin Light Chain Kinase [a]

    Lys Lys <u>Arg</u> Trp <u>Lys</u> <u>Lys</u> Asn Phe Ile Ala Val Ser Ala Ala Asn <u>Arg</u> Phe <u>Lys</u> <u>Lys</u> Ile Ser Ser Ser Gly Ala Leu Met

Smooth Muscle Myosin Light Chain Kinase (Chicken Gizzard [b] or Turkey Gizzard[c])

    Ala <u>Arg</u> <u>Arg</u> <u>Lys</u> Trp Gln <u>Lys</u> Thr Gly His Ala Val <u>Arg</u> Ala Ile Gly <u>Arg</u> Leu Ser Ser

[a]Blumenthal et al. (1985)

[b]Lukas et al. (1986)

[c]Malencik & Anderson, unpublished results (1986)

231

tion is inhibited in the complex with calmodulin (Conti & Adelstein, 1981; Malencik et al., 1982a), the phosphorylation of the fragment occurs readily in the presence of excess calmodulin and calcium. In fact, the rate of phosphorylation of the peptide under these conditions is four times greater than that determined in equivalent solutions containing EGTA instead of calcium. Apparently the peptide prepared from turkey gizzard myosin light chain kinase becomes a better substrate for cAMP-dependent protein kinase when bound to calmodulin. No reaction occurs with cGMP-dependent protein kinase, turkey gizzard myosin light chain kinase, or rabbit muscle phosphorylase kinase (Figure 5; Malencik & Anderson (1986), unpublished results).

Considering that phosphorylation has a large effect on calmodulin binding by smooth muscle myosin light chain kinase (Conti & Adelstein, 1981; Malencik et al., 1982a) and that there are similarities between the binding specificities of calmodulin and cAMP-dependent protein kinase, we had sug-

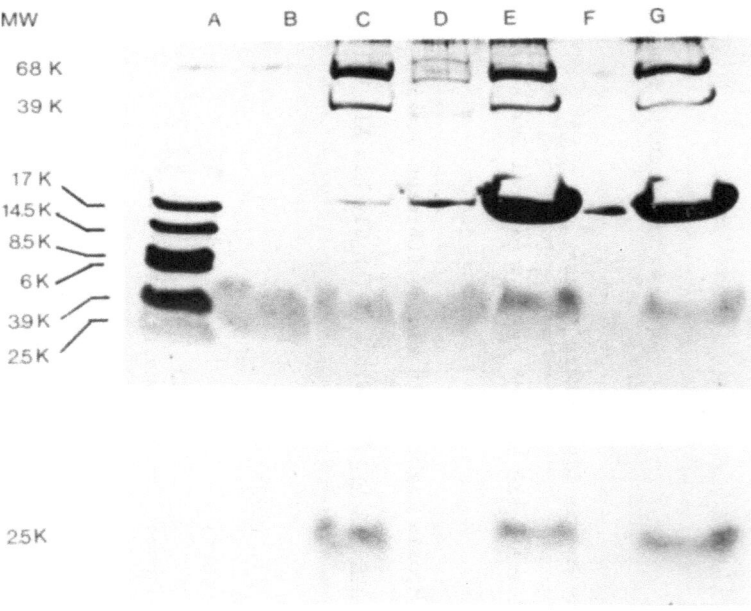

Figure 5. SDS polyacrylamide gel electrophoresis (15-30% gradient gel, unpublished) of ^{32}P-labeled fragment of myosin light chain kinase (MLCK) from turkey gizzard stained with Coomassie blue (top). The lower portion of the figure shows the section of the radioautograph corresponding to the fragment. (None of the other proteins were radioactive.) A. Marker Proteins; B. 10 μM MLCK peptide; C. B + 0.5 μM catalytic subunit of the cAMP-dependent protein kinase; D. B + 5 μM cGMP-dependent protein kinase; E. C + 1 mM Ca^{2+}, 130 μM calmodulin; F. Blank; G. C + 1 mM EGTA, 130 μM calmodulin.

gested that calmodulin and protein kinase interact with common sequences in some proteins (Malencik & Anderson, 1982; Malencik et al., 1982a,b). The inhibitory region of the 8000 Da heat-stable cAMP-dependent protein kinase inhibitor contains the following amino acid sequence (Scott et al., 1985).

Tyr-Ala-Phe-Ile-Ala-Ser-Gly-Arg-Thr-Gly-Arg-Arg-Asn-

Ala-Ile-His-Asp-Ile-Leu-Val-Ser-Ser-Ala

Several depletion peptides derived from it provided the first opportunity for direct comparison of the peptide binding specificities of calmodulin and protein kinase (Malencik et al., 1986). All of the resulting peptides known to interact effectively with either of the proteins contain the -Arg-Thr-Gly-Arg-Arg- sequence. However, the amino acid residues adjoining the N- and C-terminal ends of this basic center apparently have different roles in the binding of calmodulin and protein kinase. The peptide containing residues 4-23 of the preceding sequence forms a calcium-dependent complex with dansyl calmodulin corresponding to K_d = 70 nM and a complex with the catalytic subunit of cAMP-dependent protein kinase giving K_I = 0.8 μM. The derivative representing residues 1-19, in contrast, gives corresponding values of K_d = 5.8 μM and K_I = 0.18 μM. Key residues involved in calmodulin binding are apparently inaccessible in the native protein kinase inhibitor, which binds calmodulin poorly.

A synthetic substrate modeled after the "pseudo-substrate" site of the heat-stable protein kinase inhibitor (-Gly-Arg-Thr-Gly-Arg-Arg-Asn-Ser-Ile-His-Asp-Ile-Leu-) is readily phosphorylated at both its seryl and threonyl residues. The effect of peptide phosphorylation on calmodulin binding, with K_d increasing from 7 μM to 63 μM, is larger than that found for two other phosphorylatable peptides, but smaller than that characteristic of smooth muscle myosin light chain kinase (Malencik et al., 1986). The 20-100-fold decrease in binding occurring in the latter case may reflect conformation changes, although none have been detected, rather than immediate covalent modification of the calmodulin binding site.

DITYROSINE FORMATION DURING ULTRAVIOLET IRRADIATION OF CALMODULIN

Introduction

The results of several different kinds of physical measurement plus observations on the chemical reactivities of specific amino acid residues in calmodulin bear witness to the conformational changes that accompany calcium binding (cf. Klee & Vanaman, 1982; Forsén et al., 1986). The average fluorescence quantum yield of Tyr-99 and Tyr-138 in bovine brain calmodulin undergoes a 2.5 fold increase following interaction of the protein with calcium (Kilhoffer et al., 1981). As is true for other spectroscopic properties of amino acid residues in the C-terminal half of calmodulin (Seamon, 1980; Ikura et al., 1983), the change in fluorescence is nearly complete when 50% of the total calcium binding sites are saturated. The association of calcium with calmodulin may be sequential, with the C-terminal sites (III and IV) occupied ahead of the N-terminal sites (I and II) (cf. Forsén et al., 1986). The two tyrosine residues occur in the third (Tyr-99) and fourth (Tyr-138) calcium-binding domains. Tyr-99 faces into the third calcium binding site and is partially accessible to solvent while Tyr-138 points away from the fourth calcium-binding loop--into a hydrophobic pocket, where it makes contact with Phe-89 and Phe-141 (Babu et al., 1985).

Irradiation-Induced Changes in the Fluorescence Spectrum of Calmodulin

While performing repetitive scans of the fluorescence emission spectrum of a calcium-containing solution of bovine brain calmodulin, we were

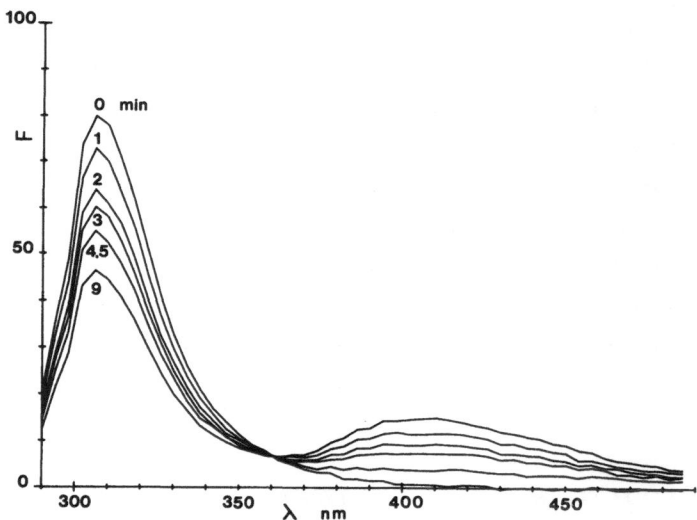

Figure 6. Rapid scans of the fluorescence emission spectrum of a solution of bovine brain calmodulin which has been irradiated at 280 nm for periods of time ranging from 0 to 9 min. Conditions: 0.5 mg/mL CaM in 10 mM Tris, 0.6 mM CaCl$_2$, pH 8.0 (25°). The emission bandpass was 4 nm. The excitation bandpass was 10 nm during irradiation and 2 nm during recording of the spectra.

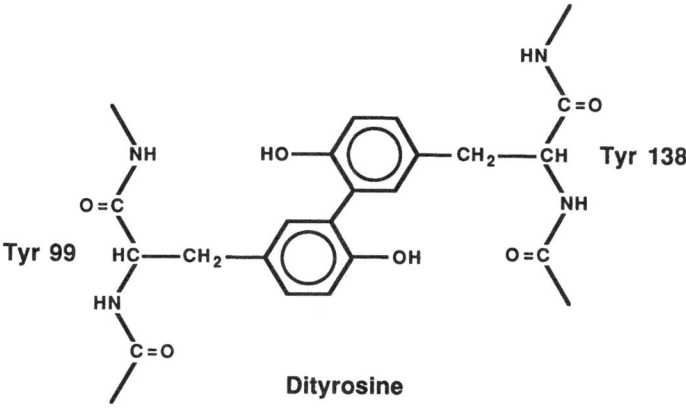

Figure 7. Structure of the dityrosine cross-link in bovine brain calmodulin.

surprised to find that 280 nm irradiation results in both a time-dependent decrease in the intrinsic tyrosine fluorescence at 300 nm and the appearance of a new emission band at 400 nm. To emphasize these effects, the excitation bandpass was increased to 10 nm and the solution illuminated, with stirring, for specified periods of time (Figure 6). The emission band at 400 nm is indicative of possible dityrosine formation in calmodulin. Dityrosine cross-links in proteins result from the phenolic coupling of phenoxy radicals (cf. review by Amadò et al., 1984). Coupling can be initiated by other radicals, such as $\cdot OH^-$, or by the photoejection of electrons during UV irradiation. Lehrer & Fasman (1967) demonstrated the formation of dityrosine during UV-irradiation of poly-L-tyrosine and its copolymers. Phenolic coupling in calmodulin is most likely to involve intramolecular cross-linking of Tyr-99 and Tyr-138 (Figure 7).

Irradiation was performed under varying conditions, both to facilitate the isolation of dityrosine-containing photoproducts of calmodulin and to elucidate the reaction. Figure 8 demonstrates the remarkable effect of calcium ion. When calcium is absent, very little change in fluorescence takes place at either 300 or 400 nm. At saturating calcium concentrations (0.5 mM), on the other hand, there is a rapid increase in intensity at 400 nm which appears to approach a plateau after 3 to 4 min of illumination. Prolonged exposure leads to a gradual decline in intensity. The effects of substoichiometric quantities of calcium--one through three mol Ca^{2+}/mol calmodulin--suggest that the reaction is most efficient when all the calcium binding sites are occupied. To determine whether calcium simply affects the

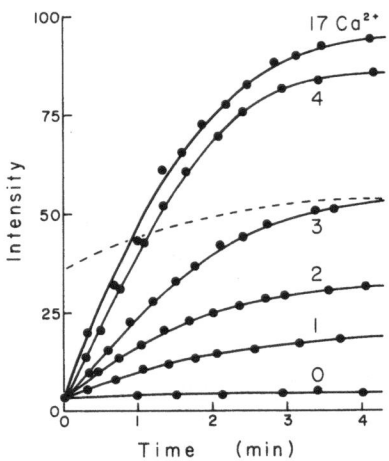

Figure 8. Calcium dependence of fluorescence changes occurring during UV irradiation of calmodulin. The fluorescence intensities at 400 nm were determined as a function of time for samples containing 0, 1, 2, 3, 4, and 17 mol Ca^{2+}/mol calmodulin. The dashed line illustrates the increase in intensity occurring when additional Ca^{2+} (0.62 mM) is added to the irradiated (5') sample containing 2 mol Ca^{2+}/mol. Conditions: 0.5 mg/mL CaM in 10 mM Tris, pH 8.0 (25°C). Excitation: 280 nm with a band width of 20 nm. Emission: 400 nm with a band width of 4 nm.

fluorescence quantum yield of the product, additional calcium (0.62 mM) was added to the irradiated (5 min exposure) sample originally containing 2 mol Ca^{2+}/mol protein. The result is an immediate ~10% increase in intensity followed by gradual changes which fall short of those obtained with samples containing excess calcium from the outset (dashed line in Figure 8). The time courses recorded after the addition of 0.5 mM calcium to irradiated calcium-free calmodulin approach those obtained with samples that had not been exposed.

The generation of the 400 nm emission band is more sensitive to calcium than it is to ionic strength or pH variation. We examined the influence of several other metal ions (Cd^{2+}, Mg^{2+}, Zn^{2+}, Fe^{3+}, Mn^{2+}, and Cu^{2+}) in order to corroborate the calcium dependence of the reaction. Diminished fluorescence intensities were obtained with millimolar concentrations of Cd^{2+}, Mg^{2+}, and Zn^{2+}, consistent with the following order of effectiveness.

$$Ca^{2+} > Cd^{2+} \gg Mg^{2+} > Zn^{2+}$$

No reaction was detected in solutions containing Mn^{2+} (1 mM), Fe^{3+} (0.1 mM), or Cu^{2+} (0.10 mM). Low concentrations of Mn^{2+} and Cu^{2+} actually inhibit the calcium-stimulated reaction (Malencik & Anderson, 1987). The behavior of the metal ions generally follows their abilities to support the calmodulin-dependent activity of cyclic nucleotide phosphodiesterase. Cadmium resembles calcium in both its ionic radius and association with calmodulin (Chao et al., 1984). The failure of Mn^{2+} to promote the fluorescence increase at 400 nm was unexpected, however. Cu^{2+} and Mn^{2+} differ from the other ions in that both are paramagnetic. This may allow them to participate in repair processes or to catalyze alternate photochemical reactions.

To further investigate the hypothesis that the 400 nm maximum in the fluorescence spectrum of irradiated calmodulin reflects the cross-linking of Tyr-99 and Tyr-138, we subjected several other proteins and polypeptides to the same reaction conditions. Of these, only poly-L-tyrosine exceeded calmodulin in the rate of fluorescence intensity change at 400 nm. However, the results obtained with this synthetic polymer are not appreciably calcium-dependent. UV irradiation of rabbit skeletal muscle troponin C, which is believed to be the protein most like calmodulin, leads to a decrease in the 300 nm emission with no change detected at 400 nm--in the presence or absence of calcium. Similar time courses were obtained with dogfish parvalbumin and the two thrombic fragments of calmodulin, containing amino acid residues 1-106 and 107-148. The failure of the reaction to occur in these examples was predicted. The sequence position of skeletal muscle troponin C homologous to Tyr-138 in calmodulin and the sequence position of parvalbumin homologous to Tyr-99 are both occupied by phenylalanine residues. Cardiac troponin C, in contrast, retains the implicated pair of tyrosine residues and exhibits the distinctive calcium-dependent increase in fluorescence at 400 nm during UV irradiation (Malencik & Anderson, 1987).

Isolation and Characterization of the Major Dityrosine-Containing Photoproduct of Calmodulin

Chromatography of a 40 mg pool of irradiated calmodulin on LKB Ultrogel AcA and phenyl-agarose columns produced several distinctive fractions. These included an intermolecularly cross-linked population, containing little dityrosine but displaying molecular weights in the range of 28,000-40,000 (representing 3.5% of the recovered protein), and a pool of more or less unmodified calmodulin (representing 83% of the recovered protein). The major 400 nm-emitting species ("X-CaM") was moderately retarded by phenyl-agarose in the presence of calcium, facilitating its separation from the other components. Pooled peak fractions, representing 2.8% of the recovered protein and 53% of the total fluorescence emission at 400 nm, were characterized in the experiments described below.

The fluorescence excitation and emission spectra of acid-hydrolyzed samples of X-CaM, adjusted to pH 9.5, are the same as those of dityrosine. Amino acid analyses of the hydrolysates confirmed the presence of dityrosine at the level of 0.59 to 0.89 mol/mol calmodulin. NaDodSO₄ electrophoresis of the unhydrolyzed derivative revealed two components of apparent molecular weights 16,000 and 14,000, accounting for 20% and 80%, respectively, of the total protein in the sample. (Native calmodulin has an apparent molecular weight of 16,000.) The increased mobility of the 14K component probably reflects incomplete folding of the cross-linked calmodulin molecule by NaDodSO₄ since no peptide bond cleavage was detectable by N-terminal determination or amino acid analysis. pH titrations, performed with solutions containing 0.184 M KCl, showed that the pK_a of the 400 nm-emitting chromophore in X-CaM varies from 7.88 in the absence of calcium to 7.59 on the addition of 3 mM $CaCl_2$ (Malencik & Anderson, 1987). Under these conditions, free dityrosine has a pK_a near 7.1. In all, our observations substantiate the proposed intramolecular cross-linking of Tyr-99 and Tyr-138.

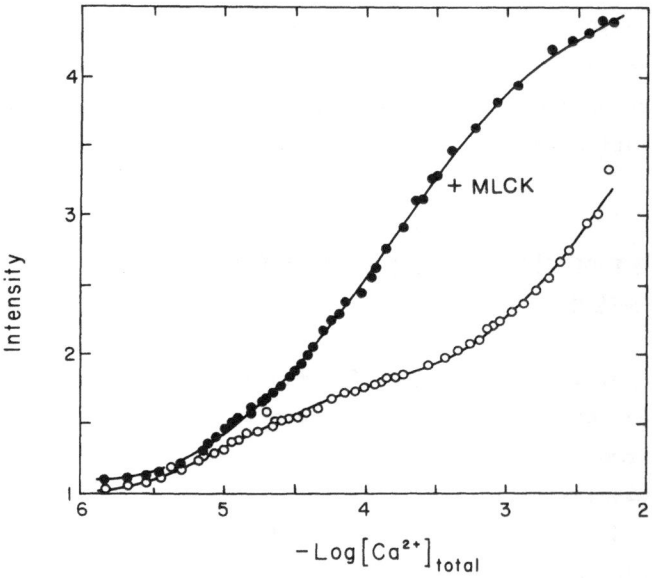

Figure 9. Fluorescence titration of the dityrosine-containing photoproduct of calmodulin with varying concentrations (M) of $CaCl_2$. Experiments were performed in both the presence (o) and absence (•) of two equivalents of smooth muscle myosin light chain kinase. Other conditions: 1.0 μM X-CaM, 50 mM Mops, pH 7.50 (25°C). Excitation: 320 nm with 2 nm band width. Emission: Schott KV 380 cut-off filter.

The interaction of calcium with the dityrosine containing photoproduct, X-CaM, was examined in measurements of both fluorescence intensity (Figure 9) and anisotropy. With a fixed pH of 7.50 and excitation wavelength of 320 nm, the intensity values are sensitive to changes in both pK_a and the fluorescence yield of the singly ionized dityrosine chromophore. The choice of wavelengths excludes any direct contribution of unmodified tyrosine. The titration reveals two stages in calcium binding but differs from comparable experiments with native calmodulin (cf. Forsén, et al., 1986) in that higher calcium concentrations, up to 10 mM or more, are needed to approach saturation. (With unmodified calmodulin, the various spectral changes are essentially complete at 0.10 mM free calcium ion.)

The shift in the equilibrium to lower calcium concentrations obtained on the addition of two equivalents of smooth muscle myosin light chain kinase (Figure 9) attests to the calcium-dependent binding of the derivative by the enzyme. A competitive displacement experiment (Figure 10) shows, however, that

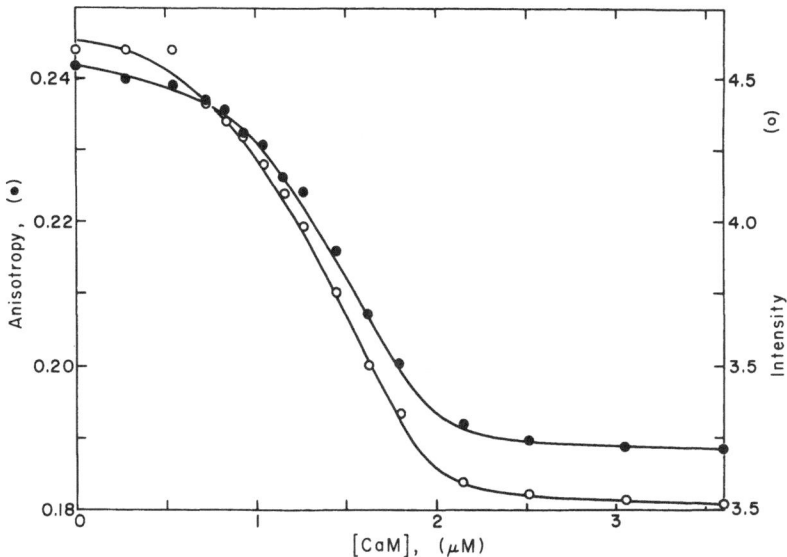

Figure 10. Competition between native calmodulin and its dityrosine-containing photoproduct in enzyme binding. A solution containing 1.0 μM X-CaM and 2.0 μM smooth muscle myosin light chain kinase was titrated with varying concentrations of native calmodulin, shown on the x-axis. The corresponding fluorescence intensities (o) and anisotropies (•) are shown on the y axis. Conditions: 5.0 mM $CaCl_2$, 50 mM Mops, pH 7.50 (25°C). Excitation: 320 nm with 2 nm band width. Emission: Schott KV 380 cut-off filter.

the ability of the cross-linked calmodulin molecule to associate with myosin light chain kinase is greatly diminished in comparison to that of the unmodified molecule. The addition of 2.0 μM calmodulin to a solution containing 5.0 mM $CaCl_2$, 2.0 μM enzyme, and 1.0 μM X-CaM results in ~92% dissociation of the latter. Random distribution of the enzyme between the native and modified populations would lead to only 33% dissociation at these concentrations. By assuming that all of the enzyme is bound, we estimate that the dissociation constant of its complex with the derivative is ~280-fold larger than that of its complex with native calmodulin. Under somewhat different conditions (0.20 M KCl plus 1.0 mM $CaCl_2$ were present), we determined that the latter value is 2.8 nM (Malencik & Anderson, 1986b).

SUMMARY

A novel calcium-dependent photochemical reaction of bovine brain calmodulin results in the intramolecular cross-linking of Tyr-99 and Tyr-138 (Malencik & Anderson, 1987). The reactivities and/or proximities of the two tyrosine residues are strongly enhanced by calcium binding, with dityrosine formation occurring most effectively when all four calcium binding sites of calmodulin are occupied. The modification results in weakened interactions of calmodulin with calcium and smooth muscle myosin light chain kinase. Cross-linking of Tyr-99 and Tyr-138 is consistent with the x-ray crystallographic studies of calmodulin, which showed that the loop regions of calcium binding sites III and IV lie side by side in an antiparallel arrangement which includes hydrogen bond formation between Il3-100 and Val-136 (Babu et al., 1985). The large changes in the ligand binding properties of calmodulin are likely to result from cross-linking rather than from tyrosine modification per se. Neither the acetylation nor the nitration of calmodulin tyrosyl residues affects the interaction with cyclic nucleotide phosphodiesterase (Walsh & Stevens, 1977; Richman, 1978; Richman & Klee, 1979).

ACKNOWLEDGEMENTS

Our work has been supported by research grants from the National Institutes of Health (AM13912), the Muscular Dystrophy Association, and the Oregon Affiliate of the American Heart Association.

FOOTNOTES

Abbreviations: Mops, 3-(N-morpholino)-propanesulfonic acid; Tris,
tris(hydroxymethyl)aminomethane; EDTA, ethylenediaminetetraacetic acid;
CaM, calmodulin; MLCK, myosin light chain kinase; NaDodSO$_4$, sodium dodecyl
sulfate.

REFERENCES

Amadò, R., Aeschbach, R., and Neukom, H., 1984, In Vitro Production and
    Characterization, Methods Enzymol., 107:377.

Anderson, S. R., and Malencik, D.A., 1986, Peptides Recognizing Calmodulin,
    in: "Calcium and Cell Function," W. Y. Cheung, ed., Academic Press,
    New York.

Babu, Y.S., Sack, J.S., Greenhough, T.G., Bugg, C.E., Means, A.R., and
    Cook, W. J., 1985, Three-Dimensional Structure of Calmodulin, Nature,
    315:37.

Banks, B.E.C., and Shipolini, R.A., 1986, Chemistry and Pharmacology of
    Honey-bee Venom, Venoms of the Hymenoptera, in: "Biochemical, Pharmaco-
    logical, and Behavioral Aspects," T. Piek, ed., Academic Press, New
    York.

Barnette, M.S., Daly, R., and Weiss, B., 1983, Inhibition of Calmodulin
    Activity by Insect Venom Peptides, Biochem. Pharmacol., 32:2929.

Berg, O.G., Winter, R.B., and von Hippel, P.H., 1981, Diffusion-driven
    Mechanisms of Protein Translocation on Nucleic Acids 1. Models and
    Theory, Biochemistry, 20:6929.

Blumenthal, D.K., Takio, K., Edelman, A.M., Charbonneau, H., Titani, K.,
    Walsh, K.A. and Krebs, E.G., 1985, Identification of the Calmodulin-
    Binding Domain of Skeletal Muscle Myosin Light Chain Kinase, Proc. Natl.
    Acad. Sci., U.S.A., 82:3187.

Brostrom, C.O., Huang, Y.C., Breckenridge, B. McL., and Wolff, D.J.,
    1975, Identification of a Calcium-Binding Protein as a Calcium-Dependent
    Regulator of Brain Adenylate Cyclase, Proc. Natl. Acad. Sci. U.S.A.,
    72:64.

Burger, D., Cox, J.A., Comte, M., and Stein, E.A., 1984, Sequential
    Conformational Changes in Calmodulin upon Binding of Calcium,
    Biochemistry, 23:1966.

Carlson, G.M., Bechtel, P.J., and Graves, D.J., 1979, Chemical and Regulatory
    Properties of Phosphorylase Kinase and Cyclic AMP-Dependent Protein
    Kinase, Adv. in Enzymol. Relat. Areas Mol. Biol., 50:41.

Chao, S.H., Suzuki, Y., Zysk, J.R., and Cheung, W.Y., 1984, Activation of Calmodulin by Various Metal Cations as a Function of Ionic Radius, Mol. Pharmacol., 26:75.

Cheung, W.Y., 1967, Cyclic 3',5'-Nucleotide Phosphodiesterase: Pronounced Stimulation by Snake Venom, Biochem. Biophys. Res. Comm., 29:478.

Cheung, W.Y., Bradham, L.S., Lynch, T.J., Lin, Y.M., and Tallant, E.A., 1975, Protein Activator of Cyclic 3':5' Nucleotide Phosphodiesterase of Bovine or Rat Brain also Activates Its Adenylate Cyclase, Biochem. Biophys. Res. Comm., 66:1055.

Cheung, W.Y., 1984, Calmodulin: Its Potential Role in Cell Proliferation and Heavy Metal Toxicity, Fed. Proc., 43:2995.

Chou, P.Y., and Fasman, G.D., 1978, Empirical Predictions of Protein Conformation, Ann. Rev. Biochem., 47:251.

Comte, M., Maulet, Y., and Cox, J.A., 1983, Ca^{++}-Dependent High-Affinity Complex Formation Between Calmodulin and Melittin, Biochem. J., 209:269.

Conti, M.A., and Adelstein, R.S., 1981, The Relationship Between Calmodulin and Phosphorylation of Smooth Muscle Myosin Kinase by the Catalytic Subunit of 3':5' cAMP-Dependent Protein Kinase, J. Biol. Chem., 256:3178.

Cox, J.A., Comte, M., Fitton, J.E., and DeGrado, W.F., 1985, The Interaction of Calmodulin with Amphilic Peptides, J. Biol. Chem., 260:2527.

Crouch, T.H., and Klee, C.B., 1980, Positive Cooperative Binding of Calcium to Bovine Brain Calmodulin, Biochemistry, 19:3692.

DeGrado, W.F., Prendergast, F.G., Wolfe, H.R., and Cox, J.A., 1985, The Design, Synthesis, and Characterization of Tight-Binding Inhibitors of Calmodulin, J. Cell. Biochem., 29:83.

Edelman, A.M., Takio, K., Blumenthal, D.K., Hansen, R.S., Walsh, K.A., Titani, K., and Krebs, E.G., 1985, Characterization of the Calmodulin-Binding and Catalytic Domains in Skeletal Muscle Myosin Light Chain Kinase, J. Biol. Chem., 260:11, 275.

Forsén, S., Vogel, H.J., and Drakenburg, T., 1986, Biophysical Studies of Calmodulin, in: "Calcium Cell Funct.", W. Y. Cheung, ed., Academic Press, New York.

Grand, R.J.A., Shenolikar, S., and Cohen, P., 1981, The Amino Acid Sequence of the Subunit (Calmodulin) of Rabbit Skeletal Muscle Phosphorylase Kinase, Eur. J. Biochem., 113:359.

Gysin, G. and Schwyzer, R., 1983, Head Group and Structure Specific Interactions of Eukephalins and Dynorphin with Liposomes: Investigation by Hydrophobic Photolabeling, Arch. Biochem. Biophys., 225:467.

Ikura, M., Hiraoki, T., Hikichi, K., Mikuni, T., Yazawa, M., and Yagi, K.,

1983, Nuclear Magnetic Resonance Studies on Calmodulin: Calcium-Induced Conformational Change, Biochemistry, 22:5273.

Kilhoffer, M.-C., Demaille, J.G., and Gerara, D., 1981, Tyrosine Fluorescence of Ram Testis and Octopus Calmodulin. Effects of Calcium, Magnesium, and Ionic Strength, Biochemistry, 20:4407.

Kincaid, R.L., and Vaughan, M., 1986, Direct Comparison of Ca^{2+} Requirements for Calmodulin Interaction with and Activation of Protein Phosphatase, Proc. Natl. Acad. Sci. U.S.A., 83:1193.

Kincaid, R.L., Vaughan, M., Osborne, J.C., Tkachuk, V.A., 1982, Ca^{++}-dependent Interaction of 5-Dimethylamino-Naphthalene-1-Sulfonyl-Calmodulin with Cyclic Nucleotide Phosphodiesterase, Calcineurin, and Troponin I, J. Biol. Chem., 257:10638.

Klee, C.B., 1977, Conformational Transition Accompanying the Binding of Ca^{2+} to the Protein Activator of 3',5'-Cyclic Adenosine Monophosphate Phosphodiesterase, Biochemistry, 16:1017.

Klee, C.B., and Vanaman, T.C., 1982, Calmodulin, Adv. Protein Chemistry, 35:213.

Krauss, E.M., and Chan, S.L., 1982, Intramolecular Hydrogen Bonding in Gramicidin S. 2. Ornithine, J. Am. Chem. Soc., 104:6953.

Lehrer, S.S., and Fasman, G.D., 1967, Ultraviolet Irradiation Effects in Poly-L-Tyrosine and Model Compounds. Identification of Bityrosine as a Photoproduct, Biochemistry, 6:757.

Levin, R.M., and Weiss, B., 1977, Binding of Trifluoperazine to the Calcium Dependent Activator of Cyclic Nucleotide Phosphodiesterase, Mol. Pharmacol., 13:690.

Lukas, T.J., Burgess, W.H., Prendergast, F.G., Lau, W., and Watterson, D.M., 1986, Calmodulin Binding Domains: Characterization of a Phosphorylation and Calmodulin Binding Site from Myosin Light Chain Kinase, Biochemistry, 25:1458.

Malencik, D.A., and Anderson, S.R., 1982, Binding of Simple Peptides, Hormones, and Neurotransmitters by Calmodulin, Biochemistry, 21:3480.

Malencik, D.A., and Anderson, S.R., 1983a, High Affinity Binding of the Mastoparans by Calmodulin, Biochem. Biophys. Res. Comm., 114:50.

Malencik, D.A., and Anderson, S.R., 1983b, Binding of Hormones and Neuropeptides by Calmodulin, Biochemistry, 22:1995.

Malencik, D.A., and Anderson, S.R., 1984, Peptide Binding by Calmodulin and Its Proteolytic Fragments and by Troponin C, Biochemistry, 23:2420.

Malencik, D.A., and Anderson, S.R., 1985, Effects of Calmodulin and Related Proteins on the Hemolytic Activity of Melittin, Biochem. Biophys. Res. Comm., 130:22.

Malencik, D.A., and Anderson, S.R, 1986a, Demonstration of a Fluorometrically

Distinguishable Intermediate in Calcium Binding by Calmodulin-Mastoparan Complexes, <u>Biochem. Biophys. Res. Comm.</u>, 135:1050.

Malencik, D.A., and Anderson, S.R., 1986b, Calmodulin-Linked Equilibria in Smooth Muscle Myosin Light Chain Kinase, <u>Biochemistry</u>, 25:709.

Malencik, D.A., and Anderson, S.R., 1986c, Association of Calmodulin with Venom Peptides, <u>Fed. Proc.</u>, 45:1692.

Malencik, D.A., and Anderson, S.R., (1987) Dityrosine Formation in Calmodulin, <u>Biochemistry</u>, 26:695.

Malencik, D.A., and Fischer, E.H., 1982, Structure, Function, and Regulation of Phosphorylase Kinase, <u>in</u>: "Calcium and Cell Function," W.Y. Cheung, ed., Academic Press, New York.

Malencik, D.A., Anderson, S.R., Bohnert, J.L., and Shalitin, Y., 1982a, Functional Interactions Between Smooth Muscle Myosin Light Chain Kinase and Calmodulin, <u>Biochemistry</u>, 21:4031.

Malencik, D.A., Huang, T.-S., and Anderson, S.R., 1982b, Binding of Protein Kinase Substrates by Fluorescently Labeled Calmodulin, <u>Biochem. Biophys. Res. Commun.</u>, 108:266.

Malencik, D.A., Scott, J., Fischer, E.H., Krebs, E.G., and Anderson, S.R., 1986, Association of Peptide Analogues of the Heat-Stable Inhibitor of cAMP-Dependent Protein Kinase with Calmodulin, <u>Biochemistry</u>, 25:3502.

Maulet, Y., and Cox, J.A., 1983, Structural Changes in Melittin and Calmodulin upon Complex Formation and Their Modulation by Calcium, <u>Biochemistry</u>, 22:5680.

McDowell, L., Sanyal, G., and Prendergast, F.G., 1985, Probable Role of Amphiphilicity in the Binding of Mastoparan by Calmodulin, <u>Biochemistry</u>, 24:2979.

Muchmore, D.C., Malencik, D.A., and Anderson, S.R., 1986, ^{1}H NMR Studies of Mastoparan Binding by Calmodulin, <u>Biochem. Biophys. Res. Comm.</u>, 137:1069.

Nakajima, T., 1986, Pharmacological Biochemistry of Vespid Venoms, <u>in</u>: "Venoms of the Hymenoptera. Biochemical, Pharmacological and Behavioral Aspects," T. Piek, ed., Academic Press, New York.

Perry, S.V., Cole, H.A., Hudlicka, O., Patchell, IV.B., and Westwood, S.A., 1984, Role of Myosin Light Chain Kinase in Muscle Contraction, <u>Fed. Proc.</u>, 43:3015.

Prozialeck, W.C., and Weiss, B., 1982, Inhibition of Calmodulin by Phenothiazines and Related Drugs: Structure-Activity Relationships, <u>J. Pharmacol. Exp. Ther.</u>, 222:509.

Richman, P.G., 1978, Conformation-Dependent Acetylation and Nitration of the Protein Activator of Cyclic Adenosine 3',5'-Monophosphate Phosphodiesterase. Selective Nitration of Tyrosine Residue 138, Biochemistry, 17:3001.

Richman, P.G., and Klee, C.B., 1979, Specific Perturbation by Ca^{2+} of Tyrosyl Residue 138 of Calmodulin, J. Biol. Chem., 254:5372.

Scott, J.D., Fischer, E.H., Demaille, J.G., and Krebs, E.G., 1985, Identification of a Inhibitory Region of the Heat-Stable Protein Inhibitor of the cAMP-Dependent Protein Kinase, Proc. Natl. Acad. Sci. U.S.A., 82:4379.

Seamon, K.B., 1980, Calcium- and Magnesium-Dependent Conformational States of Calmodulin as Determined by Nuclear Magnetic Resonance, Biochemistry, 19:207.

Seeholzer, S.H., Cohn, M., Putkey, J.A., Means, A.R., and Crespi, H.L., 1986, NMR Studies of a Complex of Deuterated Calmodulin with Melittin, Proc. Natl. Acad. Sci., 83:3634.

Small, J.V., and Sobieszek, A., 1980, The Contractile Apparatus of Smooth Muscle, Int. Rev. Cytol., 64:241.

Stull, J.T., 1980, Phosphorylation of Contractile Proteins in Relation to Muscle Function, Adv. Cyclic Nucleotide Res., 13:39.

Teale, F.W.J., 1960, The Ultraviolet Fluorescence of Proteins in Neutral Solution, Biochem. J., 76:381.

Terwilliger, T.C., and Eisenberg, D., 1982, The Structure of Melittin. II. Interpretation of the Structure, J. Biol. Chem., 257:6016.

Walsh, M., and Stevens, F.C., 1977, Chemical Modification Studies on the Ca^{2+}-Dependent Protein Modulator of Cyclic Nucleotide Phosphodiesterase, Biochemistry, 16:2742.

Weber, G., 1961, Enumeration of Component in Complex Systems by Fluorescence Spectrophotometry, Nature (London), 190:27.

Weber, G., 1965, The Binding of Small Molecules to Proteins, in: "Molecular Biophysics", B. Pullman and M. Weissbluth, ed., Academic Press, New York, N.Y.

Weiss, B., Prozialeck, W., Cimino, M., Barnette, M.S., and Wallace, T.L., 1980, Pharmacological Regulation of Calmodulin, Ann. N.Y. Acad. Sci., 356:319.

Weiss, B., Prozialeck, W.C., and Wallace, T.L., 1982, Interaction of Drugs with Calmodulin: Biochemical, Pharmacological, and Clinical Implications, Biochem. Pharmacol., 31:2217.

DEFINITION OF MONOCLONAL ANTIBODY-FLUORESCENT LIGAND INTERACTIONS

Edward W. Voss, Jr.

Department of Microbiology
University of Illinois
131 Burrill Hall
407 S. Goodwin Avenue
Urbana, IL 61801 USA

ANTIBODY HETEROGENEITY

Due to the characteristic complex cellular and genetic constituents of
the immune system, humoral antibody responses exhibit heterogeneity and
diversity at various levels. First, immune responses induced by complex
macromolecules, containing many structurally different epitopes, result in
a mixture of antibodies possessing different specificities and affinities.
Second, antibody responses to relatively simple and chemically defined
epitopes (haptens) are usually diverse as characterized in terms of the
observed range in affinities. The magnitude of affinity measured depends
on the biochemical nature and immunogenic properties of the epitope and
carrier. Finally, a third level of heterogeneity results from the multiple
immunoglobulin classes (or subclasses) resulting from the recombination of
different constant (C) genes, which dictate isotypy, with the same variable
(V) genes. This report will focus on an experimental approach to understand
the structure-function basis of antibodies differing in affinities elicited
to the fluorescent probe, fluorescein. The experimental premise to employ
a fluorescent epitope is attractive due to: 1) the sensitivity of fluorometric measurements; 2) the fact that when fluorescence is an intrinsic
property of the ligand, it provides a means by which behavior of the hapten
can be studied in complex mixtures; 3) the capacity of fluorescence to
reflect changes in the micro-environment; thus, monitoring intensity can
provide information about the types of bonds involved in the bimolecular
complex as well as the native state of the active site; 4) the potential
use of fluorescence polarization measurements to indicate molecular size

facilitating measurements of ligand-protein interactions; and 5) the availability of fluorescence quenching and solvent perturbation studies to provide useful functional information which can be correlated with resolution of the physical-chemical structure of the antibody active site.

Currently, established rules concerning structure-function relationships governing antibody specificity are virtually nonexistent. This is attributable to the fact that a systematic approach to the antibody problem has not been possible until relatively recently. Historically, immunoglobulin (Ig) heterogeneity prevented or inhibited the generation of correlative information regarding the primary, secondary, tertiary and quaternary structure of antibodies. Several significant factors have contributed to recent advances in experimentally overcoming the complexities of heterogeneity. First, the advent of hybridoma methodology (Köhler and Milstein, 1975) permitted consistent generation of homogeneous antibody populations of defined specificity. Second, advances in fluorescence theory, methodology and instrumentation have promoted the use of fluorescent probes as chemically defined immunological epitopes. Third, elucidation of affinity maturation, which is the increase in the average affinity of a heterogeneous antibody population with time, provides a mechanism to obtain antibodies with a wide range of affinities specific for the appropriate epitope. Finally, the development of the fluorescein hapten system as a site-filling epitope (Voss, 1984), which elicits production of antibodies spanning an unusually wide range in affinities (10^4-$10^{14}M^{-1}$), has been advantageous. All these factors synergistically facilitated the establishment of a systematic approach experimentally to the structure-function analyses of antibody specificity.

THE FLUORESCEIN HAPTEN SYSTEM

Fluorescein (Figure 1) has become an important hapten immunologically. Both the polyaromatic and dianionic nature of fluorescein conforms to properties generally considered suitable for immunogenicity. The size of fluorescein, as noted by a molecular weight of ~350, is important since studies (Voss et al. 1976) have shown that fluorescein approximates a site-filling ligand. The high fluorescence quantum yield (Φ) of 0.92 is beneficial in the development of sensitive fluorescent ligand binding assays, although the fluorescence lifetime of 4 nsec is not sufficiently long to measure certain molecular events. A relatively large number of studies (Voss, 1984) have shown that fluorescein is a potent hapten immuno-

248

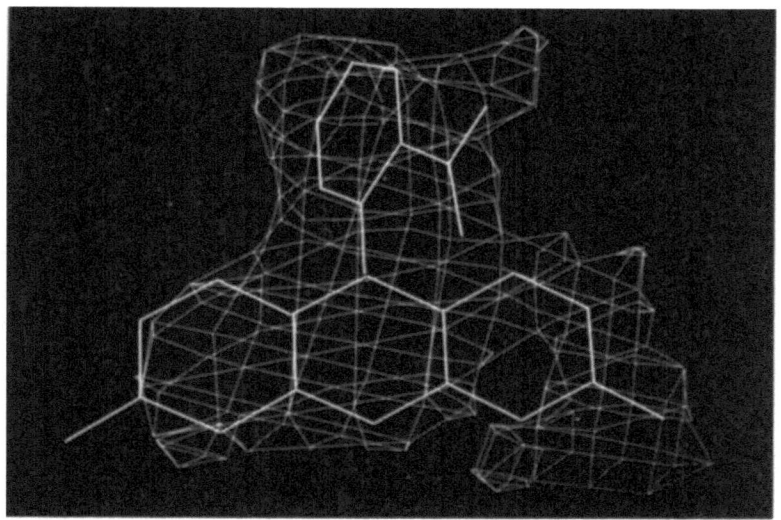

Figure 1. Cage electron density model of the fluorescein ligand obtained by Fourier Analysis at 2.7-Å resolution. (Provided by A. B. Edmundson, University of Utah, also see Edmundson et al., 1984).

logically when covalently conjugated to T-cell (lymphocyte) dependent carriers such as keyhole limpet hemocyanin (KLH). The immune response to fluorescein is significantly diverse. This phenomenon in all species studied to date (murine, lapine, avian, guinea pig and bovine) necessitated the employment of hybridoma methodology in order to dissect the response and generate homogeneous monoclonal antibody populations of defined specificity (Kranz and Voss, 1981). Table 1 represents a partial listing of monoclonal anti-fluorescein antibodies generated to the hapten as a function of two different carriers. As previously indicated, KLH is a T-cell dependent carrier while amino-ethyl Ficoll is a T-cell independent carrier. It can be seen (Table 1) that isotypy and affinity of anti-fluorescein monoclonal antibodies can be regulated through different carriers to which the fluorescyl hapten is covalently attached in the synthesis of the immunogen.

HOMOGENEITY OF MONOCLONAL ANTIBODY ACTIVE SITES

Monoclonal anti-fluorescein antibodies purified from either murine ascites fluid (in vivo) or from serum-free in vitro tissue culture (Petrossian, 1986) were judged homogeneous by isoelectric focusing and linear kinetics describing the first order dissociation rate. The latter was performed as described (Herron and Voss, 1983). The principle of the dissociation rate assay is that the fluorescence of the disodium salt of fluorescein (FDS) is significantly quenched upon binding by the anti-fluorescein monoclonal antibody (see Qmax, Table 1). In determining the affinity of each monoclonal anti-fluorescein antibody the second order association rate was assumed to be constant ($5.6 \times 10^6 M^{-1} sec^{-1}$). The latter is slower than the rate assigned to diffusion controlled reactions (Pecht, 1982) and is probably due to the fact that the fluorescein ligand is a rigid ellipsoid (Figure 1), and therefore precise geometric orientation is required before binding can occur within the antibody active site.

Table 1. Anti-Fluorescein Monoclonal Antibodies Derived from BALB/cV Mice

I. Immunogen FL-KLH[a]

A. High-Affinity Idiotype Family (4-4-20 prototype)

Clone No.	H Chain Isotype	L Chain Isotype	Keq(M^{-1})	Qmax (%)	Absorption Maximum (nm)[b]
4-4-20	IgG2a	K	1.7×10^{10}	96.4	505
9-40	IgG1	K	3.3×10^7	45.4	508
10-25	IgG1	K	5.4×10^7	55.3	508
5-14	IgG3	K	1.0×10^8	43.8	509
5-27	IgG1	K	5.3×10^8	36.4	509

B. Intermediate-Affinity Idiotype Family (9-40 Prototype)

Clone No.	H Chain Isotype	L Chain Isotype	Keq(M^{-1})	Qmax (%)	Absorption Maximum (nm)[b]
3-24	IgG1	K	2.8×10^6	38.0	504
12-40	IgG1	K	3.4×10^6	85.0	502

C. Low-Affinity Idiotype Family (3-13 Prototype)

Clone No.	H Chain Isotype	L Chain Isotype	Keq(M^{-1})	Qmax (%)	Absorption Maximum (nm)[b]
3-13	IgG1	K	3.8×10^4	90.8	ND
3-17	IgG2a	K	5.9×10^4	78.9	ND

(This family includes 8 other antibodies 3-12, 3-14, 3-29, 3-32, 3-35, 3-36, 3-43 and 3-45).

D. Other Monoclonal Anti-fluorescein Antibodies

Clone No.	H Chain Isotype	L Chain Isotype	Keq(M^{-1})	Qmax (%)	Absorption Maximum (nm)[b]
20-19-1	IgG1	K	1.8×10^8	97.9	510
20-20-3	IgG1	K	1.0×10^7	98.8	497
6-10-6	IgG1	K	7.3×10^7	98.0	506
20-4-4	IgG2b	K	5.0×10^7	49.8	513
4-6-9	IgG1	K	2.0×10^7	91.0	ND
4-6-10	IgG1	K	2.5×10^7	72.0	504
20-6-3	IgG1	K	1.9×10^7	ND	507
6-19-1	IgG2b	K	1.0×10^7	76.0	500
20-17-3	IgG1	K	5.5×10^6	89.0	506
2-3	IgG1	K	1.9×10^9	95.0	504
2-27	IgG1	K	6.0×10^8	92.3	507
2-40	IgG1	K	9.1×10^8	93.9	507

II. Immunogen Fl-Amino Ethyl Ficoll

Clone No.	H Chain Isotype	L Chain Isotype	Keq(M^{-1})	Qmax (%)	Absorption Maximum (nm)[b]
1-30-2	IgG3	K	$<1.0 \times 10^5$		
2-10-3	IgG3	K	$<1.0 \times 10^5$		
3-27-1	IgG3	K	$<1.0 \times 10^5$		
1-18-3	IgM	K	$<1.0 \times 10^5$		
1-39-2	IgM	λ	$<1.0 \times 10^5$		
2-4-4	IgM	λ	$<1.0 \times 10^5$		
2-7-6	IgM	λ	$<1.0 \times 10^5$		
2-39-2	IgM	λ	$<1.0 \times 10^5$		
2-42-1	Igm	K	$<1.0 \times 10^5$		
2-47-5	IgM	K	$<1.0 \times 10^5$		

III. Derived from F_1 (NZB x NZW)

Clone No.	H Chain Isotype	L Chain Isotype	Keq(M^{-1})	Qmax (%)	Absorption Maximum (nm)[b]
18-2-3	IgM	K	2.9×10^{10}	95.0	505

---

[a] Fluorescein covalently conjugated to keyhole limpet hemocyanin.
[b] λ_{max} of antibody bound fluorescein.

Based on a relatively constant association rate, antibody affinity is dictated by significant differences (logarithmic) in the dissociation rate.

The frequency of affinities for the characterized monoclonal antibodies elicited to the fluorescyl-KLH immunogen conform to a normal distribution with an average affinity of 4-5 x $10^8 M^{-1}$ (Std. dev. $\pm$ 4.5, Table 1).
Thus, when viewed as a population of antibodies the hybridomas collectively
simulated the classical heterogeneous hyperimmune response observed in
serum analyses (Herron, 1983).

THERMODYNAMICS OF THE ANTI-FLUORESCEIN: FLUORESCEIN INTERACTION

The thermodynamic properties exhibited by selected monoclonal antibodies listed in Table 1 have been described in detail (Herron et al.,
1986). In general curvilinear van't Hoff plots (ln K_a vs T^{-1}) were
observed for all monoclonal anti-fluorescein antibodies studied, indicating
that their standard enthalpy changes ($\Delta H°$) were temperature dependent. The
anti-fluorescein-ligand interaction was also investigated in terms of
unitary free energy ($\Delta G\mu$), unitary entropy ($\Delta S\mu$), heat capacity changes
($\Delta Cp°$) and standard volume changes ($\Delta V°$). Analyses of the thermodynamic
properties revealed two competing forces in fluorescein binding. First, a
hydrophobic effect was judged to be the thermodynamic driving force for
fluorescein binding. Second, ligand binding restricts the mobility of
active-site residues, which makes an unfavorable contribution to $\Delta G°$. The
latter forms the basis for hapten induced conformational changes.

PROPERTIES OF THE ANTI-FLUORESCYL SITE

Watt and Voss (1977) explored molecular mechanisms by which antibody
bound fluorescein undergoes a red spectral shift in its absorption spectrum
and its fluorescence is quenched. Displacement of λ_{max} (Δ10-20 nm) to the
red correlates with polarity of the microenvironment. The relative reduction in the quantum yield of fluorescein has been consistently >80% with
polyclonal anti-fluorescein antibodies but has shown a significant range
with monoclonal antibodies (35-99%). The degree of fluorescence quenching
is therefore not constant and appears dependent on many factors. Screening
of individual L-amino acids for the ability to quench the fluorescence of
fluorescein in aqueous solution resulted in the finding that L-tryptophan,
and to a lesser extent L-tyrosine, were effective quenchers (Watt and Voss,
1977). Stern-Volmer analyses and fluorescence lifetime (τ) measurements of
tryptophan quenching revealed complex fluorescein-tryptophan dynamics
(involving both static and dynamic components). In addition, titration of

252

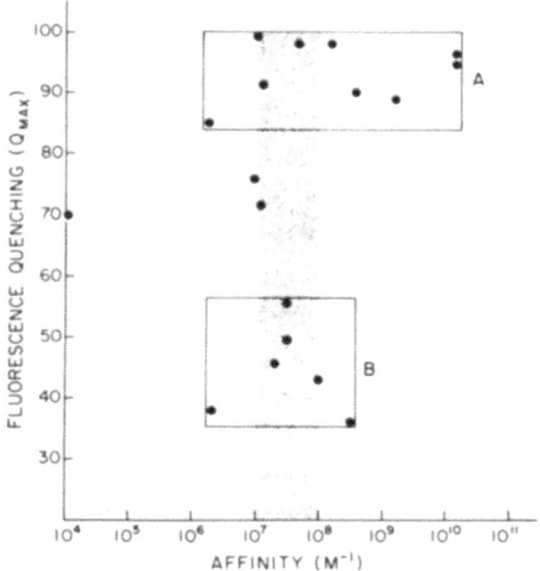

Figure 2. Correlation between Qmax and affinity of 18 monoclonal anti-fluorescyl antibodies (see Table 1). The shaded area encompasses those monoclonal antibodies in the affinity range of 10^7 to $10^8 M^{-1}$. Box A includes Ig molecules which quench the fluorescein ligand efficiently. Box B denotes monoclonal antibodies which quench inefficiently.

anti-fluorescein antibodies with fluorescyl ligand showed that the fluorescence emission of excited tryptophanyl residues within the antibody active site were quenched. Collectively, these experimental results suggested that tryptophan was involved either directly in fluorescence quenching or indirectly through energy transfer mechanisms within the antibody active site.

A plot of affinity vs Qmax (Figure 2), based on values reported in Table 1, revealed interesting correlations between binding and fluorescence quenching. The plot shows that Qmax values approximate a bimodal distribution. One group of six monoclonal antibodies shows an average quenching (Qmax) of 44.8% (Std. dev. ± 6.5). A second group of nine antibodies cluster around a mean Qmax of 93.3% (Std. dev. ± 4.5). A quenching factor of two-fold separates the two groups (Figure 2). In addition, analysis of the nine monoclonal antibodies which populate the affinity group containing the highest number of Ig molecules ($10^7 - 10^8 M^{-1}$) indicates that there is no correlation between antibodies in this intermediate affinity range and the two Qmax groupings. However, a skewed distribution is evident when one considers the six monoclonal antibodies with affinities $>10^8 M^{-1}$. Eighty-three percent of such Ig molecules are shown to display highly efficient quenching mechanisms (88 to 99%). Hybridoma 5-27 is an exception to the trend cited and therefore becomes an interesting antibody for further study. The restricted number of groupings, and lack of convincing evidence for a continuum of quenching values suggests a limited number of fluorescence quenching mechanisms within the antibody active site. Hypothetically, the hybridomas populating the inefficient quenching group (i.e. 44.8%) may possess a smaller number of bonds including specific bonds giving a finite degree of quenching, but lack other types of bonds (qualitatively and quantitatively) which the efficient quenching (i.e. 93.3% group) Ig molecules possess. Thus, in the low-moderate affinity range, fluorescence quenching is dependent qualitatively on the types of bonds participating in the ligand-protein interaction. When affinity exceeds $10^8 M^{-1}$, the multiplicity of bonds increases the chances for additional interactions some of which result in more efficient quenching of the fluorescence of bound fluorescein.

Solvent perturbation studies have been used to further characterize both polyclonal and monoclonal anti-fluorescyl active sites. Deuterium oxide, iodide and oxygen have been employed in perturbation studies (Voss et al., 1979; Watt and Voss, 1980; Kranz et al., 1981). The relative fluorescence quantum yield of fluorescein bound to specifically purified high affinity liganded polyclonal rabbit and chicken IgG antibody was significantly enhanced in deuterium oxide (D_2O) relative to complexes in aqueous (H_2O) solution. The degree of fluorescence enhancement correlated with the intrinsic affinity of polyclonal antibody for the ligand but the same correlation was not evident in examinations of individual monoclonal antibodies. Quantitatively, fluorescence enhancement in D_2O (95%) was identical for ligand bound to IgG and F(ab)'$_2$ fragments derived from the same hybridoma. The isotope enhancement effect was not due to ligand dissociation as shown by a comparison of difference spectroscopy and equilibrium dialysis results in D_2O and H_2O buffers, and by similar enhancement of fluorescence with affinity labeled antibody (Voss et al., 1980). Identical extrinsic circular dichroism of antibody bound fluorophore in D_2O and H_2O, coupled with the absence of a measurable isotope effect on the quenching of fluorescein by L-tryptophan, supported the contention that the D_2O-effect was an excited state phenomenon (Voss et al., 1979).

Comparative kinetic studies of ligand dissociation and D_2O enhancement were performed with both heterogeneous and homogeneous anti-fluorescyl

254

antibodies (Kranz et al., 1981). Polyclonal rabbit and homogeneous mouse (monoclonal) antibody preparations were purified by affinity chromatography and found to be pure by sodium dodecylsulfate-polyacrylamide gel electrophoresis and immunoelectrophoresis. Relatively high affinities of all liganded antibody preparations were determined by dissociation rate studies, demonstrating comparatively long lifetimes describing dissociation of bound fluorescein. In addition, rabbit anti-fluorescyl preparations were found to display marked heterogeneity of dissociation rates while mouse monoclonal anti-fluorescyl preparations exhibited a single linear off-rate indicating homogeneity. D_2O fluorescence enhancement studies showed non-linear kinetics with both heterogeneous and homogeneous antibody active sites. Temperature studies of ligand D_2O enhancement and dissociation rates using monoclonal anti-fluorescyl antibodies revealed similar, yet different activation energies (22.7 $\pm$ 0.8 cal and 20.2 $\pm$ 0.3 cal, respectively) for both phenomena. The studies demonstrated that the anti-fluorescein antibody active site consists of both solvent accessible and relatively inaccessible components, and that ligand binding involved both exchangeable hydrogen atoms and other unresolved interactions.

Differential accessibility of liganded high affinity rabbit polyclonal anti-fluorescyl IgG antibody combining sites to the aqueous milieu has been investigated by solvent perturbation of the fluorescence of bound fluorophore (Watt and Voss, 1979). Iodide, a dynamic quencher of fluorescein, was selected for use in these studies after examination of a number of water-soluble fluorescence quenchers. Quenching of antibody-bound fluorophore by iodide was measured with a number of liganded polyclonal anti-fluorescyl preparations, demonstrating partial solvent exposure of the fluorophore as well as heterogeneity of the high affinity antibody populations. Fluorescence quenching, lifetime, and absorption spectroscopy provided evidence that the antibody-bound fluorophore quenched by iodide interacted with it directly and that anomalous binding of the anion to the surface of the protein, resulting in ground state perturbations of the immunoglobulin, could not explain the observed results. Iodide quenching has proven useful in comparing the architecture of the anti-fluorescein sites between families of monoclonal antibodies (Table 1; Bates et al., 1985). Oxygen quenching studies with various monoclonal anti-fluorescyl antibodies (IgG) have revealed that the accessibility of bound ligand is a constant and independent of affinity.

In further studies, fluorescein isothiocyanate was reacted with purified rabbit polyclonal anti-fluorescyl antibodies under defined conditions (Watt and Voss, 1978) as a nonreversible affinity labeling reagent. Relative to the normal antibody control, significantly more ligand was conjugated to the purified antibody which was then inhibited from further binding with radioactively labeled ligand. The labeled antibody preparations were reacted with an anti-fluoroscyl antibody solid phase column to remove Ig molecules on which the fluorescein residues were not sequestered within the antibody cavity. Fluoroscyl ligand specifically conjugated to purified and absorbed antibody preparations exhibited a red spectral shift characteristic of specifically bound fluorophore. Fluorescence of ligand covalently conjugated to specific antibody was quenched about 95% relative to 90% for reversibly bound ligand. The quantum yield of the former is indicative of a thiocarbamyl linkage within the antibody active site. Fluorescence lifetime measurements by phase and modulation showed a close correlation between liganded and affinity labeled antibody relative to free ligand and fluoroscyl conjugated non-antibody proteins. The fluoroscyl affinity label was not dissociated from the antibody active site upon denaturation of the protein in 6 M guanidine-HCl. Reduction alkylation and resolution of immunoglobulin derived H (γ) and L chains in two different dispersing agents showed that the ligand remained linked to either polypeptide. Fluoroscyl ligand reversibly bound to high affinity IgG (i.e. non-affinity labeled) was quantitatively released under conditions required for H and L chain production. The fluorescence of affinity labeled ligand was enhanced in high concentrations of D_2O (Watt and Voss, 1978). Similar studies showed that the cyanuric chloride derivative of fluorescein amine was also a stable affinity labeling reagent (Watt and Voss, 1984).

ANTI-FLUORESCEIN ANTIBODY SYSTEM AS A STRUCTURE-FUNCTION MODEL

The mammalian anti-fluorescein antibody response can be characterized as diverse and displaying significant affinity maturation (Jarvis and Voss, 1984). The principle inherent in affinity maturation is significant in formulating structure-function studies as shown diagrammatically in Figure 3. Information portrayed in Figure 3 shows that upon hyperimmunization with fluorescein conjugated to a T-cell dependent carrier the anti-fluorescein response becomes progressively more heterogeneous and is characterized by a significant increase in the average affinity of the antibody population. Thus, harvesting a hyperimmune spleen for cell fusion in order to obtain monoclonal antibodies means that antibodies of many different affinities coexist. Generally the range of affinities obtained within monoclonal populations is between 10^4-10^{10} M^{-1}, although Watt et al.

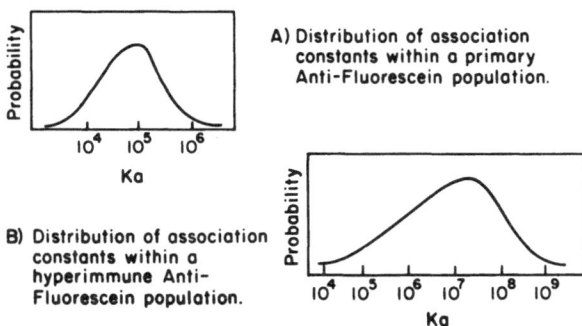

Figure 3. Purified antibodies are heterogeneous with respect to free energy of antigen-antibody interaction. Heterogeneous binding activity of purified polyclonal anti-fluorescyl antibodies. The figure illustrates important principles involved in the concept of affinity maturation characteristic of the humoral immune response to fluorescein.

(1980) showed affinities $>10^{14}M^{-1}$ in hyperimmune rabbit antisera. Thus, in terms of a structure-function model to explore antibody specificity, the anti-fluorescein system is exemplified by a non-restricted diversity which is important in procuring many different antibodies for structural studies.

NORMALIZATION FOR IDIOTYPIC RELATEDNESS

Although monoclonal anti-fluorescein antibodies obtained from the same murine inbred strain or even the same individual animal are attractive reagents for comparative structure-function studies, interpretable results are not assured. The problem is that a random comparison of anti-fluorescein antibodies may yield results too complex for conclusive analysis. In order to further isolate and normalize antibody populations more suitable for comparative structural studies, idiotype families were generated using hybridoma methodology. This was accomplished using two sequential screening assays of equivalent sensitivity. Upon sacrificing BALB/c mice hyperimmunized with Fl-KLH, the spleenocytes were fused chemically with Sp2/0-Ag-14 BALB/c derived myeloma cells using polyethylene glycol (PEG). Sp2/0 myeloma cells no longer secrete the original Ig protein due to induced mutations, thus avoiding intracellular formation of hybrid (mixed) Ig molecules. The resulting hybridoma cells were screened initially for anti-fluorescein activity using a solid phase radioimmune assay employing fluorescyl deriva- tized bovine serum albumin as the test antigen. Hybridomas exhibiting anti-fluorescein activity were subsequently screened in an anti-idiotype

Table 2. Relative Characterization of Anti-Fluorescein Antibodies Comprising the High-Affinity Idiotype Family

Clone No.	Idiotypic Relatedness (%)[a]	I_g pI[b]	Fab pI[c]	D_2O Enhancement (%)[d]	Iodide Quenching (%)[e]
4-4-20	100.0	7.4	7.1	130.7 ± 1.8	2.5
9-40	97.3	7.0	7.3	122.6 ± 0.7	15.5
10-25	98.8	7.1	7.5	134.1 ± 0.6	13.2
5-14	95.4	7.7	7.8	126.6 ± 1.8	8.4
5-27	82.7	6.8	7.1	121.6 ± 0.2	13.7

[a]   Based on values from Bates et al., (1985). Values based on normalization of the homologous 4-4-20 idiotype system to 100%.

[b]   Isoelectric point values for the complete immunoglobulin molecule.

[c]   Isoelectric point values for Fab fragments derived by papain digestion.

[d]   Increase in fluorescence of antibody-bound fluorescein in 95% D_2O relative to the quantum yield in H_2O.

[e]   Values reported as % quenching of bound fluorescein titrated with KI between 0 and 0.2M (Watt and Voss, 1979).

assay. Polyclonal anti-idiotype reagents were generated by multiple immunizations of adult albino rabbits with a murine monoclonal anti-fluorescein antibody (prototype) of defined affinity (Table 1). All anti-idiotype reagents utilized were ligand inhibitable. Idiotypy can be defined as antigenic determinants located on the Fv ($V_H + V_L$) fragment (Figure 4) of an antibody molecule (Kabat, 1984) and it has been determined that idiotypically similar antibodies are closely related structurally. Thus, idiotypy is a reasonable normalization parameter to minimize differences between monoclonal anti-fluorescein antibodies to definable levels.

Although idiotypic normalization is not a guarantee of structural homology suitable for structure-function interpretation, previous studies (Kabat, 1984; Rudikoff, 1984) suggested that selection of Ig molecules for the same specificity and the same idiotype inferred common V-regions.

Data presented in Table 2 further suggest that idiotypic normalization has resulted in the selection of Ig molecules with similar properties. Table 2 represents a comparison of five hybridomas which comprise the high affinity 4-4-20 idiotype family (see Table 1). Integrating information presented in both Tables 1 and 2, it can be seen that by selecting for antibody molecules of the same specificity and for a high degree of idiotypic cross relatedness (80% or greater), that although the intra-family affinity range was approximately 500 fold, the molecules differed most significantly in percent fluorescence quenching (Qmax) of the bound fluorescein ligand. The extent of spectral shift was similar and each antibody showed similar active site inaccessibility to iodine, similar percent fluorescence enhancement in D_2O, as well as closely related fine-specificity binding patterns for fluorescyl analogues (erythrosin, eosin and tetramethyl-rhodamine).

FRAGMENTATION OF MONOCLONAL ANTIBODIES FOR STRUCTURAL ANALYSES

In order to derive the appropriate reagent for eventual structure-function analyses, it has become necessary to isolate Fab fragments from each hybridoma molecule. Fab fragments can be derived by proteolytic digestion of monoclonal antibodies (Figure 4). The Fab fragment is an important reagent, since comparative primary sequence analyses emphasize the variable regions which convey antibody specificity. Papain digestion also eliminates Fc related microheterogeneity attributed to the attached carbohydrate moieties. In addition, in order to obtain tertiary structural information, Fab fragments preferentially crystallize relative to the intact immunoglobulin. Such a pattern in crystallization can be attributed to the relatively free and rapid motion of the Fc fragment around the hinge region. However, it is important to note that historically, even with pure Fab fragments, it has been difficult to systematically obtain x-ray diffraction information from crystallization attempts.

Papain was selected as the proteolytic enzyme based on the fact that the papain digestion sites within the hinge region are well documented. Comparative studies conducted with purified liganded and non-liganded preparations of each member of the 4-4-20 idiotype family (Table 2) revealed interesting results (Gibson and Edmundson, 1986). First, it was observed that the relative rate of digestion of monoclonal antibody in the liganded form was faster than the non-liganded form. Second, the

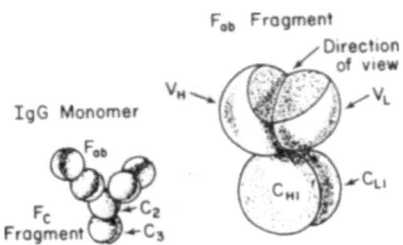

Figure 4. Schematic representation of the antibody molecule, emphasizing domain structure and interactions. The figure depicts the intact Ig molecule (left side) and the Fab fragment derived by proteolytic digestion (right side). The shaded cleft donated by "direction of view" represents the location of the antibody active site at the V_H-V_L interface but not necessarily the shape.

quantitative yield of Fab fragments was significantly higher with the liganded antibody relative to digestion of the non-liganded molecule. Thus, ligand bound to the antibody active site within the Fab fragment rendered the hinge region more accessible to the proteolytic action of papain. Bound ligand also seemed to result in constriction of the Fab domains conveying resistance to prolonged digestion and resulting in nearly theoretical yields of Fab fragments.

CRYSTALLIZATION OF ANTI-FLUORESCYL FAB FRAGMENTS

Fab fragments have been derived from each of the monoclonal anti-fluorescyl antibodies comprising the high affinity idiotype family (Table 2) by papain digestion. All liganded Fab fragments have crystallized readily, while the non-liganded fragments have not yielded crystals suitable for x-ray diffraction analyses. The rate of crystal growth by the liganded Fab fragments can be characterized as relatively fast (weeks) compared to the non-liganded fragments which has been slow (months) or inconsequential. Thus, as noted in the papain digestion studies, the presence of ligand in the antibody active site facilitated crystal development and growth. X-ray diffraction studies are currently in progress with crystals from members of the high-affinity idiotype family (Table 2). Notably the Fab fragments derived from hybridomas 4-4-20 and 9-40 have proven most suitable for diffraction studies (Edmundson, 1986). An interesting precedence does exist for the crystallographic (Fourier) analysis of fluorescein binding to an antibody active site. Edmundson et al. (1984) using crystals of the human Mcg light-chain (Bence-Jones) dimer succeeded in diffusing fluorescein into the highly aromatic binding cavity. The location and orientation of the bound ligand was studied by Fourier analysis at 2.7Å. Although the interaction of fluorescein with the Mcg dimer was found to be of very low affinity ($\sim 1 \times 10^4 M^{-1}$) in solution, an interesting and important model has been experimentally established for future comparative studies (Herron et al., 1985a).

In judging the suitability of selecting antibody molecules on the basis of idiotypic relatedness, it should be noted that each member of the high affinity (4-4-20) idiotype family crystallized at their respective isoelectric points (Table 2) in PEG 3350. This result, together with the previously mentioned active site properties, suggests that idiotypic relatedness is a suitable marker for grouping molecules into structurally related families.

FAB CONFORMATIONAL STATES

Historically it has been difficult to systematically obtain x-ray diffraction information from crystallization attempts with various immunoglobulins. The relatively low frequency of success has been difficult to explain. This report and Herron et al. (1985b) indicate the monoclonal antibody crystallization process of Fab fragments is facilitated by the presence of bound ligand. As mentioned previously, using enzymatically derived fragments from affinity purified anti-fluorescein antibodies, it has become possible to consistently crystallize univalent fragments in the liganded state. Parallel studies conducted with the non-liganded state showed insufficient crystal development. It would appear that the presence of bound ligand influences the rate, magnitude and quality of crystal growth (Herron et al., 1985b).

Such crystallographic information is consistent with the concept that the purified non-liganded Fab molecule may exist in several conformational states (Figure 5). Hsia and Little (1973) inferred stabilization of different conformations by different haptens. Studies by Lancet and Pecht (1976) and Herron (1984) further indicated different conformational states.

Lancet and Pecht (1976) using temperature-jump experiments, measured the kinetics of binding of Dnp-lys with the low affinity murine myeloma MOPC 460. Results showed two resolvable relaxation times indicating a

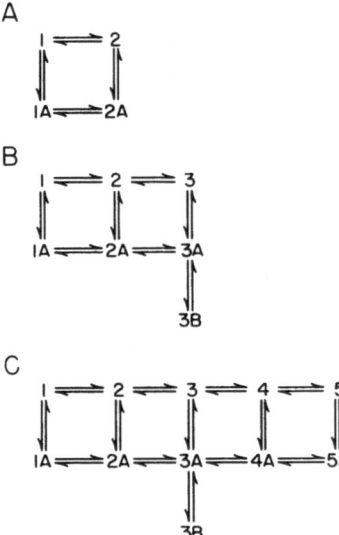

Figure 5. Selected models involving proposed variable region immunoglobulin conformational states.
A. Two-state model system proposed by Lancet and Pecht (1976).
B. Hypothetical three-state model system, including isomerization to a unique liganded state (3B).
C. Five-state model conformational system with State 3 for homologous ligand binding. In all models (A, B and C) States 1, 2, 3, etc. represent the non-liganded conformational states. 1A, 2A, 3A, etc. indicate liganded states which kinetically involve both the fast and slow steps in the forward rate. State B is the proposed unique liganded conformer.

fast bimolecular hapten-protein association followed by a relatively slow monomolecular protein isomerization. They concluded that the latter was indicative of conformational transitions. Lancet and Pecht (1976) arrived at a model consistent with the schematic equilibria (Figure 5A), hypothesizing only two non-liganded conformational states based on the fitting of relaxation time and amplitude data. The Lancet and Pecht scheme was consistent with an antigen-binding driven shift (and selection) in equilibrium for States 1A or 2A (Figure 5A). Thus, they proposed that ligand bound (but not equivalently) to either States 1 or 2 which then underwent a conformational shift consistent with a 1A<--->2A isomerization. The slow component observed in their temperature-jump experiments was considered evidence for such isomerization.

NUMBER OF CONFORMATIONAL STATES

A general scheme of two major non-liganded conformers has been

offered by Lancet and Pecht (1976) and Zidovetzki et al., (1980) for
hapten specific homogeneous murine antibodies. An assignment of two
conformational states (Figure 5A) was based on the best fit of all data to
the g matrix. Although it is conceivable that two major conformational
states which bind hapten describe immunoglobulin MOPC 460, it should not
yet be generalized that just two ligand reactive conformers are characteristic of all antibodies. Kaivarainen et al., (1981) proposed several
but limited numbers of fluctuating cavities.

Two-state models are usually presented to explain a number of structural changes in proteins, such as the allosteric effects by enzymes
(Ikegami, 1977). The two-state model, however, does not account for all
the cooperative effects of secondary bonds (hydrogen bonds, hydrophobic
interactions, and electrostatic or dipole interactions) in a protein.
Local fluctuations in structure are expected even in a highly ordered
native state. Generally, a multi-state model is more realistic based on
statistical considerations of local fluctuations in the polypeptide chain
(Ikegami, 1977). One can perhaps expect one unique conformation at 0°K
dictated by the law of minimum energy, where there is no flexibility of the
chain and entropy of the system equals zero. However, as a result of
thermal agitation, every secondary bond fluctuates in a random manner
between the bonded and unbonded states. The number of secondary bonds in
their bonded state is not constant but is continuously fluctuating. Thus,
many different conformations belong to a structure because of the flexibility of the polypeptide chain. These concepts are best expressed in a
multi-state model.

KINETIC ANALYSIS

Kinetically, it is now evident that conformational changes (continuum
between States 1, 2, and 3 in Figure 5B) take place in the 10^{-2} to 10^{-7} sec
range (Careri, et al., 1979). It is likely that ligand may be bound preferentially to one of the conformers with high affinity (e.g. State 3) or to
all 3 conformers in decreasing affinity 3>2>1. The actual rate of isomerization between non-liganded conformers was not measured by either Pecht
(1982) or Herron (1984). Upon ligand interaction, the kinetic studies
suggest that two events occur. The first event is fast and results in the
formation of 1A, 2A, etc. (Figure 5). Although the interaction of homologous ligand with each conformer may differ in affinity, the association
rate (K_{on}) can be assumed to be relatively constant (i.e. near-diffusion

controlled 1-5 x $10^6 M^{-1} sec^{-1}$). The subsequent slow step recorded by both Pecht (1982) and Herron (1984) would infer either isomerization between: 1) 1A and 2A (Figure 5A), 2) 1A, 2A and 3A (Figure 5B), 3) all 5 isomers in Figure 5C, or 4) the formation of conformer 3B. To assess these options, it is important to consider the kinetics of isomerization. As mentioned previously, the rate of isomerization of the non-liganded conformers is 10^{-2} to 10^{-7} sec., while the liganded conformers show a rate of isomerization of 10^{-2} to 10^{-4} sec. Both rates may differ with the affinity of the interaction. It is interesting to hypothesize that the rate of isomerization may be relatively faster in low affinity sites making it more difficult to stabilize with ligand. High affinity sites may display a slower isomerization rate and would be stabilized more easily. Thus, the rate of isomerization of the liganded states may have differed significantly between the studies conducted by Herron (1984) and Lancet and Pecht (1976), the latter being faster than the high affinity reaction described by Herron.

Herron's analysis of the slow step showed linear kinetics in terms of fluorescence quenching suggesting homogeneous kinetics indicative of a single event. However, the single event must be interpreted on the basis of the number of conformers interacting with the ligand. If two conformers are invoked, as in the case of Pecht (1982), then the single step is likely to be the isomerization (1A<-->2A) equilibrium. However, if the issue of multiple ligand reacting conformers is addressed then the interpretation of the slow single step is more consistent with the formation of a unique conformer 3B (Figures 5B and 5C). If 1A, 2A and 3A exist in an isomerization equilibrium, but with different rates between each step due to different affinities, Herron would not have observed linear kinetics for the slow step. Thus, it would appear that Herron (1984) may have observed the formation of the crystallizable liganded stable complex 3B. The significance of the final state 3B in a multiple conformer model is that it acts as an "equilibrium sink". It removes or reduces the concentration of the intermediate complex 3A from the multiple equilibria scheme in Figure 5 (B and C). The "equilibrium sink" is especially important in the consideration of five conformers (Figure 5C). As conformer 3A is converted to 3B, the equilibrium shifts from 1A and 2A as well as from 4A and 5A in response to the depletion of 3A. In a high affinity interaction, 3B may be viewed as exhibiting a very slow dissociation rate consistent with values reported by Watt et al (1980). The relatively slow kinetics would indicate that the final conformational change is a major transition and not easily achieved, reflecting significant intrasite amino acid side chain modulations (Romans and Dorrington, 1985). Such aromatic residue modulations have been observed

264

(Edmundson et al., 1984) and are usually motions occurring in solution over periods $\geq$ seconds (Karplus, 1985).

If the rates of transition between 1A, 2A and 3A are slow, similar to the 3A and 3B interconversion, the off-rates of the low affinity interactions 1A and 2A would result in a partial regeneration of non-liganded conformers 1 and 2. The complexity of the various equilibria based on these transitions would result in non-linear kinetics for the slow step. Herron's observation of linear kinetics suggests quite the opposite. It is likely that the ligand bound to a preferred state (e.g. State 3) triggering two events. State 3 can be hypothetically viewed as the most favorable "energetically-poised" conformer for ligand interaction. First the ligand would serve to capture the liganded state and subsequently States 1 and 2 (Figure 5) would quickly convert to State 3 and bind ligand. The second step in the binding kinetics observed by both Herron (1984) and Lancet and Pecht (1976) would be consistent with a conformation transition into a unique liganded state (e.g. State 3B in Figure 5B and 5C). It is the contention of this paper that it is the liganded State 3B (Figure 5) which crystallizes readily as a homogeneous Fab population, with most molecules fixed in the same conformational state. The major conformational states depicted in Figure 5 may differ sufficiently in geometric properties that crystal lattice formation is inhibited. Thus in effect, different conformers existing in such a mixture behave as pseudocontaminants. It is possible that minor conformational states still exist within liganded Fab states but are not of major significance to inhibit the crystallization process.

SUMMARY

Kinetic, equilibrium and thermodynamic analyses are presented which characterize the interaction of monoclonal anti-fluorescein antibodies with the fluorescent ligand.

Multiple conformational states are proposed for the immunoglobulin variable region. The existence of conformers is based on observations that bound ligand facilitates crystallization of homogeneous Fab fragments. Bound ligand is viewed as a stabilizer of one conformer from a series of non-liganded conformers. Statistical considerations related to local fluctuations due to thermal energy at the polypeptide chain level are used to support the multiple-state model.

ACKNOWLEDGEMENTS

The author is deeply indebted to Dr. Gregorio Weber for many years of advice, encouragement and friendship.

This work was supported by Grant AI 20960 awarded by the National Institutes of Health, Department of Health and Human Services.

REFERENCES

Bates, R. M., Ballard, D. W. and Voss, E. W. Jr., 1985, Comparative Properties of Monoclonal Antibodies Comprising a High Affinity Anti-fluorescyl Idiotype Family, Mol. Immuno., 22:871.

Careri, G., Fasella, P. and Gratton, E., 1979, Enzyme Dynamics: The Statistical Physics Approach, Ann. Rev. Biophys. Bioeng., 8:69.

Edmundson, A. B., 1986, Personal communication.

Edmundson, A. B., Ely, K. R. and Herron, J. N., 1984, A Search for Site-filling Ligands in the Mcg Bence-Jones Dimer: Crystal Binding Studies of Fluorescent Compounds, Mol. Immunol., 21:561.

Gibson, A. L. and Edmundson, A. B., 1986, Personal communication.

Herron, J. N. and Voss, E. W. Jr., 1983, Analysis of Heterogeneous Dissociation Kinetics in Polyclonal Populations of Rabbit Anti-fluorescyl IgG Antibodies, Mol. Immunol., 20:1323.

Herron, J. N., 1984, Equilibrium and Kinetic Methodology for the Measure-ment of Binding Properties in Monoclonal and Polyclonal Populations of Antifluorescyl IgG Antibodies, in: "The Fluorescein Hapten: An Immunological Probe", E. W. Voss, Jr., ed., CRC Press, Inc., Boca Raton.

Herron, J. N., Ely, K. R. and Edmundson, A. B., 1985a, Pressure-induced Conformational Changes in a Human Bence-Jones Protein Mcg, Biochemistry, 24:3453.

Herron, J. N., Gibson, A. L., Voss, E. W., Jr. and Edmundson, A. B., 1985b, Ligand Binding Properties of Monoclonal Antifluorescyl Antibodies in Solution and Crystals, Fed. Proc., 44:1328.

Herron, J. N., Kranz, D. M., Jameson, D. M. and Voss, E. W. Jr., 1986, Thermodynamic Properties of Ligand Binding by Monoclonal Antifluorescyl Antibodies, Biochemistry, 25:4602. (also see abstract by Herron et al. in this volume).

Hsia, J. C. and Little, J. R., 1973, Structure Properties of the Ligand Binding Sites of Murine Myeloma Proteins, FEBS Letts., 31:80.

266

Ikegami, A., 1977, Structural Changes and Fluctuations of Proteins, <u>Biophys. Chem.</u>, 6:117.

Jarvis, M. R. and Voss, E. W. Jr., 1984, Fluorescein in Kinetic Studies of Affinity Maturation, <u>in</u>: "Fluorescein Hapten: An Immunological Probe", E. W. Voss, Jr., ed., CRC Press, Inc., Boca Raton.

Kabat, E. A., 1984, Idiotype Determinants, Minigenes, and the Antibody Combining Site, <u>in</u>: "The Biology of Idiotypes", M. I. Greene and A. Nisonoff, ed., Plenum Press, New York.

Kaivarainen, A., Kaivarainen, E., Franek, F. and Olsovska, Z., 1981, Effect of Hapten Binding on Antibody-water Interaction. A concept of Fluctuating Cavities, <u>Immunol. Letts.</u>, 3:323.

Karplus, M., 1985, Dynamic Aspects of Protein Structure, <u>Annals N.Y. Acad. Sci.</u>, 439:107.

Köhler, G. and Milstein, C., 1975, Continuous Cultures of Fused Cells Secreting Antibody of Predefined Specificity, <u>Nature (London)</u>, 256:495.

Kranz, D. M. and Voss, E. W. Jr., 1981, Partial Elucidation of an Anti-hapten Repertoire in BALB/c Mice. Comparative Characterization of Several Monoclonal Anti-fluorescyl Antibodies, <u>Mol. Immunol.</u>, 18:889.

Kranz, D. M., Herron, J. N., Giannis, P. E. and Voss, E. W. Jr., 1981, Kinetics and Mechanism of Fluorescence Enhancement due to Deuterium Oxide Perturbation of Fluorescyl Ligand Bound to Purified Homogeneous and Heterogeneous Antibodies, <u>J. Biol. Chem.</u>, 256:4433.

Lancet, D. and Pecht, I., 1976, Kinetic Evidence for Hapten-induced Conformational Transition in Immunoglobulin MOPC 460, <u>Proc. Natl. Acad. Sci. USA</u>, 73:3549.

Pecht, I., 1982, Dynamic Aspects of Antibody Formation, <u>in</u>: "The Antigens", M. Sela, ed., Academic Press, New York, 6:1.

Petrossian, A., 1986, Personal communication.

Romans, D. G., and Dorrington, K. J., 1985, Binding of Hapten-liganded Antibody to Macromolecular Antigen, <u>Fed. Proc.</u>, 45:738.

Rudikoff, S., 1984, Structural Correlates of Idiotypes Expressed on Galactan Binding Antibodies, <u>in</u>: "The Biology of Idiotypes", M. I. Greene and A. Nisonoff, ed., Plenum Press, New York.

Voss, E. W. Jr., 1984, Immunological Properties of Fluorescein, <u>in</u>: "Fluorescein Hapten: An Immunological Probe", E. W. Voss, Jr., ed., CRC Press, Inc., Boca Raton.

Voss, E. W. Jr., Eschenfeldt, W. and Root, R. T., 1976, Fluorescein: A Complete Antigenic Group, <u>Immunochemistry</u>, 13:447.

Voss, E. W. Jr., Watt, R. M. and Weber, G., 1980, Solvent Perturbation of the Fluorescence of Fluorescyl Ligand Bound to Specific Antibody. Fluorescence Enhancement of Antibody Bound Fluorescein (Hapten) in Deuterium Oxide, Mol. Immunol., 17:505.

Watt, R. M., and Voss, E. W. Jr., 1977, Mechanisms of Quenching of Fluorescein by Anti-fluorescein IgG Antibodies, Immunochemistry, 14:533.

Watt, R. M., and Voss, E. W. Jr., 1978, Characterization of Affinity Labeled Fluorescyl Ligand to Specifically Purified Rabbit IgG Antibodies, Immunochemistry, 15:875.

Watt, R. M., and Voss, E. W. Jr., 1979, Solvent Perturbation of the Fluorescence of Fluorescein Bound to Specific Antibody. Fluorescence Quenching of the Bound Fluorophore by Iodide, J. Biol. Chem., 254:1684.

Watt, R. M., Herron, J. N. and Voss, E. W. Jr., 1980, First Order Dissociation Rates Between High Affinity Rabbit Anti-fluorescyl IgG Antibody and Homologous Ligand, Mol. Immunol., 17:1237.

Watt, R. M., and Voss, E. W. Jr., 1984, Affinity Labeling of Antifluorescyl Antibodies, in: "Fluorescein Hapten: An Immunological Probe", E. W. Voss, Jr., ed., CRC Press, Inc., Boca Raton.

Watt, R. M., Herron, J. N. and Voss, E. W. Jr., 1980, First Order Dissociation Rates Between High Affinity Rabbit Anti-fluorescyl IgG Antibody and Homologous Ligand, Mol. Immunol., 17:1237.

Zidovetzki, R., Blatt, Y., Glaudemans, C. P. J., Mantula, B. N., and Pecht, I., 1980, A Common Mechanism of Hapten Binding to Immunoglobulin and Their Heterologous Chain Recombinants, Biochemistry 19:2790.

INTRAMOLECULAR EXCITATION ENERGY TRANSFER IN BICHROMOPHORIC MOLECULES -

FUNDAMENTAL ASPECTS AND APPLICATIONS

Bernard Valeur

Laboratoire de Chimie Générale
Conservatoire National des Arts et Métiers
292 rue Saint-Martin
75003 Paris, France

INTRODUCTION

The bichromophoric molecules considered in the present paper consist of two chromophores covalently linked by a molecular spacer (for instance, a polymethylene chain) so that no conjugation between them is possible. Several intramolecular excited-state processes can occur in this kind of molecule (De Shryver et al., 1977a).

Intramolecular excimer formation: on excitation of one of the chromophores, which are identical in this case, the excitation energy may be delocalized over the whole molecule which leads to an emission band characteristic of the excimer (Zachariasse et al., 1985).

Intramolecular exciplex formation: the same phenomenon may occur with unlike chromophores (Okada et al., 1972).

<u>Intramolecular photochemical reaction</u>: the excitation energy is utilized for a change in the chemical structure either reversibly or irreversibly: photocyclomerization (Desvergne et al., 1983).

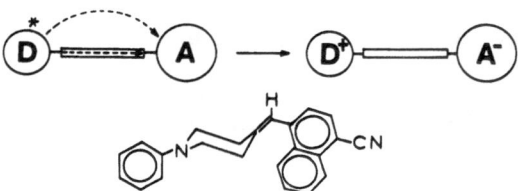

A necessary (but non-sufficient) condition to observe these phenomena is that the molecular spacer is flexible enough to permit close approach of the two chromophores. The aromatic rings ought to be face-to-face for excimer and exciplex formation. In contrast, this condition of flexibility is not required for intramolecular transfer of electron or excitation energy.

<u>Intramolecular electron transfer</u> is indeed a through-space or a through-bond process (Mes et al., 1984).

<u>Intramolecular excitation energy transfer (EET)</u>: the electronic excitation energy of the donor is partially or totally transferred to the acceptor which emits its own fluorescence. The molecular spacer may be flexible, rigid (i.e. molecular frame) or partially rigid.

This intramolecular process has been a subject of considerable interest because of its implications in numerous fields: photophysics, photo-chemistry, biology, polymers,... Bichromophores can be indeed used as models for biomolecules or for mimicking biological systems. End-capped bichromophoric polymers allow the study of static and dynamic conformations. Practical applications are also of interest: laser dyes, frequency conversion of light (scintillators, fluorescent solar concentrators, ...).

MECHANISMS OF ELECTRONIC ENERGY TRANSFER

Only the main features of these mechanisms will be presented. The reader is referred to the reviews by Wilkinson (1968) and Turro (1978).

Radiative transfer

The fluorescence photons emitted by the donor may be absorbed by the acceptor if the emission spectrum of the donor overlaps the adsorption spectrum of the acceptor. This process does not require any interaction between the two chromophores and occurs in two steps:

$$D^* \rightarrow D + h\nu_D$$

$$A + h\nu_D \rightarrow A^*$$

This leads to a distortion of the donor emission spectrum but the lifetime of donor remains unchanged unless the two chromophores are identical; in this case, multiple reabsorption and reemission leads to a lengthening of the lifetime. This reabsorption phenomenon is easily observed with molecules exhibiting strong overlap between absorption and emission spectra (such as chlorophyll or porphyrin molecules).

The efficiency of radiative transfer (often called "trivial" transfer) depends not only on spectral overlap but also on the volume of the vessel used and on the concentration.

Non-radiative transfer

In radiationless transfer of electronic energy, deactivation of the donor and excitation of the acceptor occur simultaneously without emission of photons during this process. Some interaction between the donor and the acceptor is required but the interaction energy can be very small so that long-range energy transfer is possible. The resonance condition to be fulfilled implies that the energy of some emission transitions of the donor corresponds to the energy of absorption transitions of the acceptor (Figure 1). This condition is fulfilled in the region where the donor emission spectrum overlaps the acceptor absorption spectrum.

Two kinds of interaction are involved in non-radiative transfer. The total interaction energy can be indeed divided into two terms, a coulombic term and an exchange term. The first one is often approximated by a dipole-dipole interaction between the transition moments of the donor and the acceptor. This is a long-range interaction which can extend up to 60-80 Å. In contrast, the exchange interaction which represents the electrostatic interaction between the charge clouds, can be observed only at short distances ($\lesssim 10$ Å).

By a quantum mechanical treatment of resonance transfer via dipole-dipole interaction between two well-separated molecules, Förster (1949) obtained the following expression for the transfer rate constant:

$$k_T^{D-D} = \frac{1}{\tau_D^\circ} \left(\frac{R_\circ}{R}\right)^6 \qquad (1)$$

where τ_D° is the lifetime of the donor in the absence of transfer, R is the distance between the donor and the acceptor (which is assumed to remain unchanged during the lifetime of the donor), and $R_\circ$ is the critical distance at which transfer and spontaneous decay of the excited donor are equally probable. $R_\circ$, which can be determined from spectroscopic data (see tables in Berlman, 1973), is given by

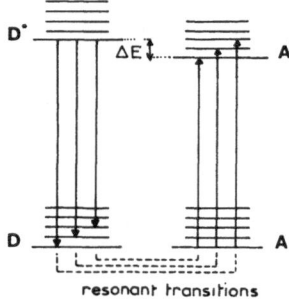

Figure 1. Energy level diagram for a donor D and an acceptor A (DE represents the difference in electronic energy.

$$R_o^6 = \frac{9000 \; (\ln 10)}{128 \pi^5 N_A} \; \tilde{n}^{-4} \kappa^2 \Phi_D^o \; J \tag{2}$$

where $\tilde{n}$ is the refractive index, N_A is Avogadro's number, Φ_D^o is the donor quantum yield in the absence of transfer, J is the overlap integral:

$$J = \int_o^\infty F_D(\lambda) \varepsilon_A(\lambda) \lambda^4 d\lambda \tag{3}$$

in which the extinction coefficient, ε_A, of the acceptor is expressed in units of $l \cdot mol^{-1} \cdot cm^{-1}$, whereas the spectral distribution of the donor fluorescence $F_D(\lambda)$ is normalized so that

$$\int_o^\infty F_D(\lambda) d\lambda = 1. \tag{4}$$

κ^2 is an orientation factor for the chromophores which is given by

$$\kappa = \cos \Theta_{DA} - 3 \cos \Theta_A \cos \Theta_D \tag{5}$$

where Θ_{DA} is the angle between the donor and acceptor transition moments, and Θ_D and Θ_A are the angles between these respectively and the separation vector. When the molecules are free to rotate at a rate which is much larger than the deexcitation rate of the donor (dynamic averaging), the average value of κ^2 is 2/3. In a rigid medium, the square of the average of κ is 0.476 for an ensemble of acceptors which are statistically randomly distributed about the donor with respect to both distance and orientation (this case is often called "static isotropic average"). The case of donor-acceptor pairs at fixed separations and with isotropically distributed orientations has been investigated by Dale (1978); some bichromophoric molecules are relevant to this case.

For a given interchromophoric distance, the transfer efficiency is related to the ratio R/R_o:

$$\Phi_T = \frac{1}{1 + (R/R_o)^6} \tag{6}$$

It should be emphasized that the inverse sixth power dependence on distance is characteristic of the dipole-dipole mechanism of energy transfer. In contrast, an exponential dependence is to be expected from the exchange mechanism: according to Dexter's theory (Dexter, 1964), the rate constant for transfer can be written in the form

$$k_T^{ex} = \frac{2}{\hbar} KJ'\exp(-2R/L) \tag{7}$$

where

$$J' = \int_0^\infty F_D(\lambda)\varepsilon_A(\lambda)d\lambda \tag{8}$$

with the normalization conditions

$$\int_0^\infty F_D(\lambda)d\lambda = \int_0^\infty \varepsilon_A(\lambda)d\lambda = 1 \tag{9}$$

R is the interchromophoric distance and L is the average Bohr radius. Since K is a constant which is not related to any spectroscopic data, it is difficult to characterize the exchange mechanism by experiments.

In fluid solutions, translational diffusion of the chromophores may play a role in the energy transfer process and, ultimately, the rate for transfer may become diffusion controlled (k = 8 RT/3000η where η is the viscosity of the medium). Various modifications of Förster's formula have been suggested (Birks, 1970) to take into account the diffusion effects.

In a mixture of chromophores, intermolecular energy transfer can be observed only at relatively high concentration. There is a competition between radiative and non-radiative processes, the latter being predominant at high concentration. In contrast, intramolecular energy transfer in bichromophoric molecules is concentration independent and only non-radiative transfer can be observed. The efficiency of intramolecular radiative transfer is indeed very low (Conrad and Brand, 1968; Mugnier et al., 1985a).

REVIEW OF BICHROMOPHORIC MOLECULES EXHIBITING INTRAMOLECULAR EXCITATION ENERGY TRANSFER

Several (but incomplete) reviews are devoted to this subject (Birks, 1970; Berlman, 1973; De Schryver et al., 1977b; Speiser, 1983). Since the interchromophoric distance and the relative orientation of the chromophores

274

are of major importance, a distinction should be made according to the link between the two chromophores: a flexible link permits variations in distance and orientation whereas a rigid link imposes restrictions and distance and/or orientation. The bichromophores of biological interest will be reviewed separately.

1. Chromophores linked by a flexible chain

The first observation of intramolecular EET was reported by Weber and Teale (1958) in compounds 1. Both emit a single fluorescence band typical

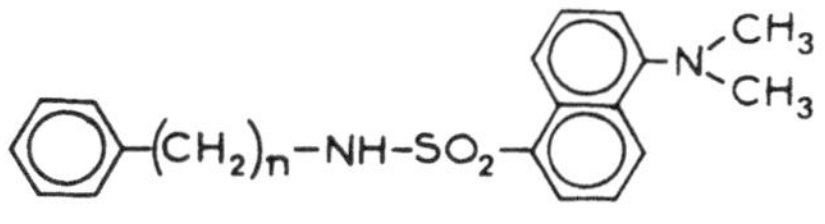

1

of 1-dimethylamino-naphthalene-5-sulfonamide and no fluorescence was observed in the region where the phenyl moiety is expected to emit. This result was interpreted in terms of resonance energy transfer between the two chromophores.

Schnepp and Levy (1962) found that the efficiency of intramolecular EET from the naphthyl chromophore to the anthryl chromophore in compounds 2 was 30% whatever the number of methylene groups separating the two moieties.

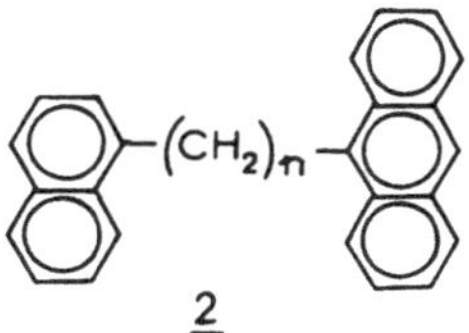

2

No naphthalene emission was observed and only the anthracene fluorescence spectrum was recorded.

Intramolecular singlet and triplet EET was studied (Leermakers et al., 1963; Lamola et al., 1965) in compound 3. Intersystem crossing in

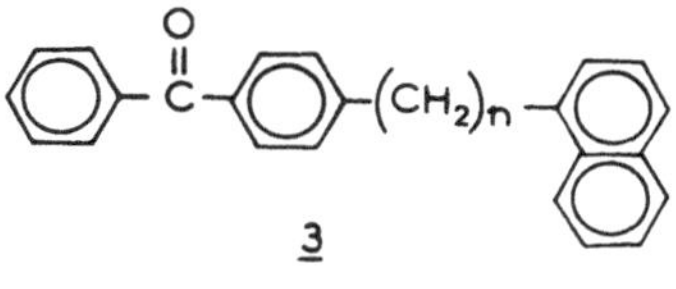

3

benzophenone is very rapid and triplet energy is subsequently transferred to naphthalene with an efficiency of 100% in all three compounds; the rate of triplet EET is greater than $10^{10} s^{-1}$. An exchange mechanism was proposed to occur; the possibility of complex formation is to be considered in view of the significant bathochromic and hyperchromic effects noted in the absorption spectra. Upon excitation of the naphthalene moiety, singlet energy is transferred to benzophenone with an efficiency of about 90% for n = 1, 75% for n = 2 and 85% for n = 3; the rate of singlet EET is about 10^7 - 10^9 s^{-1}.

Complete transfer of triplet energy from phthalimide or carbazole to naphthalene (compounds 4 and 5) was observed in a rigid glass by Breen and Keller (1968). Charge transfer from naphthalene to phthalimide accounts

for additional absorption and emission bands not characteristic of either chromophore. Transfer of singlet energy can occur from naphthalene to phthalimide or carbazole.

Bichromophores of type 6 were studied by Conrad and Brand (1968) in order to explore the potential of energy transfer experiments for calculating distances between fluorescent residues on polypeptides and proteins.

Transfer of singlet EET occurs from tryptophan to the dansyl group. From

276

the values of transfer efficiency, the average interchromophoric distances
can be calculated by means of Förster's equation. These distances were
found to increase consistently with the number of methylene groups separating the two chromophores and were in agreement with estimates based on
theoretical end-to-end distances of methylene chains.

Cowan and Baum (1970) observed triplet EET from acetophenone to
trans-β-methylstyryl (compounds $\underline{7}$) followed by trans-cis isomerization of

$$\underline{7}$$

the olefin chromophore. The rate constants for triplet EET decreases by a
factor of 20 in going from n = 2 to n = 4. The exchange mechanism assisted
by collisions between the donor and acceptor accounts for this result.

Tamaki (1973) investigated the intramolecular singlet EET from phenol
to indole, these chromophores being linked by a chain consisting of methylene groups and amide groups (system $\underline{8}$). Fluorescence and phosphorescence

R_1	R_2
$(CH_2)_2$	CH_3
$(CH_2)_3$	CH_3
$(CH_2)_3$	H
$CH_2-CO-NH-CH_2$	CH_3
$CH_2-CO-NH-CH_2$	H
$(CH_2)_2-NH-CO-(CH_2)_2$	CH_3

arises only from indole and the transfer efficiency varies from 0.48 to
0.78 in ethanol. Weak intramolecular hydrogen bonding of the indole
chromophore to the phenolic oxygen or the π orbitals of the phenol ring
accounts for the distortion of the absorption spectra. The low fluorescence yields of compounds in which the link is $-(CH_2)_3$ was explained in
terms of exciplex involving the indole group in the excited state and the
phenol group in the ground state.

In order to increase the efficiency of flashlamp-pumped dye lasers,
Schäfer and co-workers (1978) suggested to use bichromophoric laser dyes
exhibiting rapid intramolecular EET. A broader coverage of the flashlamp
spectrum is thus achieved. A decrease in threshold energy was indeed
observed in bichromophores consisting of p-terphenyl (the donor) and
dimethyl-POPOP (the acceptor) linked together by a $-CH_2-$ or a $-CH_2-S-CH_2-$

bridge (compound 9). The transfer efficiency is close to 100% and the

9

observed fast rise of the fluorescence suggest a transfer time shorter than
10^{-12} s in agreement with theoretical estimates (Kopainsky et al., 1978).
Further spectroscopic studies of these compounds were carried out by
Ketskemety et al. (1982).

Triplet-triplet absorption of laser dyes can be efficiently prevented
by intramolecular triplet EET to a triplet quencher linked to the laser dye
by a short saturated chain, as reported by Liphardt et al. (1981) in
compounds 10. Thanks to triplet EET to trans-stilbene, the efficiency of
dimethyl-POPOP is increased by a factor of 500 for n = 1 and 200 for n = 2.

10

A Förster-type mechanism for transfer seems to be more likely than an
exchange mechanism of Dexter-type and the measured energy transfer rates
show a strong dependence on the length of the bridge between the two
chromophores (Schäfer et al., 1982). Differences in behavior of bichromophores 9 and 10 as regards intramolecular EET can be explained on the basis
of the energy level of the triplet states (Liphardt et al., 1982).

Bichromophoric molecules consisting of two fluorescent dyes linked by
a saturated bridge are well suited for efficient frequency conversion of
light and can thus be used in various applications such as wavelength
shifters, scintillators, fluorescent solar concentrators etc., as suggested
by Valeur and co-workers (1982) with compound 11. The effects of various

11

parameters (temperature, viscosity, nature of the solvent, number of methylene groups separating donor and acceptor moieties) on the efficiency of energy transfer were examined (Mugnier et al., 1985a). The rate constants of intramolecular EET were measured by multifrequency phase modulation fluorometry (Mugnier et al., 1985b; Mugnier et al., unpublished results). The methodologies and the results of these static and dynamic studies will be presented later in the present article.

The methods of supersonic jet spectroscopy allow one to observe intramolecular EET in an isolated molecule. Zewail and co-workers (1982) reported the first observation of time resolved formation of a product of intramolecular EET in compound 12. The anthracene chromophore was excited

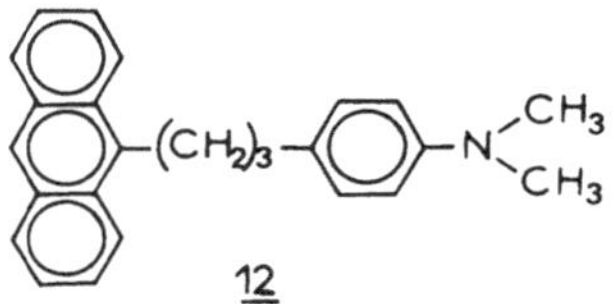

12

by picosecond pulses and the transfer of energy was probed by observing the emission from the exciplex formed between anthracene and dimethylaniline. A supersonic free jet was also used by Ebata et al. (1984) for the study of intramolecular EET from o- or m-xylene moiety in compound 13. The rate constant of energy transfer was found to be 10^8 - 10^9 s^{-1}.

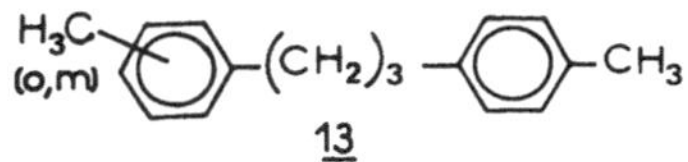

13

Reversible singlet energy transfer has been recently observed by DeSchryver and co-workers (1985) with compound 14, in addition to the

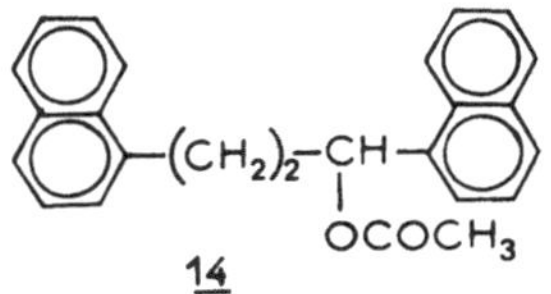

14

existence of two excimers. Fluorescence decay measurements allowed them to determine the two rate constants characterizing the reversible energy transfer. The authors pointed out the importance of this process in excimer-forming molecules.

In these studies of chromophores linked by a flexible chain, triplet EET (when it is possible) always occurs with very high efficiency via an exchange mechanism. In most cases a dipole-dipole mechanism (Förster-type) is responsible for singlet EET whose efficiency is often high, but this is not a general rule. In some cases, complex formation may be involved in the energy transfer processes.

2. Chromophores linked by a rigid or partially rigid molecular frame

In order to hold the two chromophores at a relatively fixed distance, Latt et al. (1965) used a bisteroid bridge (compound 15). From the measured

transfer efficiency, the interchromophoric distance can be calculated from Förster's equation assuming random donor-acceptor orientations. The distance was found to be 21.3 $\pm$ 1.6 Å for 15 a which is in agreement with the distance measured on molecular models. The somewhat lower value obtained for 15 b (16.7 $\pm$ 1.4 Å) can be explained by restricted movements of naphthalene and anthracene about their attachment sites.

A rigid steroid bridge was further used by several authors. Keller and Dolby (1967) found that, in compounds 16, an efficient transfer of

singlet energy from benzophenone or carbazole to naphthalene occurs according to a Förster mechanism. The interchromophoric distances calculated (14
Å for 16 a and 15 Å for 16 b) are consistent with the distance measured on
molecular models (~ 15 Å). It should be noted that the transfer of triplet
energy in these compounds is not very efficient (39% for 16 a and 30% for
16 b) in contrast to the complete transfer observed with the same chromophores linked by a flexible methylene chain (compounds 3 and 5) allowing
shorter interchromophoric distances.

Rauh et al. (1969) examined the transfer of singlet energy in four
indole alkaloids in which the indole chromophore (the donor) is separated
from naphthalene (the acceptor) by a saturated fused-ring network, but
rotations of the naphthalene moiety are allowed. A transfer efficiency of
100% was observed; such a high efficiency is not predicted by Förster's
mechanism in view of the interchromophoric distances and restricted orientations involved and exchange modes are assumed to operate as well.

In contrast to the preceding examples, no rotation of the donor
(N-methyl-indole) and the acceptor (ketone) is possible in system 17 since
these chromophores are fused to a rigid steroid that separates them by 10.2

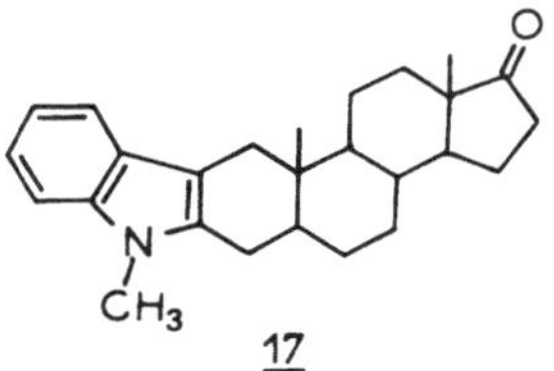

17

Å (Haugland et al., 1971). The overlap integral J was varied over a
40-fold range by altering the solvent. The transfer rate was found to be
proportional to J, as predicted by Förster's theory.

Keller (1968) investigated bichromophores (18) in which the donor
(naphthalene) and the acceptor (anthrone) are held relatively rigidly at
right angles to each other by a spiro linkage. Complete transfer of both

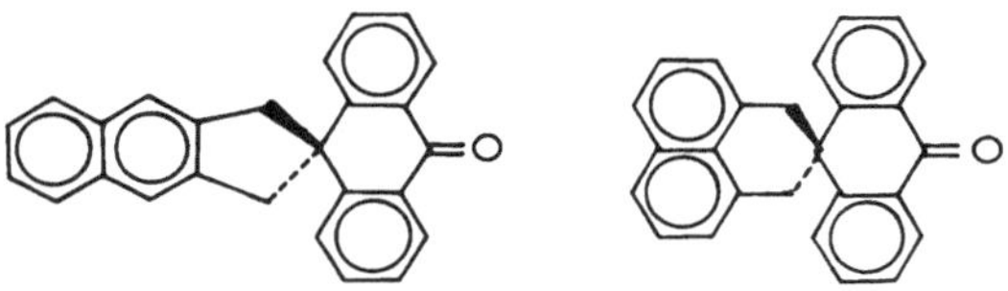

18

singlet and triplet energy transfer was observed. However the assertion of
perpendicularity between naphthalene and anthrone chromophores is question-
able, as noticed by De Member and Filipescu (1968). Rigid perpendicular
orientation between p-dimethoxybenzene and fluorene is achieved in compound
19 and no transfer could be detected (De Member and Filipescu, 1968)
consistent with the orientation requirements for electric dipole energy
transfer. However, Lamola (1969) found entirely different results using
the same bichromophore. Since the absorption spectrum of 19 is significantly different from the sum of the absorption spectra of the model
compounds, the question of intramolecular EET is somewhat ambiguous.

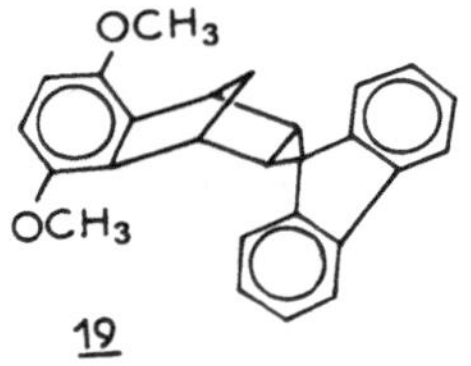

19

Lamola emphasized that Förster's theory is valid strictly for large interchromophoric distances where the coulombic interaction between the donor
and acceptor oscillators predominates and where the point-dipole approximation is valid. At distances of less than 10 Å, these requirements are not
fulfilled.

Efficient but not total triplet EET between tetralin-1,4-dione and
fluorene (compound 20) was reported by Filipescu et al. (1969). The

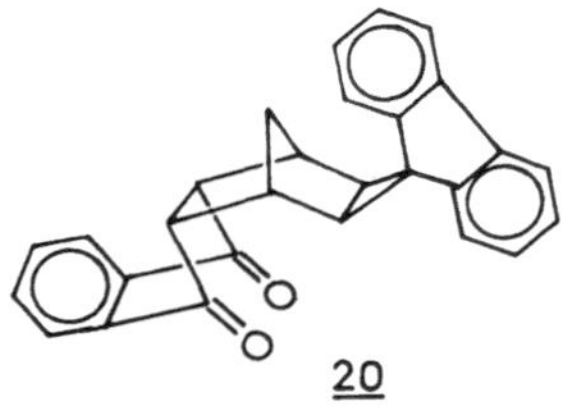

20

observed transfer efficiency is compatible with a theoretical estimation of
the exchange-transfer probability.

Singlet and triplet EET between cyclopentenone and phenanthrene
attached to a rigid spiro-cyclopropane-norbonane molecular frame (21) was
studied by Bunting and Filipescu (1970). Their results are consistent with
efficient (but not total) singlet EET from phenanthrene to cyclopentanone,
followed by back-transfer of triplet energy from cyclopentanone to phenan-

21

threne. From the comparison between experimental and calculated rate
constants for singlet and triplet transfer, it can be concluded that
transfer occurs by coulomb and exchange interaction at the singlet level
and by exchange mechanism at the triplet level.

Another possibility of keeping the donor and acceptor chromophores at
fixed distance is to use a rod-like organic spacer composed of bicyclo[2.2.2]
octane units. Various bichromophores of this type (22, 23, 24) were

studied by Zimmerman and McKelvey (1971) and Zimmerman et al., (1980). A
dual fluorescence (i.e. from both donor and acceptor) was observed. The
transfer efficiency decreases with increasing distance between chromophores;
however the dependence on distance is greater than the inverse sixth power
relationship derived from Förster's theory. This result was interpreted in
terms of an especially efficient overlap-delocalization mechanism when only
one bicyclo[2.2.2] octane unit separates the two chromophores.

Chromophores linked by two polymethylene chains were investigated by
Speiser and co-workers (Speiser et al., 1980; Getz et al., 1980a; Getz et
al., 1980b; Hassoon et al., 1983; Speiser and Katriel, 1983; Hassoon et
al., 1984). In all compounds 25 to 28, the overlap between the donor
emission spectrum and the acceptor absorption spectrum is very small; the

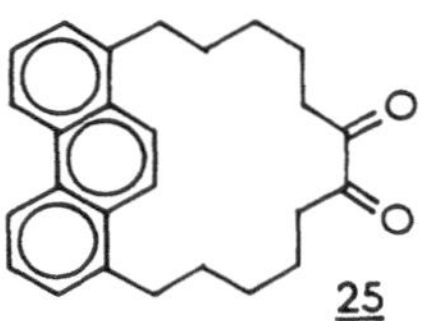

25

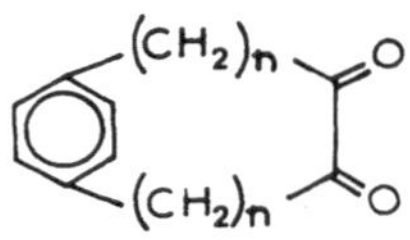

(CH₂)ₙ / (CH₂)ₙ — n=4,5,6 — 26

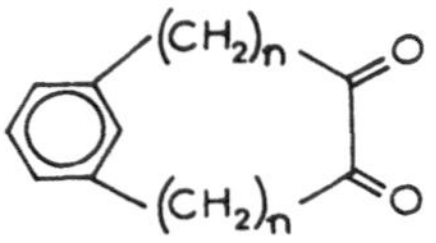

n=3,4 — 27

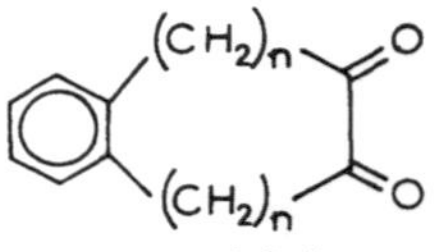

n=2,3,4 — 28

consequent poor transfer efficiency leads to the observation of dual fluorescence. It was demonstrated that in compound 25 energy transfer occurs from higher vibrational states of the phenanthrene moiety or higher energy conformers of this compound. Bichromophores 26, 27 and 28 are well suited to the study of short range intramolecular EET. The results show that a Dexter-type exchange mechanism is responsible for singlet EET since the transfer efficiency is strongly structure dependent. For instance, almost complete energy transfer is observed for compound 26 with n = 4 whereas no transfer occurs in this compound with n = 6.

3. Bichromophores of biological interest

In addition to the works of Conrad and Brand (1968) and Haugland et al. (1971), described in the last two sections, the investigation of Stryer and Haugland (1967) clearly demonstrates that the energy transfer process via dipole-dipole resonance interaction can serve as a spectroscopic ruler in biological systems. An α-naphthyl group (the donor) and a dansyl group (the acceptor) can be held at fixed distances ranging from 12 to 46 Å by means of spacers consisting of oligomers of poly-L-proline whose conformation is a trans helix (29). The observed efficiency of energy transfer as

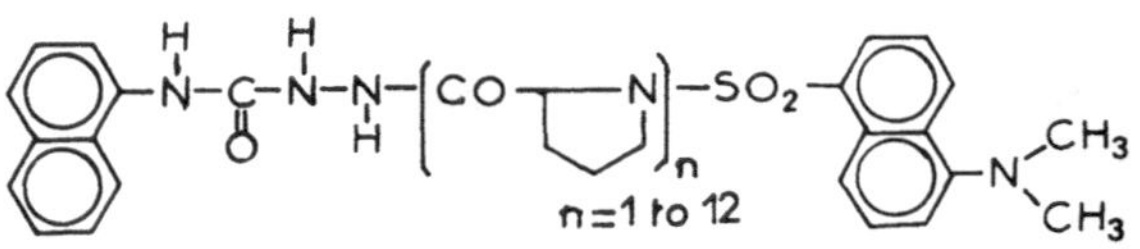

29

a function of distance is in excellent agreement with the theoretical R^{-6} dependence predicted by Förster's theory. Such a spectroscopic ruler has been successfully used in many investigations (Steinberg, 1971; Stryer, 1978). The limitations of this method were discussed by Eisinger et al. (1969).

Energy transfer between two chromophores linked by oligopeptides, oligonucleotides or other small biopolymers can provide information about distributions of accessible molecular conformations, as suggested by Cantor and Pechukas (1971) and Grinvald et al. (1972). The distribution of end-to-end distances of oligomers of 2-hydroxyethyl-L-glutamine was studied by Steinberg and co-workers (1975) in this way (compound 30). The donor

$$
\text{[naphthyl]}-CH_2CH-NH-\left[\begin{array}{c}O\\||\\C-CH-NH\end{array}\right.\left.-SO_2-\text{[naphthyl]}-N\begin{array}{c}CH_3\\CH_3\end{array}\right.
$$

with side chains:
C=O, NH, (CH_2)_2, OH on the donor unit;
(CH_2)_2, C=O, NH, (CH_2)_2-OH on the repeating unit, n, $n = 4$ to 9

30

and acceptor moieties linked at each end were the same as in compound 29. The fluorescence decay of the donor is not a single exponential owing to the distribution of distances which results in a large number of rate constants for energy transfer. The parameters of previously proposed expressions to describe end-to-end distribution in polymers can be determined by numerical analysis of the fluorescence decay curves. Information on the brownian motion of the ends of the chain can also be obtained when changes in distance occur during the lifetime of the donor (Haas et al., 1978). Energy transfer is indeed enhanced by translational diffusion of the donor and acceptor moieties toward each other. Values of the diffusion coefficient were derived from the emission kinetics of the donor in solvents of varying viscosity.

Several other studies of end-to-end energy transfer in oligopeptides and polypeptides have been reported (Guillard et al., 1975; Guillard and Englert, 1976; Leclerc et al., 1978; Chiu and Bersohn, 1977; Sisido et al., 1979; Sisido et al., 1985).

The first investigation of intramolecular EET between linked nucleotides was reported by Weber (1957) in dihydro diphosphopyridine nucleotide; he found that 30 per cent of the photons absorbed by the adenine appear as nicotinamide fluorescence and he suggested the existence of an intramolecular complex. Dinucleotides and other model systems in which the bases are joined by a short linkage (generally -(CH_2)_3-) offer the simplest way to study interactions between the bases of nucleic acids. It has been shown by ORD and NMR experiments that, in these systems, the stacking tendency of the bases is strong so that their relative positions can be expected to resemble those found in nucleic acids. An extensive review of the studies on excited-state interactions between bases in these systems has been presented by Eisinger and Lamola (1971). The large red-shift in the fluorescence spectrum with respect to that of the constituent bases is ascribed to the formation of an exciplex between the bases. The phosphorescence spectra show that, in most cases, emission occurs from the chromophore which possesses the lowest triplet level thus indicating triplet energy transfer. Recently, Gryczynski et al. (1985) have studied adenine--$(CH_2)_n$-Y_t systems ($\underline{31}$). The stacking interactions between the two

$\underline{31}$

nucleotides were evaluated by measuring the hypochromic effect which was found to be stronger for n = 3. From the comparison between corrected fluorescence excitation spectra and absorption spectra, the efficiencies of energy transfer from adenine to the Y_t base were determined as 0.41, 0.81, 0.44 and 0.17 for systems with n = 2, 3, 5 and 6, respectively. These variations as a function of n are consistent with different degrees of stacking as revealed by hypochromicity measurements.

Interactions between nucleotides and organic molecules can be studied in bichromophoric compounds in which the base is linked to a dye by a polymethylene chain. Lee et al. (1977) reported intramolecular EET from uracil or thymine to a dansyl group ($\underline{32}$) or a NBD group ($\underline{33}$). The efficiencies of transfer to the dye range from ~30% in $\underline{32}$ and ~59% and 95% in $\underline{33\ b}$ and $\underline{33\ a}$ respectively. It is remarkable that the efficiency of transfer in uracil derivatives (<u>32 a</u> and <u>33 a</u>) is about the same as the efficiency of quenching of uracil photohydration, thus indicating that the excited-state precursor for both fluorescence sensitization and photohydration is a singlet excited-state of uracil.

In order to get more insight into interaction processes and photoreactions between nucleic acids and coumarins, Wenska and S. Paszyk (1984) prepared bichromophores containing adenine or thymine linked by a polymethylene chain with 7-oxycoumarin (<u>34</u>, <u>35</u>). A hypochromic effect was

observed in the adenine derivative (<u>34</u>) but not in the thymine derivative (<u>35</u>). In both systems, the fluorescence of coumarin was partially quenched. Energy transfer from adenine or thymine to the coumarin moiety was demonstrated; an efficiency of 60% was found in compound <u>35</u>. Intermolecular dimerization of coumarin residues was observed upon irradiation of compound <u>34</u>.

It is of interest to understand DNA binding with bifunctional intercalators because of their biological properties and antitumoral activities. An example relevant to the present review by compound <u>36</u> (Gaugain et al.,

1978a) in which energy transfer from the acridine moiety to the phenanthridinium moiety was observed by Gaugain et al. (1978b). Only the phenanthridinium ring is intercalated in DNA; the efficiency of energy transfer depends on the square of the adenine-thymine base pair content.

Gust and Moore (1985) proposed synthetic systems mimicking the energy transfer and charge separation occurring in natural photosynthesis. Carotenoid antenna function consists of the absorption of light and the transfer of singlet excitation to chlorophyll or, in the case of 37, to the porphyrin moiety. In this compound, the transfer efficiency between the

37

carotenoid and porphyrin moieties was found to be 25% (Moore et al., 1980). NMR studies revealed that the two chromophores are in a stacked conformation, their relative distance being ~4 - 5 Å. At such a short distance, Förster's theory is no longer valid and exchange interactions are expected to occur. Moreover, distortion of the absorption spectra led the authors to consider additional mechanism involving stronger coupling. This study reflects the complex nature of the energy transfer mechanism in photosynthetic membranes. Photoprotection from singlet oxygen damage is explained by intramolecular triplet EET from porphyrin to carotenoid which presumably occur via an electron-exchange mechanism (Gust et al., 1985).

Other bichromophoric systems, studied as models for photosynthesis consist of anthracene carboxylate or phenylbenzoate (the donor) to pyropheophorbide α (the chlorophyll-like chromophore) covalently linked to each other (Sarkar et al., 1982). The energy transfer mechanism was shown to be of the Förster-type, the acceptor transition dipole being represented by the Soret band of pyropheophorbide α moiety.

Covalently linked metalloporphyrins have been studied by several authors because dimer formation and exciton interactions were found to play an important role in the charge separation process in the reaction centers of photosynthetic bacteria and plants (Schwarz et al., 1972; Anton et al., 1978; Selensky et al., 1981; Mialocq et al., 1984; Kaizu et al., 1986). In these studies, special attention was paid to intramolecular EET.

Flavins belong to another class of chromophores of photobiological importance. Since short interflavin distance is required for efficient photoreduction, it is of interest to study dimeric flavin models in which the two chromophores are linked by a polymethylene chain (compound 38)

38

(Visser et al., 1983). Information on the relative orientation of the two chromophores and on the rate for intramolecular singlet EET was obtained from time-dependent fluorescence anisotropy carried out in a polymethyl-methacrylate matrix. The rate constant for transfer was found to be 9 x $10^8 s^{-1}$ and $1.6 \times 10^9 s^{-1}$ for n = 3 and 6 respectively.

Intramolecular EET has been studied in other various bichromophoric molecules relevant to biology or medicine: neuroleptics (Thiery, 1970), bilirubins (McDonagh and Lightner, 1985), nitrobenzoxadiazole derivatives of polyene antibiotics (Peterson, 1985), ...

STEADY-STATE AND TIME-DEPENDENT STUDY OF INTRAMOLECULAR ENERGY TRANSFER: METHODOLOGY AND APPLICATION TO SOME COUMARIN BICHROMOPHORES

The following methodology can be applied to any bichromophoric molecule.

1. Ground-state interaction?
   If the absorption spectrum of the bichromophore superimposes the sum of the spectra for the donor and acceptor model compounds, there is no ground-state interaction.

2. Exciplex formation?
   If there is no distortion of the emission spectrum (apart from a decrease in donor fluorescence intensity and an increase in acceptor fluorescence intensity) exciplex formation is unlikely.

3. Randomicity of mutual orientation or preferred-orientation?
   The answer is given by steady-state fluorescence polarization experiments in a rigid medium.

4. Determination of transfer efficiency
   Three methods can be used:

Steady-state experiments

a.   Comparison between the absorption spectrum
     of the bichromophore and its excitation
     spectrum.
b.   Enhancement of acceptor fluorescence.
c.   Decrease in donor fluorescence.

5. <u>Information on dynamics of transfer</u>
    Pulse or phase fluorometry: shape of the decay | Time-dependent
    of donor emission; determination of rate cons- | experiments
    tants for transfer.

As an example of investigation using this methodology, the results we
have obtained with some coumarin bichromophores (Mugnier et al., 1985a;
Mugnier et al., 1985b; Mugnier, unpublished results) will be now presented.
These bichromophores consist of two different coumarin dyes linked by a
short polymethylene chain. These coumarins have been chosen so that the

n	compound
3	D_3A
4	D_4A
8	D_8A
12	$D_{12}A$

donor emission spectrum strongly overlaps the acceptor absorption spectrum.
An efficient intramolecular non-radiative energy transfer is thus to be
expected, which is important for practical applications (dye lasers,
frequency conversion of light, ...) (Bourson et al., 1982). Furthermore,
from a fundamental point of view, an investigation of the effects of
various parameters (number of methylene groups separating the donor and
acceptor moieties, nature of the solvent, temperature, viscosity) should
provide a better understanding of intramolecular EET in the case of high
transfer efficiency and short distances.

The model compounds which have been chosen for the donor (D_m) and the
acceptor (A_m) are the following

290

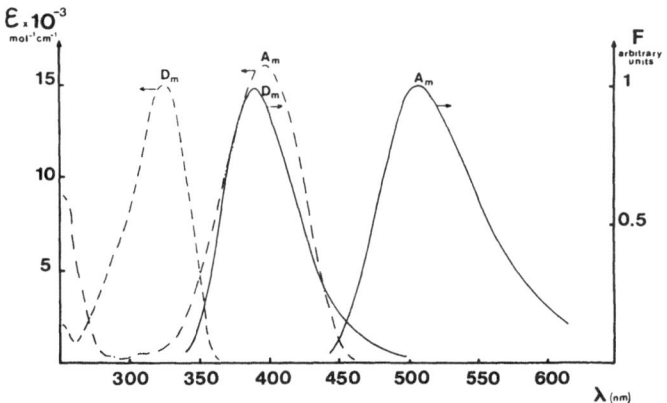

Figure 2. Absorption and corrected emission spectra of the model compounds D_m and A_m for the donor and the acceptor (in propylene glycol at 25°C).

Their absorption and fluorescence spectra are shown in figure 2. The sum of the absorption spectra of these models matches very well the absorption spectra of the bichromophores, indicating the absence of ground-state interaction. Moreover, the emission spectra do not reveal any exciplex formation. Examples of emission spectra are given in figure 3.

In the following considerations, the subscripts D and A refer to the donor and acceptor respectively, and the superscripts b and m refer to the bichromophores and the models respectively. Φ represents the quantum yield, α the absorption coefficients (proportional to absorbances) and Φ_T the transfer efficiency.

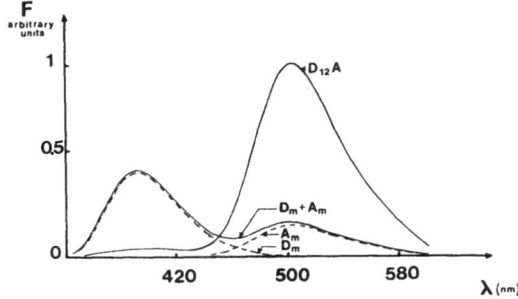

Figure 3. Corrected emission spectra of D_m, A_m and $D_{12}A$, and of an equimolar mixture of D_m and A_m ($D_m + A_m$). (In propylene glycol at 25°C; concentration: $2 \times 10^{-6} M$; excitation wavelength: 324 nm).

Mutual orientation of donor and acceptor

Assuming that only the first singlet states of the donor and the acceptor are involved in the energy transfer, the fundamental emission anisotropy, r_o (anisotropy is defined as $r = (I_{\shortparallel}/I_{\perp})/(I_{\shortparallel}+2I_{\perp})$ where $I_{\shortparallel}$ and $I_{\perp}$ are the polarized intensities respectively parallel and perpendicular to the electric vector of the polarized incident beam) of the bichromophore dissolved in an isotropic rigid medium [31] is given by:

$$r_{oA}^{b} = f_1 \, r_{oD}^{m} \, \frac{3 \, \overline{\cos^2\Theta_{DA}} - 1}{2} + f_2 \, r_{oA}^{m} \tag{6}$$

where f_1 and f_2 are the fraction of fluorescence intensity emitted by the donor and acceptor, respectively:

$$f_1 = \frac{\alpha_D^m \Phi_T}{\alpha_A^m + \alpha_D^m \Phi_T} \qquad\qquad f_2 = \frac{\alpha_A^m}{\alpha_A^m + \alpha_D^m \Phi_T}$$

The term $(3 \, \overline{\cos^2\Theta_{DA}} - 1)/2$ which involves the average cosine square of the angle Θ_{DA} between the donor and acceptor transition moments, represents the transfer depolarization factor. In the case of donor-acceptor pairs at fixed separations and with isotropically distributed orientations (in the ensemble), Dale (1978) has shown that, in a rigid medium, the transfer depolarization factor ranges from 0 to 0.04 depending on the transfer efficiency.

In the case of our coumarin bichromophores, this factor is expected to be around 0.005 for the observed transfer efficiencies (0.92 - 0.94) in propylene glycol at -60°C (vide infra). Therefore, significant departure from 0.005 would reveal some possible preferred orientation of the acceptor with respect to the donor. Fluorescence polarization measurements lead to experimental values compatible with the theoretical value of 0.005 within the standard errors. Moreover, it should be emphasized that there is neither a significant difference between D_3A and $D_{12}A$ nor a significant trend as n increases; therefore, since a preferred orientation is unlikely in $D_{12}A$ (because of the flexibility of a $-(CH^2)_{12}$-chain), a non-randomicity of mutual orientations in D_3 and consequently for any value of n, can only be very small if any.

Transfer efficiency

Two solvents have been used: propylene glycol (PG) and dimethyl-

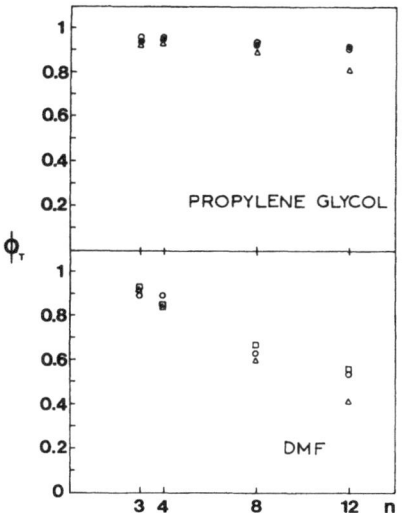

Figure 4. Variations in transfer efficiency versus n at -60°C (•) 25°C
(o), 10°C (□) and 60°C (Δ). Accuracy: 6-9%.

formamide (DMF). The transfer efficiency Φ_T was determined in PG by
three methods (vide supra): the consistency of the results indicates
that the emission of the donor cannot arise from residual uncoupled donor
molecules and that there is no additional radiationless channel for
deactivation of the donor in the presence of the acceptor. Comparison
between absorption spectrum and emission spectrum is the only possible
method for the determination of Φ_T in DMF, because the fluorescence of
the donor in the presence of transfer is undetectable.

The results are reported in figure 4. In PG, transfer efficiency is
always high and is only slightly dependent on n whereas, in DMF, there is
a significant decrease in Φ_T with increasing n value. In both solvents,
the effect of temperature is weak.

At first sight, it is surprising that no residual fluorescence of
the donor is detected in DMF despite incomplete energy transfer, whereas
dual fluorescence is observed in PG (see figure 3) and higher transfer
efficiencies are found. The fact that the quantum yield of the donor is
much lower in DMF (0.01) at 25°C) than in PG (0.11 at 25°C) accounts for
this observation.

Special attention should be paid now to the dependence of transfer
efficiency upon n. If we assume that Förster's theory applies to the

present bichromophores, the critical distance R_o can be calculated according to equation 2. For the sake of clarity, we consider only the extreme cases of a fluid solvent (DMF at 60°C, η = 0.7 cP) and a rigid solvent (PG at -60°C). The orientation factor κ^2 is taken to be 2/3 in the former case and is calculated using Dale's theory (Dale, 1978) in the latter. From the calculated R_o values (15 Å in DMF at 60°C and ~22 Å in PG at -60°C) it can be concluded that the only reason why the transfer efficiency is more sensitive to n in DMF than in PG is that R_o is lower in DMF owing to the lower quantum yield of the donor in this solvent.

From the value of R_o and transfer efficiencies, it is possible to calculate the interchromophoric distance (equation 4) or, more exactly, the average distance $\langle R^6 \rangle^{1/6}$ because the flexibility of the polymethylene chain allows a distribution of distances between donor and acceptor. A necessary but nonsufficient condition for the Förster mechanism to be valid is that the interchromophoric distance calculated in this way is shorter than the center-to-center distance of maximum extension measured on Dreiding models. This is always observed for n = 8 and 12. Moreover the calculated average distances are very similar in DMF (60°C) and in PG (-60°C) for n = 8 (14 Å) and for n = 12 (15 - 16 Å). Two conclusions can be drawn from this observation:

- Viscosity has no significant effect; in other words the diffusion phenomena play a minor role. The very low lifetime of the donor in DMF (125 ps at 25°C) accounts for this observation because the mean diffusion length of the donor is of the order of 1 - 2 Å (for free diffusion).

- Similar distances are obtained despite that the parameters involved in the expression of R_o are quite different. Therefore, a Förster mechanism is likely to be predominant in the case of D_8A and $D_{12}A$.

On the other hand, the interchromophoric distances in D_3A and D_4A can be of the order of 10 Å or less; therefore the Dexter mechanism (exchange interaction) can operate, but in view of the large spectral overlap, dipole-dipole interaction is still highly probable. However, for short distances, Förster's theory is no longer valid because the point dipole approximation is not fulfilled.

Rate constant for transfer

Information on rate of excitation energy transfer is generally obtained from the changes in the fluorescence decay of the donor.

However, when very efficient energy transfer occurs and/or when the quantum yield of the donor is very low, analysis of time-resolved fluorescence of the donor is difficult. Nevertheless, information on transfer rate is contained in time-dependent acceptor fluorescence via donor excitation. Pulse fluorometry can be used for the determination of the rise of the acceptor fluorescence following pulse excitation of the donor. Alternately, phase fluorometry experiments are based on the measurement of the phase shift between the acceptor fluorescence in the presence of transfer (i.e. excitation via the donor) and the acceptor fluorescence in the absence of transfer at various frequencies (Mugnier et al., 1985b). This kind of experiment has been carried out with the above-depicted coumarin bichromophores.

In a first approach, we consider a simple kinetic scheme involving a single rate constant k_t for transfer from donor to acceptor. In the present case, no back transfer occurs (the acceptor emission spectrum does not overlap the donor absorption spectrum). k_D and k_A are the emission rate constants of the donor and the acceptor, respectively.

If one could find an excitation wavelength at which only the donor absorbs, we have shown that the tangent of the phase shift ($\Phi_A - \Phi_A^o$) between the acceptor fluorescence in the presence of transfer and the acceptor fluorescence in the absence of transfer is given by:

$$\tan (\Phi_A - \Phi_A^o) = \omega/(k_D + k_T) \tag{7}$$

It is remarkable that this expression is identical to the tangent of the phase shift of the donor in the presence of transfer

$$\tan \Phi_D = \omega/(k_D + k_T) \tag{8}$$

In practice, one has to take into account some absorption of the acceptor at the wavelength of maximum absorption of the donor. Then, equation (7) should be replaced by the following:

$$\tan (\Phi_A - \Phi_A^\circ) = \omega/(k_D + k_T + Y((k_D + k_T)^2 + \omega^2)/k_T) \qquad (9)$$

where Y is the ratio of the absorbances of the acceptor and the donor at the excitation wavelength.

In the present application, the modulation frequency ranges from 1 to 200 MHz and the transfer rate constant is on the order of $10^{10}s^{-1}$ as expected from the values of transfer efficiencies; therefore $(k_D + k_T)^2 \gg \omega^2$, and, according to equation (9) the variations in $\tan (\Phi_A - \Phi_A^\circ)$ versus ω are expected to be linear. Such a behavior is indeed observed and, from the average slope, the following values of k_T have been obtained:

in PG at 25°C ($k_D = 1.31 \times 10^9 \text{ s}^{-1}$)

$k_T = 1.7 \times 10^{10}s^{-1}(n = 3)$; $1.9 \times 10^{10}s^{-1}$ (n = 4);

$1.5 \times 10^{10}s^{-1}$ (n = 8); $1.4 \times 10^{10}s^{-1}$ (n = 12).

in DMF at 25°C ($k_D - 8.0 \times 10^9 \text{ s}^{-1}$)

$k_T = 4.1 \times 10^{10}s^{-1}$ (n = 3); $3.6 \times 10^{10}s^{-1}$ (n = 4);

$1.8 \times 10^{10}s^{-1}$ (n = 8).

All these values are in the range predicted by Förster which means that, even at short distance, dipole-dipole interactions are still highly probable as a consequence of the large spectral overlap between donor emission and acceptor absorption. However, it should be re-emphasized that Förster's theory is not valid at short distance.

In a second approach, a distribution of rate constants is to be considered because of the distribution of distances resulting from the flexibility of the polymethylene chain linking the two moieties. In the most general case, the differential tangent can be expressed as follows (Mugnier et al., unpublished results):

$$\tan (\Phi_A - \Phi_A^\circ) = \omega \frac{\displaystyle\sum_{i=1}^{N} \frac{f_i k_{Ti}}{(k_D + k_{Ti})^2 + \omega^2}}{\displaystyle\sum_{i=1}^{N} \frac{f_i k_{Ti}(k_D + k_{Ti})}{(k_D + k_{Ti})^2 + \omega^2} + Y} \qquad (10)$$

where f_i is the weighting factor of the rate constant $k_{Ti} (\sum_i f_i = 1)$. If $(k_D + k_{Ti})^2 >> \omega^2$, tan $(\Phi_A - \Phi_A^\circ)$ is still proportional to ω. Therefore, if a linear dependence on frequency is observed, this does not necessarily mean that the rate constant is unique; rather, only an average of the distribution is experimentally attainable considering the values of k_D, k_T and the available range of modulation frequencies. Increasing the frequency should help to detect a non-linear dependence resulting from a distribution of rate constants; however, the lifetime of the acceptor ought to be short enough, otherwise a low modulation ratio of the acceptor at high frequencies would lead to inaccurate measurements of $\Phi_A - \Phi_A^\circ$.

ACKNOWLEDGEMENTS

The results presented in the fourth part of this paper are the fruit of the excellent collaboration with Dr. Jean Bourson, Jacques Mugnier, Jacques Pouget and Professor Enrico Gratton, who are all gratefully acknowledged.

REFERENCES

Anton, J. A., Loach, P. A., and Govindjee, 1978, Transfer of Excitation Energy Between Porphyrin Centers of a Covalently-Linked Dimer, Photochem. Photobiol., 28:235.

Berlman, I., 1973, Energy Transfer Parameters of Aromatic Molecules, Academic Press, NY.

Birks, J. B., 1970, Photophysics of Aromatic Molecules, Wiley, NY.

Bourson, J., Mugnier, J., and Valeur, B., 1982, Frequency Conversion of Light by Intramolecular Energy Transfer in Bifluorophoric Molecules, Chem. Phys. Letters,92:430.

Breen, D. E., and Keller, R. A., 1968, Intramolecular Energy Transfer Between Triplet States of Weakly Interacting Chromophores. I. Compounds in Which the Chromophores are Separated by a Series of Methylene Groups, J. Am. Chem. Soc., 90:1935.

Bunting, J. R., and Filipescu, N., 1970, Intramolecular Energy Transfer in Rigid Model Compounds. Singlet and Triplet Transfer Between Cyclopentanone and Phenanthrene, J. Chem. Soc. B 1750.

Cantor, C. R., and Pechukas, P., 1971, Determination of distance distribution functions by singlet-singlet energy transfer, Proc. Nat. Acad. Sci. U.S.A., 68:2099

Chiu, H. C., and Bersohn, R., 1977, Electronic Energy Transfer Between Tyrosine and Tryptophan in the Peptides Trp-(Pro)$_n$-Tyr, Biopolymers, 16:277.

Conrad, R. H., and Brand, L., 1968, Intramolecular Transfer of Excitation From Tryptophan to 1-Dimethylaminonaphthalene-5-Sulfonamide in a Series of Model Compounds, Biochemistry, 7:5777.

Cowan, D. O., and Baum, A. A., 1970, Intramolecular Triplet Energy Transfer, J. Am. Chem. Soc., 92:2153.

Dale, R. E., 1978, Fluorescence Depolarization and Orientation Factors for Excitation Energy Transfer Between Isolated Donor and Acceptor Fluorophore Pairs at Fixed Intermolecular Separations, Acta Phys. Polon., A 54:743.

De Member, J. R., and Filipescu, N., 1968, Intramolecular Energy Transfer Between Nonconjugated Chromophores. Effect of Rigid Perpendicular Orientation, J. Am. Chem. Soc., 90:6425.

De Schryver, F. C., Boens, N., and Put, J., 1977a, Excited-State Behavior of Some Bichromophoric Systems, Adv. Photochem., 10:359.

De Schryver, F. C., Boens, N., and Put, J., 1977b, Excited-State Behavior of Some Bichromophoric Systems, Adv. Photochem., 10:379.

Desvergne, J. P., Bitit, N, Castellan, A., and Bouas-Laurent, H., 1983, Study of Non-Conjugated Bichromophoric Systems Part 3. The Photo-cyclomerization of 9-(1-naphthylmethoxymethyl) anthralene and 9-(2-Furylmethoxymethyl)anthracene. Interest of the CH_2-O-CH_2 Link, J. Chem. Soc., Perkin Trans. II:109.

Dexter, D. L., 1964, A Theory of Sensitized Luminescence in Solids, J. Chem. Phys., 21:836.

Ebata, T., Suzuki, Y., Mikami, N., Miyashi, T., and Ito, M., 1984, Intramolecular Electronic Energy Transfer of Bichromophoric Molecules in a Supersonic Free Jet, Chem. Phys. Letters, 110:597.

Eisinger, J., Feuer, B., and Lamola, A. A., 1969, Intramolecular Singlet Excitation Transfer. Applications to Polypeptides, Biochemistry, 8:3908.

Eisinger, J., and Lamola, A. A., 1971, in: "Excited States of Proteins and Nucleic Acids", ch. 3, R. F. Steiner and I. Weinryb, ed., Plenum Press, New York.

Felker, P. M., Syage, J. A., Lambert, W. R., and Zewail, A. H., 1982, Direct Observation of Intramolecular Energy Transfer by Selective Picosecond Laser Excitation of a Single Chromophore in Jet-Cooled Molecules, Chem. Phys. Letters, 92:1.

Filipescu, N., de Member ,J. R., and Minn, F. L., 1969, Intramolecular Triplet-Triplet Energy Transfer Between Nonconjugated Chromophores

With Fixed Orientation, *J. Am. Chem. Soc.*, 91:4169.

Förster, Th., 1949, Experimentelle und Theoretische Untersuchung des Zwischen-molekularen Ubergangs von Elektronenan Regungsenergie, *Z. Naturforsch*,. 49:321; 1959, Transfer Mechanisms of Electronic Excitation, *Discuss. Faraday Soc.*, 27:7.

Gaugain, B., Barbet, J., Oberlin, R., Roques, B. P., and Le Pecq, J. B., 1978a, DNA Bifunctional Intercalators. 1. Synthesis and Conformational Properties of an Ethidium Homodimer and of an Acridine Ethidium Heterodimer, *Biochemistry*, 17:5071.

Gaugain, B., Barbet, J., Capelle, N., Roques, B. P., and Le Pecq, J. B., 1978, DNA Bifunctional Intercalators. 1. Synthesis and Conformational Properties of an Ethidium Homodimer and of an Acridine Ethidium Heterodimer, *Biochemistry*, 17:5078.

Getz, D., Ron, A., and Speiser, S., 1980a, Intramolecular Energy Transfer From a Vibronic State of a Bichromophoric Molecule, *J. Mol. Struct.*, 61:61.

Getz, D., Ron, A., Rubin, M. B., and Speiser, S., 1980b, Dual Fluorescence and Intramolecular Electronic Energy Transfer in a Bichromophoric Molecule, *J. Phys. Chem.*, 84:768.

Grinvald, A., Haas, E., and Steinberg, I. Z., 1972, Evaluation of the distribution of distances between energy donors and acceptors by fluorescence decay, *Proc. Acad. Sci. U.S.A.*, 69:2273.

Gryszynski, I., Kawski, A., Paszyc, S., Skalski, B., and Tempczyk, A., 1985, Hypochromic Effect and Electronic Excitation Energy Transfer in $Y_t-(CH_2)_n$-Adenine Systems, *J. Photochem.*, 30:153.

Guillard, R., Leclerc, M., Loffet, A., Leonis, J., Wilmet, B., and Englert, A., 1975, Dimensions of Oligopeptides by Singlet-Singlet Energy Transfer and Theoretical Calculations. I. Influence of Glycine on the Dimensions of Tetrapeptides, *Macromolecules*, 8:134.

Guillard, R., and Englert, A., 1976, Interpretation of Energy-Transfer Experiments by Theoretical Studies of Model Compounds Using Semiempirical Potential Functions. I. Three Linked Aromatic Peptide Units, *Biopolymers*, 15:1301.

Gust, D., and Moore, T. A., 1985, A Synthetic System Mimicking the Energy Transfer and Charge Separation of Natural Photosynthesis, *J. Photochem.* 29:173 and references cited therein.

Gust, D., Moore, T. A., Bensasson, R. V., Mathis, P., Land, E. J., Chachaty, C., Moore, A. L., Liddell, P., and Nemeth, G. A., 1985, Stereodynamics of Intramolecular Triplet Energy Transfer in Carotenoporphyrins, *J. Am. Chem. Soc.*, 107:3631.

Haas, E., Wilchek, M., Katchalski-Katzir, E., and Steinberg, I. Z., 1975,

Distribution of End-to-End Distances of Oligopeptides in Solution as Estimated by Energy Transfer, <u>Proc. Nat. Acad. Sci. U.S.A.</u>, 72:1807.

Haas, E., Katchalski-Katzir, E., and Steinberg, I. Z., 1978, Brownian Motion of the Ends of Oligopeptide Chains in Solution as Estimated by Energy Transfer Between the Chain Ends, <u>Biopolymers</u> 17:11.

Hassoon, S., Lustig, H., Rubin, M. B., and Speiser, D., 1983, Molecular Structure Effects in Intramolecular Electronic Energy Transfer, <u>Chem. Phys. Lett.</u>, 98:345.

Hassoon, S., Lustig, H., Rubin, M. B., and Speiser, S., 1984, The Mechanism of Short-range Intramolecular Electronic Energy in Bichromophoric Molecules, <u>J. Phys. Chem.</u>, 88:6367.

Haugland, R. P., Ygeurabide, J., and Stryer, L, 1971, Dependence of the Kinetics of Singlet-Singlet Energy Transfer on Spectral Overlap, <u>Proc. Nat. Acad. Sci.</u>, 63:23.

Kaizu, Y., Maekawa, H., and Kobayashi, H., 1986, Upper Excited-State Emission of a Covalently Linked Porphyrin Dimer, <u>J. Phys. Chem.</u>, 90:4234.

Keller, R. A., and Dolby, L. J., 1967, Rate Constants and the Mechanism for the Transfer of Triplet Excitation Energy, <u>J. Am. Chem. Soc.</u>, 89:2768; 1969, Intramolecular Energy Between Triplet States of Weakly Interacting Chromophores III Compounds in Which the Chromophores are Separated by a Rigid Steroid Bridge,91:1293.

Keller, R. A., 1968, Intramolecular Energy Transfer Between Triplet States of Weakly Interacting Chromophores. II. Compounds in Which the Chromophores are Perpendicular to Each Other, <u>J. Am. Chem. Soc.</u>, 90:1940.

Ketskemety, I., Farkas, E., Toth, Zs., and Gati, L., 1982, Intramolecular Energy Transfer in Laser Active Bichromophoric Molecules, <u>Acta Phys. Chem.</u>, 28:3.

Kopainsky, B., Kaiser, W., and Schäfer, F. P., 1978, Ultrafast Energy Transfer Within Bifluorophoric Molecules, <u>Chem. Phys. Lett.</u>, 56:458.

Lamola, A. A., 1969, Intramolecular Excitation Transfer in 1,4-dimethoxy-5,8-methano-6,7-exo-[fluorene-9'-spiro-1"-cyclopropane]naphthalene, <u>J. Am. Chem. Soc.</u>, 91:4786.

Lamola, A. A., Leermakers, P. A., Byers, G. W., and Hammond, G. S., 1965, Intramolecular Electronic Energy Transfer Between Nonconjugated Chromaphores in Some Model Compounds, <u>J. Am. Chem. Soc.</u>, 37:2322.

Latt, S. A., Cheung, H. T., and Blout, E. R., 1965, Energy Transfer. A System With Relatively Fixed Donor-Acceptor Separation, <u>J. Am. Chem. Soc.</u>, 87:995.

Leclerc, M., Premilat, S., and Englert, A., 1978, Nonradiative Energy

Transfer in Oligopeptide Chains Generated by a Monte-Carlo Method Including Long-Range Interactions, Biopolymers, 17:2459.

Lee, Y. J., Summers, W. A., and Burr, J. G., 1977, Energy Transfer in Fluorescent Derivatives of Uracil and Thymine, J. Am. Chem. Soc., 99:7679.

Leermakers, P. A., Byers, G. W., Lamola, A. A., and Hammond, G. S., 1963, Intramolecular Electronic Energy Transfer in 4-(1-Naphthylmethyl)-Benzophenone, J. Am. Chem. Soc., 85:2670.

Liphardt, B., Liphardt, B., and Lüttke, W., 1981, Laser Dyes With Intra-molecular Triplet Quenching, Opt. Comm.,38:207.

Liphardt, B., Liphardt, B.,Lüttke, W., and Ouw, D., 1982, Energy Transfer Processes in Two Different Bifluorophoric Laser Dyes, Appl. Phys., B 29:73.

McDonagh, A. F., and Lightner, D. A., 1985, Intramolecular Energy Transfer in Bilirubins, in: "Primary Photo-processes in Biology and Medicine, R. V. Bensasson, G. Jori, E. J. Land and T. G. Truscott, eds., Plenum Publishing Corp., New York.

Mes, G. F., de Jong, B., Van Ramesdonk, H. J., Verhoeven, J. W., Warman, J. M., de Haas, M. P., and Horsman-van den Dool, L.E.W., 1984, Excited-State Dipole Moment and Solvatochromism of Highly Fluores-cent Rod-Shaped Bichromophoric Molecules, J. Am. Chem. Soc., 106:6524.

Mialocq, J. C., Gianotti, C., Maillard, P., and Momenteau, M., 1984, Energy Transfer in "Covalently Linked" and "Face-to-Face" Bisporphyrins, Chem. Phys. Letters, 112:87.

Moore, A. L., Dirks, G., Gust, D., and Moore, T. A., 1980, Energy Transfer From Carotenoid Polyenes to Porphyrins: A Light-Harvesting Antenna, Photochem. Photobiol., 32:691.

Mugnier, J., Puget, J., Bourson, J., and Valeur, B., 1985a, Efficiency of Intramolecular Electronic Energy Transfer in Coumarin Bichromophoric Molecules, J. Lumin., 33:273.

Mugnier, J., Pouget, J., Bourson, J. and Valeur, B. to be published.

Mugnier, J., Valeur, B., and Gratton, E., 1985b, Rate of Intramolecular Electronic Energy Transfer in Coumarin Bichromophoric Molecules. An Investigation by Multifrequency Phase-Modulation Fluorometry, Chem. Phys. Letters,119:217.

Okada, T., Fiyita, T, Kubota, M., Masaki, S., and Mataga, M., 1972, Intrachain Reaction of a Pair of Reaction Groups Attached to Polymer Ends .8. Static and Dynamic Studies on the End-to-end Intrachain Energy Transfer on a Polysarcosine Chain, Chem. Phys. Letters, 14:563.

Petersen, N. O., 1985, Intramolecular Fluorescence Energy Transfer in

Nitrobenzoxadiozole Derivatives of Polyene Antibiotics, <u>Can. J. Chem.</u>, 63:77.

Rauh, R. D., Evans, T. R., and Leermakers, P. A., 1969, Intramolecular Electronic Energy Transfer in Some Indole Alkaloids and Related Donor-Acceptor Systems, <u>J. Am. Chem. Soc.</u>, 90:6897; 1969, 91:1868.

Sarkar, H. K., Song, P.-S., Park, S. C., and Lee, J., 1982, Model for Photosynthetic Light Harvesting System: Energy Transfer in Anthracene- and Biphenyl-Chlorophyll Derivatives, <u>J. Lumin.</u>, 26:347.

Schäfer, F. P., Bor, Zs., Lüttke, W., and Liphardt, B., 1978, Bifluorophoric Laser Dyes With Intramolecular Energy Transfer, <u>Chem. Phys. Lett.</u>, 56:455.

Schäfer, F. P., Zhang, F.-G., and Jethwa, J., 1982, Intramolecular TT-Energy Transfer in Bifluorophoric Laser Dyes, <u>Appl. Phys.</u> B 28:37.

Schnepp, O., and Levy, M., 1962, Intramolecular Energy Transfer in a Naphthalene-Anthracene System, <u>J. Am. Chem. Soc.</u>, 84:172.

Schwarz, F. P., Gouterman, M., Muljiani, Z., and Dolphin, D. H., 1972, Energy Transfer Between Covalently Linked Metal Porphyrins, <u>Bioinorg. Chem.</u>, 2:1.

Selensky, R, Holten, D., Windsor, M. W., Pain III, J. B., Dolphin, D. H., Gouterman, M., and Thomas, J. C., 1981, Excitonic Interactions in Covalently-Linked Porphyrin Dimers, <u>Chem. Phys.</u>, 60:33, and references cited therein.

Sisido, M., Imanishi, Y., and Higashimura, T., 1979, Molecular Weight Distribution of Polysarcosine Obtained by NCA [N-carboxy-d-amino acid anhydride] Polymerization, <u>Macromolecules</u>, 12:975.

Sisido, M., Egusa, S., Yagyu, K., and Imanishi, Y., 1985, Intramolecular End-to-End Energy Transfer on Polypeptide Chains. Effects of Chain Length, Temperature, and Chain Stiffness, <u>Polymer J.</u>, 17:587.

Speiser, S., 1983, Novel Aspects of Intermolecular and Intramolecular Electronic Energy Transfer in Solution, <u>J. Photochemistry</u>, 22:195.

Speiser, S., Katraro, R., Welner, S., and Rubin, M. B., 1980, Intramolecular Electronic Energy Transfer in 1,8(6',7'-dioxodecamethylene)phenanthrene, <u>Chem. Phys. Lett.</u>, 61:199.

Speiser, S., and Katriel, J., 1983, Intramolecular Electronic Energy Transfer Via Exchange Interaction in Bichromophoric Molecules, <u>Chem. Phys. Lett.</u>, 102:88.

Steinberg, I. Z., 1971, Long-range Nonradiative Transfer of Electronic Excitation Energy in Proteins and Polypeptides, <u>Ann. Rev. Biochem.</u>, 40:83.

Stryer, L., 1978, Fluorescence Energy Transfer as a Spectroscopic Ruler, Ann. Rev. Biochem., 47:819.

Stryer, L., and Haugland, R. P., 1967, Energy Transfer: A Spectroscopic Ruler, Proc. Nat. Acad. Sci. U.S.A., 58:719.

Tamaki, T., 1973, Intramolecular Interaction Between the Phenol and the Indole Chromophores, Bull. Chem. Soc. Jap., 46:2527.

Thiery, C., 1970, Intramolecular Excitation Transfer: A Spectroscopic Study of a Series of Neuroleptic Derivatives of Para-Fluorobutyrophenone, Mol. Photochem., 2:1.

Turro, N. J., 1978, Modern Molecular Photochemistry, Benjamin/Cummings, Menlo Park, California, chapter 9.

Vandendriessche, J., Collart, P., De Schryver, F. C., Zhou, Q. F., and Xu, H. J., 1985, Reversible Singlet Energy Transfer in 1,3-Di(2-naphthyl) propyl acetate, Macromolecules, 18:2321.

Visser, A.J.W.G., Santema, J. S., and von Hoek, A., 1983, Energy Transfer in Biflavinyl Compounds as Studied with Fluorescence Depolarization, Photobiochem. Photobiophys., 6:47.

Weber, G., 1957, Intramolecular Transfer of Electronic Energy in Dihydrodiphosphopyridine Nucleotide, Nature (London), 180:1409.

Weber, G., and Teale, F.W.J., 1958, Fluorescence Excitation Spectrum of Organic Compounds in Solution. Part 1. Systems With Quantum Yield Independent of the Exciting Wavelength, Trans. Far. Soc.,54:640.

Wenska, G., and Paszyk, S., 1984, Coumarin-Nucleotide Base Pairs Ultraviolet Absorption, Fluorescence, and Photochemical Study, Can. J. Chem., 62:2006.

Wilkinson, F., 1968, in: "Luminescence in Chemistry", E. J. Bowen, ed., Van Ostrand, London, chapter 8.

Zachariasse, K. A., Duvenek, G., and Kühnle, W., 1985, Double Exponential Decay in Intramolecular Excimer Formation: 1,3-di(2-Pyrenyl)Propane, Chem. Phys. Letters, 113:337.

Zimmerman, H. E., and McKelvey, R. D., 1971, Electron and Energy Transfer Between Bicyclo[2.2.2]octane Bridgehead Moieties, J. Am. Chem. Soc., 93:3638.

Zimmerman, H. E., Goldman, T. D., Hirzel, T. K., and Schmidt, S. P., 1980, Rod-like Organic Molecules. Energy Transfer Studies Using Single Photon Counting, J. Org. Chem., 45:3934.

RESOLUTION AT 2 GHz OF LIFETIMES AND CORRELATION TIMES OF HIGHLY PURIFIED SOLUTIONS OF HUMAN HEMOGLOBIN

Enrico Bucci, Henryk Malak, Clara Fronticelli, Ignacy Gryczynski and Joseph R. Lakowicz

Department of Biological Chemistry
University of Maryland at Baltimore
School of Medicine
660 West Redwood Street
Baltimore, MD 21201 USA

INTRODUCTION

As discussed by Weber, the presence of the heme severely quenches the emission of the tryptophyl residues in hemoglobin and myoglobin (Weber and Teale, 1959). However, emission signals are still detectable in purified solutions of hemoglobin (Alpert et al., 1980; Hirsch et al., 1981; Hirsch and Nagel, 1981), and their characteristics justify the hypothesis that they are produced by tryptophan.

The question may be posed whether the emission results from impurities in the solutions, represented by either spurious proteins, by denatured forms of hemoglobin, or by heme-free hemoglobin (or myoglobin) components due to spontaneous dissociation of the heme from the proteins. It should be stressed that a 50 to 100-fold quenching of the tryptophans, as produced by the heme, would bring the residual fluorescence of hemoglobin down to values comparable to those produced by 1-2% heme-free impurities present in the various solutions.

Measurements of the lifetimes of emission may assist in identifying the emitting probes, because quenched tryptophans will show very short lifetimes, probably in the picosecond region, while longer lifetimes, in the nanosecond region are expected from non-quenched residues. If long-lived (nsec) emissions are observed, then one must consider the presence of impurities, or suggest models which account for the failure of energy transfer.

Time-resolved decays of myoglobin and hemoglobin have been reported (Szabo et al., 1984; Albani et al., 1985; Hochstrasser and Negus, 1984). The measurements resolved the presence of lifetime components in both the picosecond and nanosecond regions. The form of the decays was found to depend upon the ligation state of the hemoglobin (Szabo et al., 1984), and was different for the isolated α and β-subunits (Albani et al., 1985). Szabo and co-workers suggested that the decay times, ranging from 70 psec to 6.5 nsec, are due to the presence of multiple conformational states of the hemoglobin. Those conformations which favor energy transfer are presumably those showing the shortest decay times. However, the fractional intensities of these shorter-lived components was found to account for about 50% of the total emission, with the remainder displaying decay times of 1.5 to 6.5 nsec. Hence, the fractional intensity of each component is substantial, and there can be little doubt that these components are present in the emission. However, it is not yet clear whether these components are due to native heme proteins, as suggested by Szabo, or due to other non-heme proteins present at a level near 1% of the total protein.

We used an alternative method to study the time-dependent fluorescence from highly purified (HPLC) preparations of hemoglobin. Recently, the technique of frequency-domain fluorometry was extended to 2 GHz, which

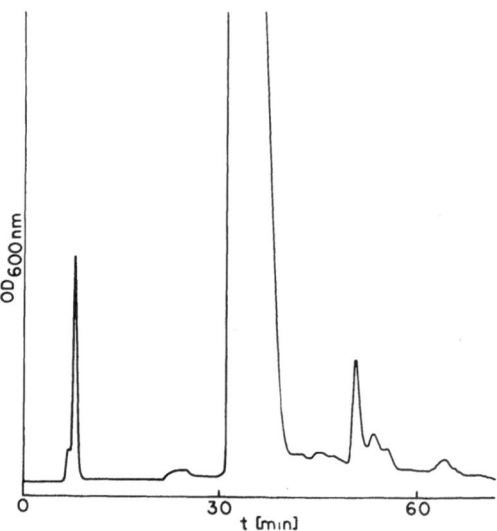

Figure 1. HPLC elution profile for hemoglobin on a DEAE PW5 column (Waters). The gradient was obtained with a 0.015 M Tris Buffer at pH 8.0 and a 0.015 M Tris buffer at pH 7.7 in 0.2 M Na acetate.

allows lifetimes and correlation times near 20 psec to be accurately measured. This instrument (Lakowicz et al., 1986) obtains the modulated light from the harmonic content of a train of psec pulses from a dye laser. We used this instrument to explore the lifetimes of emission of highly purified hemoglobin solutions. We also measured the correlation times associated with the detectable lifetimes, which were not previously reported. The data suggest that only the lifetime detectable in the psec region belong to the hemoglobin system.

MATERIALS AND METHODS

Human hemoglobin was prepared from washed red cells obtained from fresh blood samples donated by the local blood bank. They were hemolyzed in 0.07 M phosphate buffer at pH 7.0 and the stroma were eliminated by filtration through a 0.45 μ pellicon cassette. The protein was concentrated by ultrafiltration through a 1000- MW pellicon cassette, dialyzed against water, recycled through a mixed bed resin cartridge for removing polyphosphates and other ions, and stored at -90°C.

The protein was purified as oxyhemoglobin by HPLC chromatography through a DEAE PW5 preparative column (Waters) using a gradient formed by 0.015 M Tris-acetate at pH 8, and 0.015 M Tris-acetate at pH 7.7 in 0.2 M Na acetate. The procedure lasted less than 1 h at room temperature. Figure 1 shows the chromatographic profile.

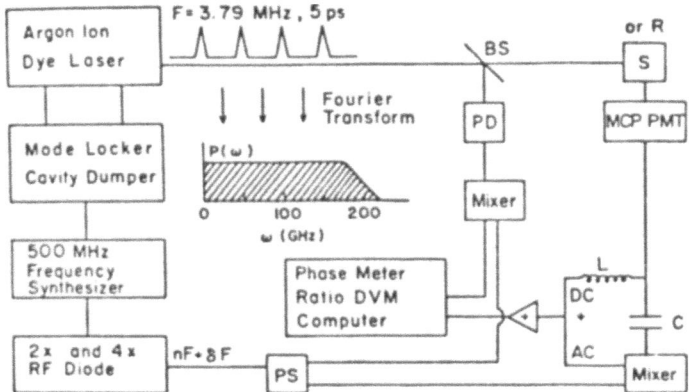

Figure 2. Diagram of the 2 GHz instrument frequency-domain instrumentation.

Deoxygenation of the hemoglobin samples was obtained by diluting the protein with nitrogen-bubbled buffers, in a closed cuvette with no liquid-gas interface. Na$^+$ dithionite was added with a syringe through a serum stopper, to a final concentration of 0.01 mg/ml. Spectrophotometric analyses indicated the complete deoxygenation of the samples.

All measurements were conducted at a protein concentration near 0.2 mg/ml in 0.01 M phosphate buffer at pH 7.0 at 4°C. Protein concentration was measured spectrophotometrically using E = 0.868 cm^2 mg^{-1} for the carbonmonoxy derivative at 540 nm. Optical densities were measured using a Cary 14 spectrofluorometer.

Lifetimes and correlation times were measured using a fluorometer operating between 4 to 1000 MHz (Lakowicz et al., 1986). The modulated excitation was provided by the harmonic content of a laser pulse train with a repetition rate of 3.76 MHz and a pulse width of 5 psec, from a synchronously pumped and cavity dumped dye laser. The emitted signals were observed with a microchannel photomultiplier, and the cross-correlation detection was performed outside the PMT. A schematic of the instrument is shown in Figure 2. Analyses of the data for estimating lifetimes and correlation times were conducted using the theory and program described elsewhere (Lakowicz and Maliwal, 1985; Lakowicz et al., 1985; Maliwal and Lakowicz, 1986).

Steady state measurements were performed with a SLM 8000 photon counting spectrofluorometer.

RESULTS

Figure 3 shows the emission spectra, upon excitation at 295 nm, of solutions of oxyhemoglobin before and after insertion of a cut-off and a broad band filters in the emission path. The filters were used for eliminating contribution from the Raman scattering and for suppressing the emission above 400 nm, which showed an increasing intensity with wavelength. On this basis all subsequent measurements were performed using the two filters in the optical path and observing the resultant emission centered at 365 nm.

In an attempt to detect the presence of heme-free hemoglobin, samples of oxyhemoglobin were titrated with hemin chloride, while monitoring the

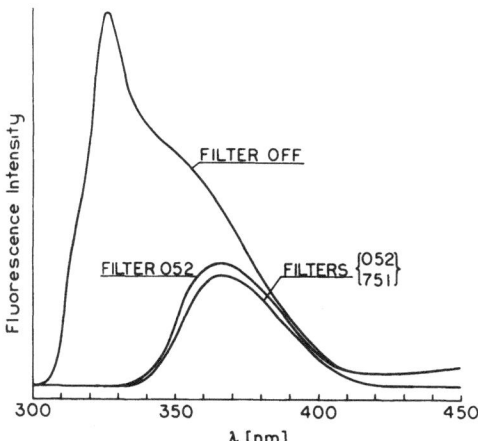

Figure 3. Fluorescence emission spectrum of oxyhemoglobin solutions in a 1 cm path cuvette, square geometry. The protein concentration was 0.2 mg/ml in 0.01 M phosphate buffer at pH 7.0, at 4°C. Excitation was at 295 nm. Where indicated Corning Filters were used in the emission path. The filter 0-52 is a cut-off at 340 nm, the filter 7-51 selects a broad band between 340 to 410 nm approximately.

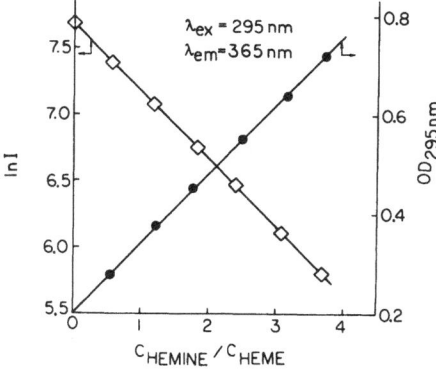

Figure 4. EFFECT OF HEMIN. Effect of increasing concentrations of hemin chloride on the emission intensity of oxyhemoglobin solutions, excitation at 295 nm, emission at 365 nm. Both filters were used in the emission path. Left, logarithmic transformation of the emission intensity in the arbitrary units; Right, total optical density of the samples at 295 nm. Other conditions as in Figure 3.

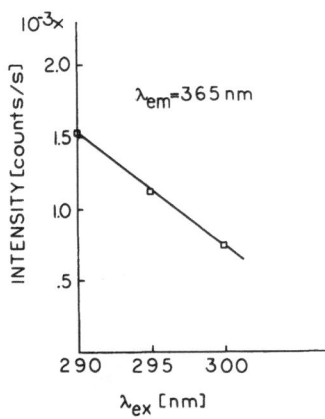

Figure 5. Excitation wavelength dependence of the emission intensity at 365 nm of oxyhemoglobin samples. Conditions as in Figure 3.

emission at 365 nm. If heme-free hemoglobin (apohemoglobin) were responsible for the emission then one expects a dramatic decrease in intensity upon addition of hemin chloride. Figure 4 shows that the decreasing emission followed a simple exponential function, as expected for the increasing optical density of the solutions. The absence of a dramatic quenching effect argues against the presence of apohemoglobin as a significant component of the emission.

The weak signals detected in the steady state measurements prevented a full investigation of the excitation and polarization spectra of our samples. Figure 5 shows that by increasing the excitation wavelength from 295 to 300 nm a decrease of emission intensity was observed, consistent with the absorption characteristics of tryptophan. As expected for tryptophan emission, the steady state anisotropy of the solutions increased with increasing wavelength.

Based on the previously published results (Szabo et al., 1984; Albani et al., 1985) we expected these samples to display decay times near 80 psec and 6 nsec. Hence, we reasoned that addition of a collisional quencher would selectively quench the longer-lived emission, and thereby increase the fraction of the emission due to the short-lived component. If this residual emission was due to tryptophan its anisotropy could not exceed about 0.32 (Lakowicz, 1983). If the residual emission were due to scattered light the anisotropy could be larger than this value. Figure 6 shows that increasing concentrations of acrylamide quenched the emission

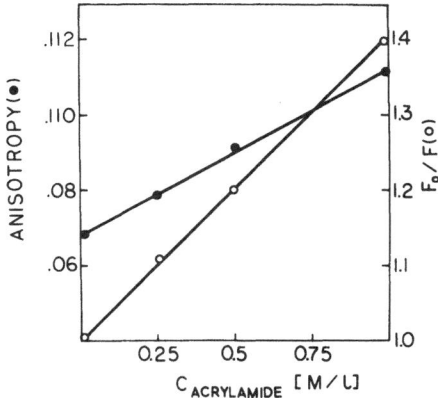

Figure 6. Static anisotropy of the emission at 365 nm of oxyhemoglobin in the presence of increasing quantities of acrylamide. o) Quenching of fluorescence, •), static anisotropy. Excitation at 295 nm, emission at 365 nm. Other conditions as in Figure 3.

intensity of the protein, and increased its anisotropy, without revealing the large anisotropies expected from scattered light. These experiments indicated that the emission signals were in major part due to fluorescence response of tryptophan fluorophores. It is surprising that the anisotropies are as low as 0.1, which may be the result of excitation at 295 nm, where the fundamental anisotropy of tryptophan is low (Lakowicz, 1983), or due to motional freedom of the tryptophan residues. It should be noted that the motions are only likely to have a significant effect on the longer-lived components of the emission.

Figures 7 and 8 show the data and the best-fit curves for the phase shift and demodulation ratios of oxy and deoxy hemoglobin samples respectively. Table 1 shows the values of lifetimes and intensities recovered from the least squares analysis. Visual inspection of the data and the χR^2 values indicate very good statistics for the three component fits. The standard deviations show the absence of overlap of the values of the various parameters. Analyses performed using only one or two lifetime components gave significantly larger values of χR^2.

Figures 9 and 10 show the data and the respective analyses for the time dependent anisotropies of oxy and deoxy hemoglobin, respectively. The parameters from the analyses are listed in Table 2. In both cases 2 correlation times gave the best fits, introduction of a third decay component did not produce better statistics. The fit for deoxyhemoglobin was not satisfactory, and introduction of additional components did not improve

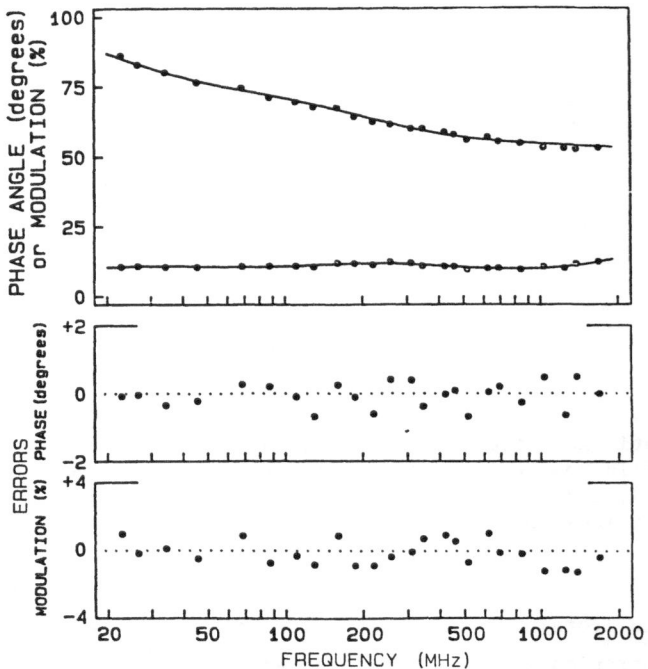

Figure 7. Frequency-resolved intensity decay of oxyhemoglobin at 4°C. Excitation at 300 nm, emission 365 nm. Other conditions as in Figure 3.

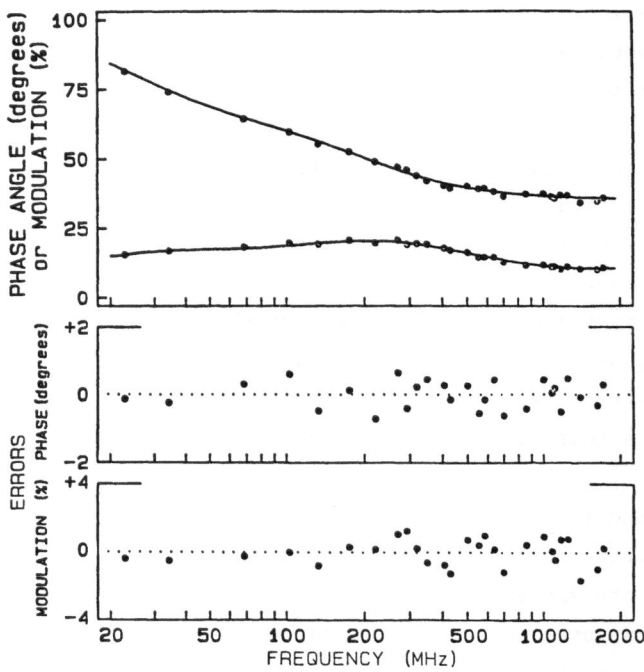

Figure 8. Frequency-resolved emission of deoxyhemoglobin. Same conditions as in Figure 7.

Table 1. Decay Times of the Intrinsic Tryptophan Emission from Hemoglobin

Sample	τ_i(ns) (sd)	α_i	f_i	χR^2
Oxy Hb	0.016 (.001)	.992	.536	
	0.820 (.028)	.007	.190	
	8.306 (.240)	.001	.274	2.88
Deoxy Hb	0.009 (.001)	.991	.349	
	0.901 (.025)	.008	.226	
	7.479 (.270)	.001	.385	3.50

Table 2. Correlation Times for the Intrinsic Tryptophan Emission from Hemoglobin.

Sample	Φ_i(ns) (sd)	$r_0 g_i$	χR^2
Oxy Hb	0.053 (.005)	0.352	
	4.595 (1.70)	0.029	2.02
Deoxy Hb	0.060 (.007)	0.298	
	19.456 (19.0)	0.044	6.85

the simulations, also introducing negative amplitudes in the estimated parameters. Moreover, the sum of the initial anisotropies of the components listed in Table 2 gave an initial total anisotropy near 0.38, distinctly higher than the frozen anisotropy of N-acetyl-tryptophanamide which was measured to be 0.30 in 100% glycerol at 5°C. This suggests that the model used for the decay law was incorrect. The model assumed that all of the detectable correlation times were equally distributed over all of the lifetimes components of the system.

DISCUSSION

The lifetimes detectable in the time dependent emission were well separated from each other, allowing a very good resolution of the parameters. It should be stressed that the parameters did not overlap even with a range of two standard deviations (95% confidence limits).

One may question the origin of the fluorescent signals. Due to the strong overlap of the emission spectrum of tryptophan with that of the absorption of the heme, the distance for half transfer of energy between

the two chromophores is large, approaching 60-70 Å. The average distance
of the tryptophans of hemoglobin from the heme is 30 Å. As the energy
transfer is inversely proportional to the 6th power of the distance, a
quenching of 98-99% is expected.

There are three different tryptophans in hemoglobin, 1 in the α and 2
in the β subunits, which are expected not to have identical emission
characteristics. It seems likely that the lifetime in the psec region
originated from the tryptophans quenched by the heme. As these tryptophans
display decayed times near 20 psec, it was no longer possible to resolve
their individual lifetimes.

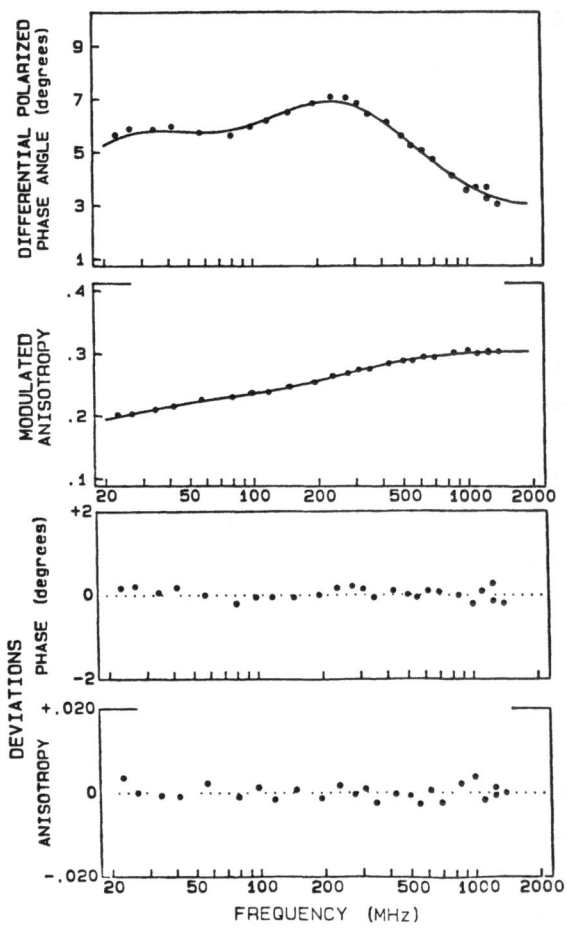

Figure 9. Frequency-resolved time-dependent anisotropy of oxyhemoglobin
samples. Conditions as in Figure 7. The solid line shows the best fit to
the data using two correlation times. The lower panels show the deviation
between the data and this model.

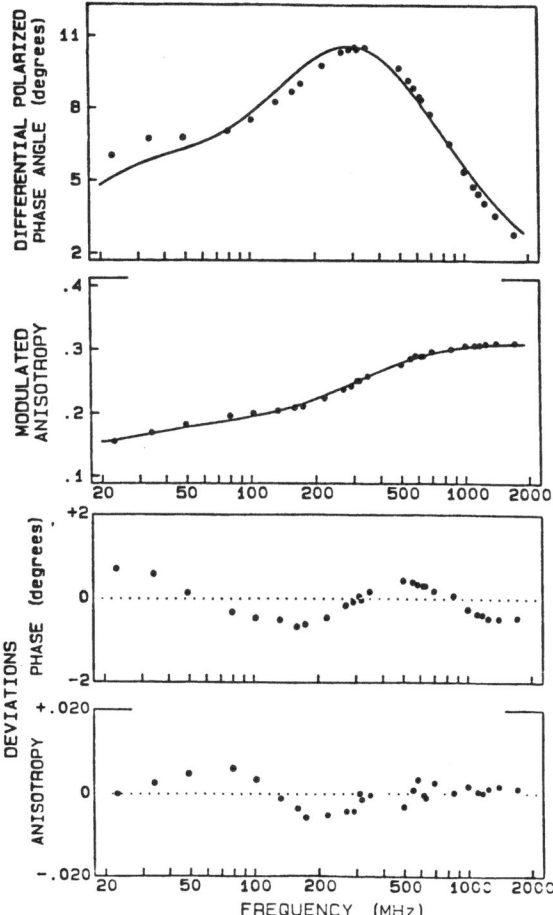

Figure 10. Frequency-resolved time-dependent anisotropy of deoxyhemo-globin samples. Conditions as in Figure 7. The solid line shows the best fit to the data using two correlation times. The lower panels show the deviation between the data and this model.

Lifetimes in the nsec region are expected from non-quenched proteins (Grinvald and Steinberg, 1976; Ross et al., 1981). Their amplitudes were very small, consistent with the presence of impurities which did not exceed 0.01% of the total protein, proving the high degree of purification obtained with HPLC techniques. These small amplitudes were also consistent with some residual tryptophan fluorescence, due to rare, very improbable orientations of the residues, so escaping the quenching (Szabo et al., 1984; Albani et al., 1985). The two hypotheses are not mutually exclusive.

At present, the first proposition (presence of impurities) seems to be more appropriate. This hypothesis is supported by the observation that going from oxy to deoxy hemoglobin the only lifetimes which changed were those in the psec region. Also, the modification of the correlation times upon removal of oxygen probably reflected an increasing amount of impurities, due to the additional manipulations necessary for the deoxygenation. Moreover the poor fits of time dependent anisotropy data to even a three correlation time model suggests that the chosen model for the decay law was inappropriate.

Our anisotropy models assume that all of the correlation times of the system were equally distributed over all of the detectable lifetimes. This was not necessarily true. For example, the sample may contain different populations of tryptophan residues, each of which display a characteristic intensity decay and a characteristic anisotropy decay. Such systems have been called associative by Brand and co-workers (Knutson et al., 1986; Davenport et al., 1986), and are known to result in unusual frequency-domain anisotropy data (Lakowicz, unpublished observations). It seems probable that our inability to fit the data for deoxy hemoglobin even with three decay times could be the result of our present inability to account for association between the intensity and anisotropy decay components. However, this is speculative, and our argument is not presently supported by quantitative analyses. Additionally, the 50-60 psec correlation times should be regarded or tentative, because these were obtained using a model which may not be appropriate for the system under study.

FOOTNOTES

This work was supported in part by the following grants: PHS-HL-13164, PHS-HLB-33629 (EB) and NSF DMB-8511065 and DMB-8502835 (JRL). Computer time and facilities were supported by the computer network of the University of Maryland at College Park and its branch at UMAB. Also used were the computer system and computer software developed in the laboratory of Dr. J. Lakowicz.

Abbreviations Used: Tris = Tris(hydroxymethyl)aminomethane; PMT = photomultiplier; Hb = hemoglobin

REFERENCES

Albani, J., Alpert, B, Krajcarski, D.T., and Szabo, A.G., 1985, A
  Fluorescence Decay Time Study of Tryptophan in Isolated Hemoglobin
  Subunits, FEBS Letters, 182:302.

Alpert, B., Jameson, D.M., and Weber, G., 1980, Tryptophan Emission From
  Human Hemoglobin and its Isolated Subunits, Photochem. Photobiol.,
  31:1.

Davenport, L., Knutson, J.R., and Brand, L., 1986, Anisotropy Decay
  Associated Fluorescence Spectra and Analysis of Rotational Hetero-
  geneity. 2. 1,6-Diphenyl-1,3,5-Hexatriene in Lipid Bilayers,
  Biochemistry, 25:1811.

Grinvald, A., and Steinberg, I.Z., 1976, The Fluorescence Decay of Tryptophan Residues in Native and Denatured Proteins, Biochim. Biophys. Acta,
  427:663.

Hirsch, R. E., Zukin, R. S., and Nagel, R. L., 1980, Intrinsic Fluorescence
  Emission of Intact Oxy Hemoglobins, Biochem. Biophys. Res. Comm.,
  93:532.

Hirsch, R. E., and Nagel, R. L., 1981, Conformational Studies of Hemoglobins Using Intrinsic Fluorescence Measurements, J. Biol. Chem.,
  256:1080.

Hochstrasser, R.M., and Negus, 1984, Picosecond Fluorescence Decay of
  Tryptophans in Myoglobin, Proc. Natl. Acad. Sci., 81:4399.

Knutson, J.R., Davenport, L., and Brand, L., 1986, Anisotropy Decay
  Associated Fluorescence Spectra and Analysis of Rotational Heterogeneity. 1. Theory and Applications, Biochemistry, 25:1805.

Lakowicz, J. R., 1983, Protein Fluorescence, in: Principles of Fluorescence
  Spectroscopy, Plenum Press, NY.

Lakowicz, J.R., Cherek, H., Maliwal, B., and Gratton, E., 1985, Time-
  Resolved Fluorescence Anisotropies of Diphenylhexatriene and Perylene
  in Solvents and Lipid Bilayers Obtained from Multifrequency Phase-
  Modulation Fluorometry, Biochemistry, 24:376.

Lakowicz, J.R., and Maliwal, B.P., 1985, A 2 GHz Frequency Domain
  Fluorometer, Biophys. Chem., 21:61.

Lakowicz, J. R., Laczko G., and Gryczynski, I., 1986, Construction and
  Performance of a Variable-Frequency Phase-Modulation Fluorometer, Rev.
  Sci. Instrum., 57:2499.

Maliwal, B.P., and Lakowicz, J.R., 1986, Resolution of Complex Anisotropy
  Decays by Variable Frequency Phase-Modulation Fluorometry: A Stimulation Study, Biochem. Biophys. Acta, 873:161.

Ross, J.A.B., Rousslang, K.W., and Brand, L., 1981, Time-Resolved

Fluorescence and Anisotropy Decay of the Tryptophan in Adrenocortico-
tropin-(1-24), <u>Biochemistry</u>, 20:4361.

Szabo, A. G., D. Krajcarski, M., and Zuker, M., 1984, Conformational
Heterogeneity in Hemoglobin as Determined by Picosecond Fluorescence
Decay Measurements of the Tryptophan Residues, <u>Chem. Phys. Lett.</u>,
108:145.

Weber, G., and Teale, F.J.W., 1959, Electronic Energy Transfer in Haem
Proteins, <u>Disc. Faraday Soc.</u>, 27:134.

TIME-RESOLVED FLUORESCENCE STUDIES ON FLAVINS

Antonie Visser

Department of Biochemistry
Agricultural University
Wageningen, The Netherlands

INTRODUCTION

Flavins are ubiquitous in nature. They mediate in electron transfer
and oxidation reactions in both bacterial, mammalian and plant systems (for
a review see Müller, 1983). The natural cofactors are flavin adenine
dinucleotide (FAD) and flavin mononucleotide (FMN), which occur in protein-
bound form (Figure 1). Both FMN and FAD are biosynthesized from the
precursor riboflavin (vitamin B2). Flavins are redox carriers and can
exist in oxidized, one-electron reduced (radical) or two-electron reduced
state. The characteristic yellow color of flavins arises from the oxidized
state absorbing blue light from the visible spectrum. Excited state
properties of flavins have recently been reviewed (Visser, 1987).

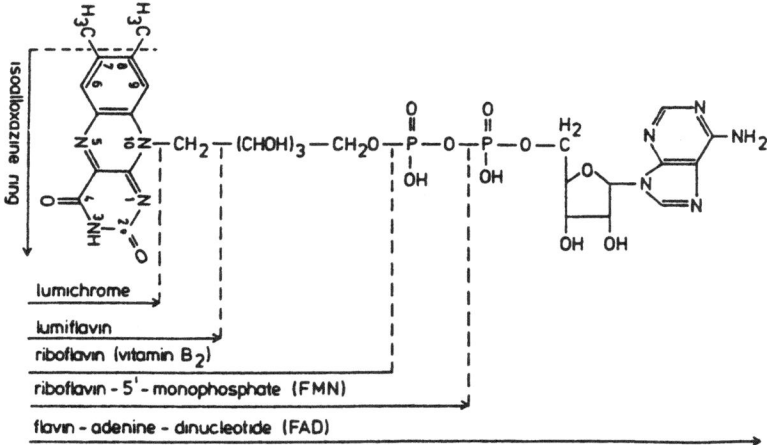

Figure 1. Structure of natural flavins (FAD, FMN and riboflavin).

Riboflavin and FMN in aqueous solution fluoresce in the green spectral region with rather high quantum yield (Q = 0.26, Weber and Teale, 1957). FAD fluorescence is much weaker (Q = 0.03, Spencer and Weber, 1972). As already noted in 1952 the majority of flavoproteins is characterized by strongly quenched flavin fluorescence (Weber, 1952). Some exceptions are lipoamide dehydrogenase and transhydrogenase (Visser et al., 1974, Veeger et al., 1976). The fact that flavin fluorescence in flavoproteins is often rather weak or apparently absent severely limits its potential use as reporter group for dynamic and structural information.

Understanding the appearance of flavin fluorescence at a molecular level has been the merit of Gregorio Weber (Weber, 1948 and 1950). Quenching of riboflavin fluorescence by purines was shown to involve a static mechanism in which the isoalloxazine forms a ground-state complex with the quencher (Weber, 1950 and 1966). The mean life of the complex could be determined and was found to be within an order of magnitude of the lifetime of the unquenched fluorescence. The quenching of the isoalloxazine fluorescence in FAD was ascribed to intramolecular complexation with adenine. Conversely, quenching of riboflavin fluorescence by potassium iodide or oxygen is caused by a collisional mechanism (Weber, 1950; Lakowicz and Weber, 1973). The lifetime of the (unquenched) excited state of the flavin was indirectly evaluated from quenching curves or by means of the degree of fluorescence polarization and amounted to about 10 ns (Weber, 1950).

The first direct, accurate determination of the lifetime of flavin fluorescence came from phase fluorometry yielding 4.7 ns for FMN in aqueous solution at room temperature (Spencer and Weber, 1969). The average lifetime of FAD fluorescence is shorter than that of FMN and together with static fluorescence measurements detailed dynamic and thermodynamic information on the intramolecular complex was obtained (Spencer and Weber, 1972). The effect of pressure upon the stability of complexes between flavin and aromatic residues has been studied with fluorescence methods (Weber et al., 1974; Visser et al., 1977a). The complexes are formed with a decrease in volume in the order of -4.0 to -5.0 ml/mol, but significantly smaller (-1.8 ml/mol) when the approach of the two aromatic rings is restricted because of a short linkage (Visser et al., 1977a).

Both aromatic residues like tryptophan, tyrosine and adenine and sulfur-containing compounds (Falk and McCormick, 1976) are very efficient quenchers of flavin fluorescence. The mechanism of quenching can be understood as follows. One out of four nitrogen atoms of isoalloxazine is

unique, since it bears less charge than a normal pyridine nitrogen
(Platenkamp et al., 1987). This atomic position of the isoalloxazine is
liable to accept electrons or undergo nucleophilic attack. In the excited
singlet state there is no distinct geometry change (Platenkamp et al.,
1980) nor change in solvation (Weber et al., 1974). Fluorescence quenching
by electron-rich molecules through a partial electron transfer mechanism
seems very well possible. Such a mechanism has been supported by
picosecond absorption spectroscopy where a transient charge transfer state
involving laser-excited flavin and indole could be detected on picosecond
timescale (Karen et al., 1983). The flavin fluorescence quenching in
flavoproteins can be attributed to the same mechanism. The presence of
aromatic and sulfur-containing amino acids in the flavin binding region has
been demonstrated in various flavoproteins by x-ray crystallography.

Owing to the availability of picosecond and high frequency CW lasers
and very sensitive detection systems considerable progress has been made
with fluorescence dynamics of flavins and flavoproteins. Several examples
will be presented in a section dealing with applications, which is preceded
by a section on methodological aspects.

METHODOLOGY

Instrumentation

A mode-locked Argon ion laser with available wavelengths in the blue,
green and ultraviolet spectral region (458, 473, 477, 496, 502 and 514 nm;
351 and 364 nm) constitutes a very appropriate light source for flavin
fluorescence because the excitation is right within the first absorption
band of oxidized (blue) or fully reduced (ultraviolet) flavins (see e.g.
Ghisla et al., 1974). Details of the mode-locking and extra-cavity modulation (to reduce the pulse repetition rate from 76 MHz to 600 or 300 kHz)
have been presented elsewhere (Visser and van Hoek, 1979; van Hoek and
Visser, 1981 and 1985). The pulse width is about 100 ps FWHM (full width
at half maximum). In principle pulse compression with at a least a factor
of ten is possible by coupling the laser light into an optical fiber (see
Fleming, 1986, for a survey). The experimental set-up for the polarized
fluorescence decay measurements has been described in previous publications
(Visser et al., 1985; van Hoek and Visser, 1985). The exciting laser light
is 100% vertically polarized. In the detection light path a sheet of
polaroid (HN38) was placed to separate the parallel and perpendicular

components of fluorescent intensity. The compartment can accommodate a
regular thermostated cell holder, a liquid helium cryostat or an oven for
experiments in the vapour phase (Eweg et al., 1980). Either interference
filters or two Jarrell-Ash 0.25 m monochromators in tandem can be put in
the detection light path for isolation of the fluorescence. The detection
method is that of time-correlated single photon counting (O'Connor and
Phillips, 1984). The mode-locked Argon ion laser pulse width as measured
by the detection system amounted to about 500 ps. The broadening from 100
to 500 ps is due to the limited time response of the photomultiplier used
(Philips PM2254B). It is expected to reduce this figure at least by
several factors with a microchannel plate detector (Bebelaar, 1986).
Photon counts were stored in selected memory subgroups of a multichannel
analyzer (ND66, Nuclear Data). Every 10 seconds (controlled by the micro-
processor of the multichannel analyzer) the polarizer was rotated 90° from
its original position in 0.3 seconds and the two polarization components
were stored in two different subgroups (1024 channels each) of the multichannel analyzer. The multichannel analyzer is linked to a MicroVAX II
computer for data analysis.

Data Analysis

The parallel, $I_{\shortparallel}(t)$, and the perpendicular $I_{\perp}(t)$, polarized components
of the fluorescence are measured. From these components, the total
fluorescence can be calculated:

$$S(t) = I_{\shortparallel}(t) + 2I_{\perp}(t) \qquad (1)$$

S(t) is a convolution of the actual fluorescence, s(t), with the impulse
response function, P(t):

$$S(t) - P(t) * s(t) = \int_{o}^{t} P(t') \, s(t-t')dt' \qquad (2)$$

The function s(t) can often be represented by a sum of exponential terms:

$$s(t) = \sum_{i} \alpha_i \exp(-t/\tau_i) \qquad (3)$$

The τ_i's are the fluorescence lifetimes and the α_i's are the corresponding
amplitudes. The parameters are α_i and τ_i and they have to be determined
from S(t). The most frequently used method is that of least square fitting.

322

In this procedure the weighted sum of squared residues, WSSR, is minimized:

$$\text{WSSR} = \sum_{k=1}^{n} w_k [S(t_k) - S_c(t_k)]^2 \tag{4}$$

$S_c(k)$ is the calculated fit of $S(k)$ (k refers to the kth time channel of the multichannel analyzer). The weighting factors are given by the inverse of the variance of $S(k)$ (Phillips et al., 1985; Vos et al., 1987). The weighting factor can be approximated by the reciprocal of the number of counts in the particular time channel. Application of proper weighting factors both for polarized intensity components and background emission is very important in the fitting procedure (Vos et al., 1987). A measurement of the impulse response function P(t) using a scatterer will lead to erroneous results because of the wavelength dependence of the response function. A number of authors has proposed the use of a reference compound to obtain the impulse response profile (Wahl et al., 1974; Wijnaendts van Resandt et al., 1982; Zuker et al., 1985; Löfroth, 1985). With the fluorescence response $S_r(t)$ of a reference compound having a single fluorescence lifetime τ_r, the decay parameters can be obtained from:

$$S(t) = \sum_i \alpha_i S_r(t) + S_r(t) * \{\sum_i \alpha_i (\frac{1}{\tau_r} - \frac{1}{\tau_i})\exp(-t/\tau_i)\} \tag{5}$$

In the weighting factors for the residues both the poissonian noise in S(t) and $S_r(t)$ must be accounted for. Details are given by Vos et al. (1987). Minimization of the weighted residuals can be performed with τ_r fixed, if an accurate value is known for it, or with τ_r as an extra iteration variable. A limitation of the method is that τ_r may not be equal to one of the lifetime components τ_i because the term $1/\tau_r - 1/\tau_i$ of equation 5 would cancel.

The anisotropy r(t) of the fluorescence is defined as follows:

$$r(t) = \frac{i_{\shortparallel}(t) - i_{\perp}(t)}{i_{\shortparallel}(t) + 2i_{\perp}(t)} \tag{6}$$

where $i_{\shortparallel}$ and $i_{\perp}$ are the undistorted intensity components to be distinguished from $I_{\shortparallel}$ and $I_{\perp}$. Like s(t), r(t) can be analyzed as a sum of exponentials:

$$r(t) = \sum_j \beta_j \exp(-t/\Phi_j) \tag{7}$$

where Φ_j are the rotational correlation times. Experimentally one obtains $I_{,,}(t)$ and $I_{\perp}(t)$. $I_{,,}(t)$ and $I_{\perp}(t)$ can be combined to give the following convolution products:

$$I_{,,}(t) = P(t) * \{\tfrac{2}{3} r(t)\, s(t) + \tfrac{1}{3}\, s(t)\} \tag{8a}$$

$$I_{\perp}(t) = P(t) * \{\tfrac{1}{3}\, s(t) - \tfrac{1}{3}\, r(t)\, s(t)\} \tag{8b}$$

As was shown by Gilbert (1983) and Cross and Fleming (1984) the latter equations can be simultaneously analyzed to yield the optimized parameters β_j and Φ_j of $r(t)$. The advantage of this procedure is that the appropriate weighting factors for the residuals can be used. When a reference compound is used the expressions for $I_{,,}(t)$ and $I_{\perp}$ are (Löfroth, 1985; Vos et al., 1987):

$$I_{,,}(t) = \tfrac{1}{3}(\sum_i \alpha_i + 2\sum_{ij}\alpha_i\beta_j)S_r(t) + \tfrac{1}{3} S_r(t) * \sum_i\{\alpha_i(\tfrac{1}{\tau_r} - \tfrac{1}{\tau_i})\exp(-t/\tau_i)\} +$$

$$\tfrac{2}{3}S_r(t) * \sum_{ij}\{\alpha_i\beta_j(\tfrac{1}{\tau_r} - \tfrac{1}{\tau_i} - \tfrac{1}{\Phi_j})\exp(-t/\tau_i - t/\Phi_j)\} \tag{9a}$$

$$I_{\perp}(t) = \tfrac{1}{3}(\sum_i \alpha_i - \sum_{ij}\alpha_i\beta_j)S_r(t) + \tfrac{1}{3} S_r(t) * \sum_i\{\alpha_i(\tfrac{1}{\tau_r} - \tfrac{1}{\tau_i})\exp(-t/\tau_i)\} +$$

$$\tfrac{1}{3}S_r(t) * \sum_{ij}\{\alpha_i\beta_j(\tfrac{1}{\tau_r} - \tfrac{1}{\tau_i} - \tfrac{1}{\Phi_j})\exp(-t/\tau_i - t/\Phi_j)\} \tag{9b}$$

The weighting factors for the residuals have been given by Vos et al., 1987.

In principle two rotational models can be distinguished, namely associative and non-associative rotation (Dale et al., 1977). In the associative case each lifetime component is associated with a particular correlation time. In practice it implies that some of the cross terms in

324

the s(t)r(t) product have to be omitted. In the non-associative model each lifetime component is associated with the same rotational component.

Two computer programs were developed for the analysis of polarized fluorescence decay data. In the first one S(t) can be fitted to a sum of exponentials to yield parameters α_i and τ_i. The fitting routine is provided by the subroutine ZXSSQ from the International Mathematical and Statistical Library (IMSL Inc., Houston). The second program uses α_i and τ_i as input parameters and simultaneously fits $I_{\shortparallel}(t)$ and $I_{\perp}(t)$ to yield optimized parameters β_j and Φ_j. The program also uses ZXSSQ. The quality of the fit is judged with visual inspection and with the help of various statistical functions and parameters, namely: the weighted residuals, autocorrelation of the weighted residuals, standard deviations, the number of zero passages in the autocorrelation function (Ameloot and Hendrickx, 1982), the reduced χ^2 value and the Durbin-Watson parameter (O'Connor and Phillips, 1984).

APPLICATIONS

Riboflavin (vitamin B₂)

In Figure 2 the fluorescence decay of riboflavin is presented. The excitation wavelength is 457.9 nm and the polarized emission is monitored at 531 nm. A very suitable reference compound at this fluorescence wavelength is erythrosin B. The fluorescence lifetime of erythrosin B is extremely short. F. Perrin, already more than 50 years ago, has reported a fluorescence lifetime of 84 ps from polarization of fluorescence (Perrin, 1929). In the analysis of the total fluorescence the lifetime of riboflavin and that of the reference compound have been fitted simultaneously. Riboflavin in water (pH 7.0) can be considered as a fluorescence standard yielding a value of 4.72 ns at 20°C. Erythrosin B in water pH 7.0 gave a lifetime of 60 ps.

The fluorescence anisotropy decay of riboflavin can be analyzed as well. The results are presented in Figure 2. It is clear that the anisotropy is created and decays within the pulse profile. Rigorous deconvolution is required to recover the rotational correlation time (see also Bootsma et al., 1985). The reproducibility of the correlation time turned out to be very good. The temperature dependence of the correlation time is proportional with the expected η/T (η = viscosity, T = temperature) as observed for the closely related lumazine molecule (Visser et al., 1985). The time

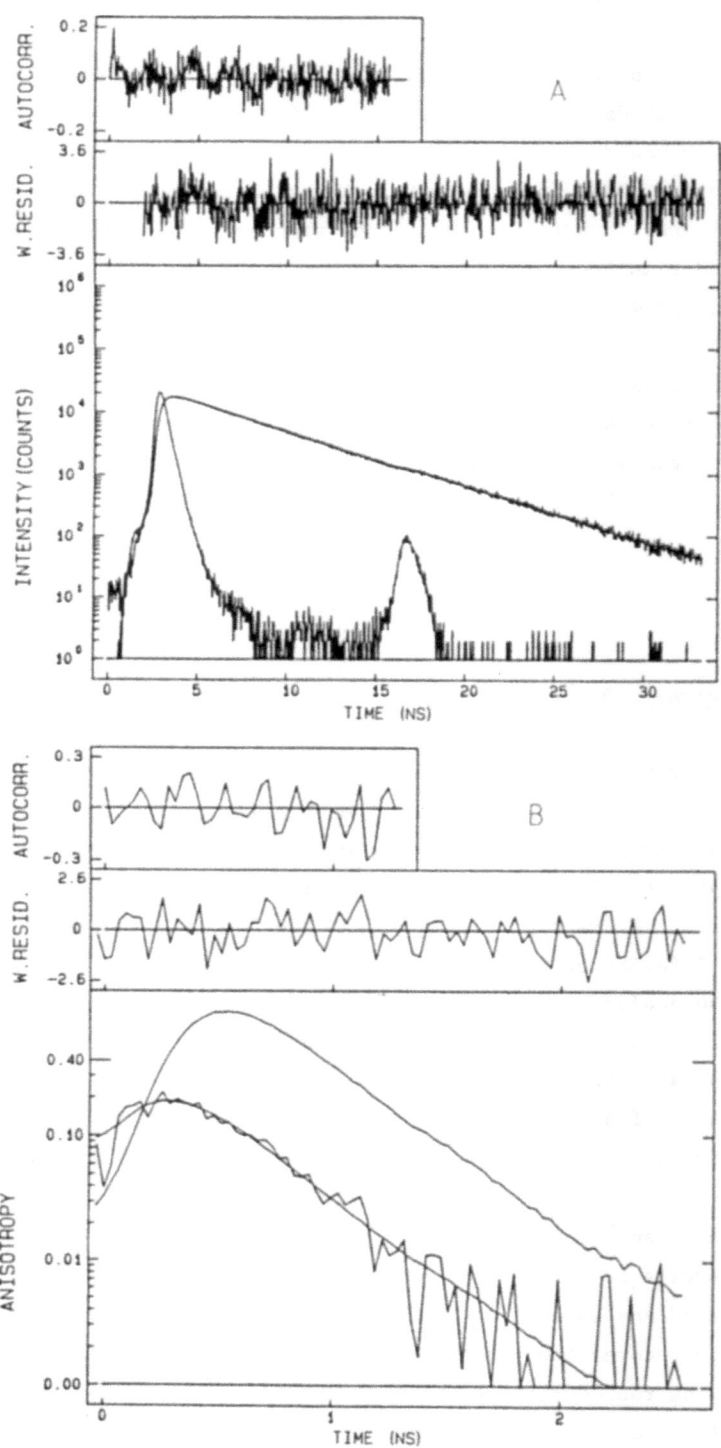

resolution of the experiment will improve considerably by the use of a microchannel plate detector (Rigler et al., 1985). Such a modification is currently being implemented in our equipment.

Flavin Adenine Dinucleotide (FAD)

The conformation of FAD in neutral aqueous solution is consistent with the concept of an equilibrium between open and closed forms (Spencer and Weber, 1972; Barrio et al., 1973). Both conformers are interconvertible in ground and excited states (Figure 3). The fluorescence decay of FAD has been shown to be nonexponential (Visser, 1984). The fluorescence describing the cyclic scheme can be modeled according to a biexponential decay law. The rate constant for folding in the excited state and the fluorescence lifetime of the intramolecular complex can be evaluated from the data. The intensity of the fluorescence is composed of the excited state concentrations of both forms and obeys the following relationship (Visser, 1984):

$$I(t) \sim (\frac{1}{\tau_2} - M)\exp(-t/\tau_1) + (M - \frac{1}{\tau_1})\exp(-t/\tau_2) \qquad (10a)$$

$$M = \frac{k_1 + k_2 \ K}{K + 1} \qquad (10b)$$

$$\frac{1}{\tau_1} \cdot \frac{1}{\tau_2} = k_1 k_2 + k_2 k_F + k_1 k_R \qquad (10c)$$

Figure 2. A. Fluorescence decay analysis of riboflavin (1.0 µM) in water pH 7.0 at 20 C with erythrosin B (10 µM) in water as reference compound. Wavelengths of excitation and emission were 458 and 531 nm, respectively. Shown are the fluorescence response curves of erythrosin B and riboflavin and the calculated riboflavin fluorescence response curves of erythrosin B and riboflavin fluorescence with a lifetime of 4.715 ± 0.004 ns. The quality of the fit is indicated by the weighted residuals and autocorrelation function on top of the curves. χ^2=1.16, Durbin-Watson parameter (Visser et al., 1985) DW=1.91 and number of zero passages in the autocorrelation function (Ameloot and Hendrickx, 1982) ZP=228 (1024 channels, time equivalence per channel = 0.032467). The reference lifetime was fitted to 60 ± 1 ps.

B. Anisotropy decay analysis of riboflavin. Analysis was carried out over the whole decay curve of parallel and perpendicular components (1024 channels each). Shown is a replot over a limited time range of the fitted and experimental anisotropy. The correlation time amounted to 128 ± 3 ps, the initial anisotropy was fixed to 0.38. The fluorescence response curve of erythrosin B is shown to illustrate the rise and decay of the anisotropy. χ^2=1.11, DW=1.96, ZP($I_{||}$)=241, ZP($I_\perp$)=259.

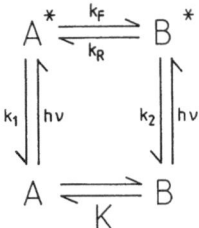

Figure 3. Scheme illustrating stacked (B) and unstacked (A) forms of FAD in ground and excited states.

$$\frac{1}{\tau_1} + \frac{1}{\tau_2} = k_1 + k_2 + k_F + k_R \qquad (10d)$$

The expression simplifies considerably if the rate constants of folding and unfolding are much smaller than the fluorescence rate constants as is probably the case for NADH.

Results of decay analysis of FAD (Visser, 1984) and NADH (Visser and van Hoek, 1981) fluorescences are collected in Table 1. The main conclusion from the results is that, owing to the short fluorescence lifetime of the closed form of FAD, excited state equilibrium is never attained. The fluorescence lifetime of FMN does not show up in the analysis. The lifetime of the complex can be calculated to be 13 ns from k_R and the equilibrium constant in the ground state (taken from Spencer and Weber, 1972). In case of NADH the fluorescence lifetimes of the two forms are so short that no interconversion in the excited state takes place. Therefore, the amplitudes are measures of the percentages stacked and unstacked according to the following relationship:

$$I(t) \sim \exp(-k_1 t) + K\exp(-k_2 t) \qquad (11)$$

Flavoproteins

Apart from riboflavin bound to lumazine apo-protein from _Photobacterium leiognathi_ (Kulinski et al., 1987) all flavoproteins hitherto studied exhibit heterogeneous fluorescence decay. Examples are D-amino acid oxidase (Nakashima et al., 1980; Yagi et al., 1983), lipoamide dehydrogenase (Visser et al., 1980; de Kok and Visser, 1984) and p-hydroxybenzoate hydroxylase (Visser et al., 1984a). The nature of the quenching has been

328

Table 1. Fluorescence Decay Parameters, Stacking Rate and Equilibrium Constants of FAD, FMN, NADH and NMNH at $^{\circ}C^{a)}$

Sample	α_1	τ_1 (ns)	α_2	τ_2 (ns)	k_F $(10^9 5^{-1})$	
FAD	0.72	2.82	0.28	0.31	4.0	5.1[b)]
FMN	1.00	4.70	-	-	-	-
NADH	0.82	0.25	0.18	0.69	-	0.22
NMNH	1.00	0.29	-	-	-	-

a) From Visser (1984) and Visser and van Hoek (1981)
b) From Spencer and Weber (1972)

discussed above. The analysis of the fluorescence decay of the flavin with a sum of exponentials is arbitrary. The flavin microenvironment in a flavoprotein is changing continuously because of fluctuations in structure. A distribution of lifetimes may be a better representation of the experimental fluorescence decay. Three examples of time-resolved fluorescence of flavoproteins will be shown. Two of them have FAD as prosthetic group, namely glutathione reductase and lipoamide dehydrogenase. The third example is the FMN-containing flavodoxin isolated from Desulfovibrio vulgaris.

Glutathione reductase (M_r = 100,000) catalyzes the reduction of oxidized glutathione by NADPH. Reduced glutathione has an important function in red blood cells to maintain cysteine residues in blood proteins like hemoglobin in the reduced state. The 3-dimensional structure of glutathione reductase is known (Thieme et al., 1981). The functional form of the protein is dimeric. Both subunits are involved in binding glutathione. Both FAD and NADPH are bound in an extended conformation. A disulfide bridge between two cysteines is in close contact with FAD. Electrons supplied by NADPH are transferred from reduced FAD to the disulfide bridge and then to oxidized glutathione. Because of the close vicinity of sulfur atoms the isoalloxazine fluorescence is expected to be strongly quenched. This is indeed observed. The static fluorescence is very weak and the fluorescence decay is dominated by a short component.

The fluorescence decay is complex. The results can only be fitted with four exponentials. In Table 2 the results of the analysis are

Table 2. Analysis of Fluorescence Decay of FAD in Glutathione Reductase From Human Erythrocytes.

α_1	τ_1 (ns)	α_2	τ_2 (ns)	α_3	τ_3 (ns)	α_4	τ_4 (ns)	χ^2	DW[a]	ZP[b]
1.00	1.32	-	-	-	-	-	-	242	0.009	0
0.93	0.26	0.07	3.20	-	-	-	-	21	0.096	0
0.930	0.084	0.058	1.06	0.012	4.43	-	-	2.3	0.97	63
0.947	0.049	0.039	0.67	0.011	2.35	0.003	5.38	1.3	1.89	205

a) Durbin-Watson parameter (see Visser et al., 1985)

b) Number of zero passages in the autocorrelation function (Ameloot and Hendrickx, 1982)

shown when the number of exponentials increases from one to four. The strong heterogeneity indicates a multitude of microenvironments of the flavin. Quenching groups like sulfhydryl residues diffuse towards the flavin and collide to deactivate the fluorescence. The diffusion is probably hindered because of spatial constraints imposed by the poly-peptide chain. A lifetime distribution may be a better representation, although the distribution is expected to be broad. The fluorescence anisotropy decay can be fitted to a single exponential (Figure 4). The model chosen was the non-associative one. The correlation time of 3.6 ns is too short to account for rotation of the whole protein. One would expect a rotational correlation in the order of 38 ns at 20°C on the basis of an empirical formula (Visser et al., 1983a):

$$\Phi = 3.84 \times 10^{-4} M_r \text{ (ns)} \tag{12}$$

for $\bar{v} = 0.735$ cm³/g, h = 0.2 cm³/g, $\eta = 1$ cP and T = 293 K.

The shorter correlation time reflects the flexibility of the flavin prosthetic group. Depolarization of the fluorescence owing to energy transfer from one FAD to the other within the dimeric protein can be excluded, since the anisotropy decays to zero at long times. When energy transfer among rigidly held chromophores takes place, the anisotropy decays to a constant anisotropy (cf. Visser et al., 1983b).

Lipoamide dehydrogenase (E3) is a constituent of the pyruvate dehydro-

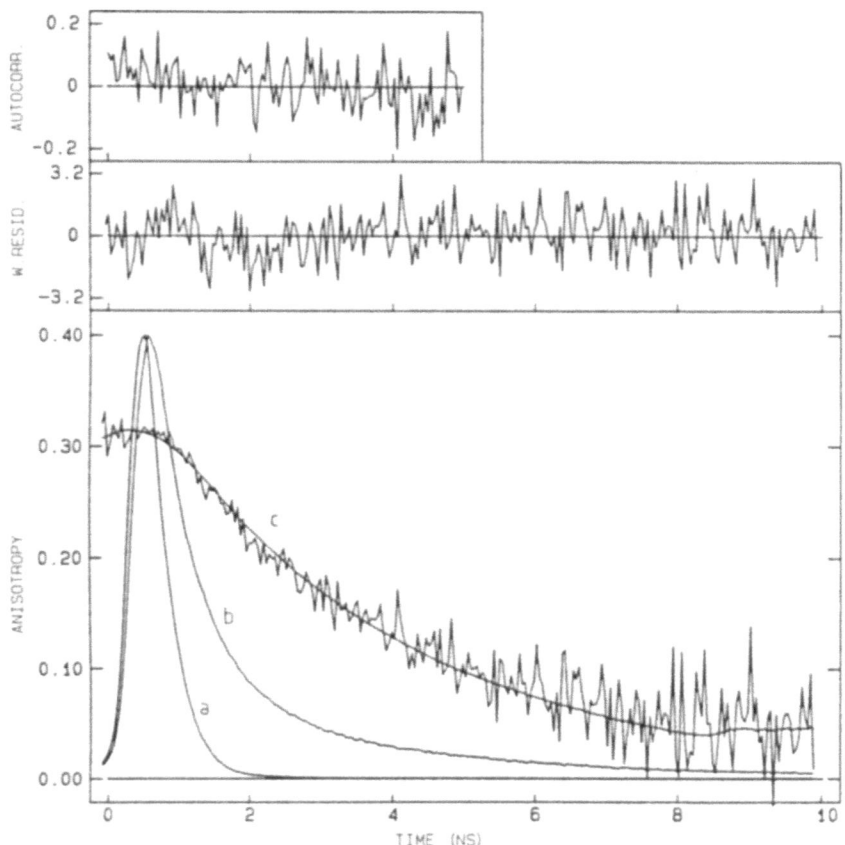

Figure 4. Fluorescence anisotropy decay (c) of FAD in glutathione reductase from human erythrocytes at 20 C. The reference (a) and fluorescence (b) responses are shown in the same figure. See Figure 2 for details of analysis. The correlation time $\Phi=3.55 \pm 0.05$ ns and the initial anisotropy is 0.322 ± 0.001.

genase multienzyme complex (Reed and Cox, 1970; Bosma et al., 1982). The other enzymes are pyruvate decarboxylase (E1) and lipoyl acetyltransferase (E2). The overall reaction of the complex is the oxidative decarboxylation of pyruvate and the formation of acetylcoenzyme A. The Escherichia coli multienzyme complex has been extensively studied. Lipoamide dehydrogenase catalyzes the oxidation of reduced lipoic acid residues attached to E2. Electron microscopy indicates that the complex is formed around a core of 24 E2 molecules in octahedral symmetry. The overall molecular weight of the complex is 4.2×10^6. Lipoamide dehydrogenase isolated from the complex is a dimeric flavoprotein ($M_r = 110,000$). Its active site is related to the one of glutathione reductase in the sense that a disulfide bridge is in the vicinity of the flavin. The lipoamide dehydrogenase from Azotobacter vinelandii has been crystallized (Schierbeek et al., 1983).

We have studied the dynamic fluorescence properties of E. coli and A. vinelandii lipoamide dehydrogenases in and out of the complex (Grande et al., 1980; de Kok and Visser, 1984 and 1987). The fluorescence decay of FAD in the E. coli enzyme is heterogeneous (Table 3). The average lifetime is distinctly longer than that of glutathione reductase, while the fluorescence quantum yield is larger. For an optimum fit a triple exponential function had to be used. The same arguments as for glutathione reductase hold for this flavoprotein. A distribution of lifetimes may be a better description than three lifetime components ranging from 0.2 to 3.5 ns. As in glutathione reductase the initial anisotropy decay can be interpreted more unambiguously (Figure 5). The anisotropy decays exponentially with a correlation time of 25 ns. This correlation time is too short to account for a spherical rotor with molecular weight 110,000 (42 ns at 20°C). A biexponential decay model consisting of a fixed correlation time of 42 ns (Φ_2) and a shorter (free floating) correlation time (Φ_1) has been tried with equally good results (collected in Table 3):

$$r(t) = \beta_1 \exp(-t/\Phi_1) + \beta_2 \exp(-t/\Phi_2) \tag{13}$$

Table 3. Fluorescence and anisotropy decay parameters of E. coli lipoamide dehydrogenase in and out of the pyruvate dehydrogenase multienzyme complex at 20°C.

Lifetimes

	α_1	τ_1 (ns)	α_2	τ_2 (ns)	α_3	τ_3 (ns)
free	0.34	0.17	0.27	1.69	0.39	3.48
complex	0.38	0.21	0.28	1.50	0.34	3.34

Correlation times

	β_1	Φ_1 (ns)	β_2	Φ_2 (ns)
free	0.31	25	-	-
	0.05	10	0.26	42
complex	0.31	36	-	-
	0.13	12	0.18	10^4

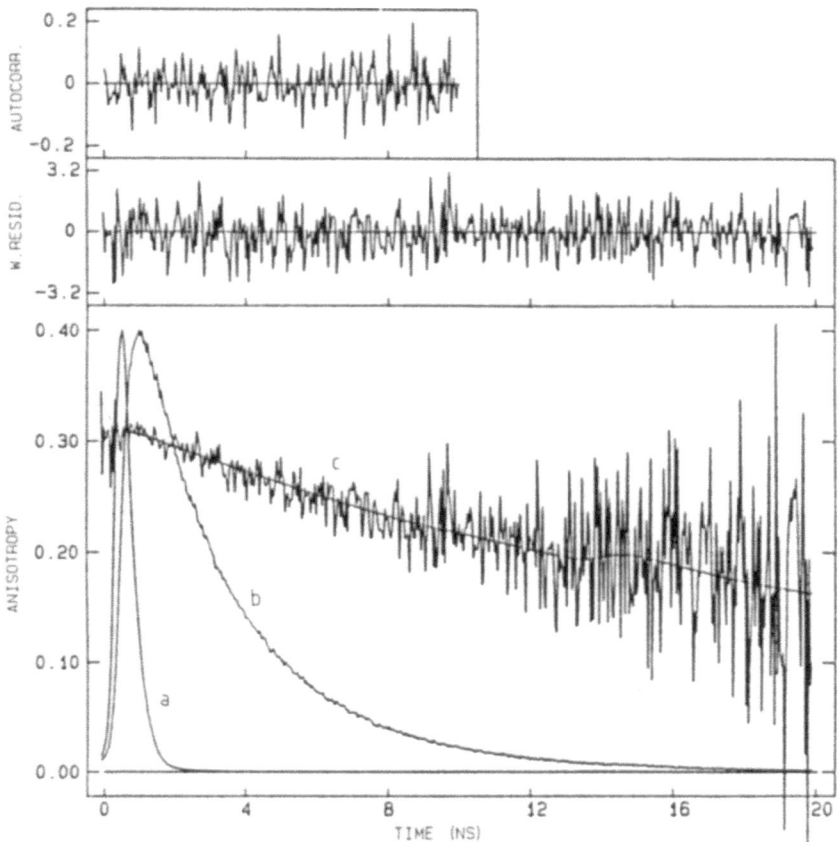

Figure 5. Fluorescence anisotropy decay of FAD (c) in E. coli lipoamide dehydrogenase at 20°C. The reference (a) and fluorescence (b) responses are shown in the same figure. See Figure 2 for details of analysis. The correlation time Φ-25.0 ± 0.5 ns and the initial anisotropy is 0.312 ± 0.001.

The correlation time $\Phi_1 = 10$ ns may be tentatively interpreted by a restricted subunit motion. The amplitude of this motion follows from the preexponential factor of the longer correlation time according to the following relation (Lipari and Szabo, 1980):

$$\beta_2/(\beta_1 + \beta_2) = \beta_2/r(o) = [\tfrac{1}{2}\cos\Theta_o(1 + \cos\Theta_o)]^2 \qquad (14)$$

From the data in Table 3 we found $\Theta_o = 20°$.

The fluorescence decay results for lipoamide dehydrogenase in the multienzynme complex are hardly different from that of the isolated enzyme. This indicates that the microstructure of the flavin is not drastically altered. A surprising result is the anisotropy decay of the enzyme within

333

the complex. The anisotropy decays during the time scale of the experiment (correlation time 30 ns) indicating flexibility of E3 within the complex. The complex itself rotates with a correlation time of about 10 ns (Visser et al., 1981). The initial anisotropy decay could be equally fitted to the following model:

$$r(t) = \beta_1 \exp(-t/\Phi_1) + r(\infty) \tag{15}$$

The results are summarized in Table 3. The angle of this motion can be estimated from $r(\infty)/r(0)$ (see equation 14 with $\beta_2 = r(\infty)$) giving 34°, which corresponds with a considerable displacement. The independent motion of E3 is probably functional in the catalysis of the multienzyme complex Grande et al., 1980). In addition to the spatial freedom of the lipoic acid residues the restricted motion of E3 may be used in efficient transfer of reducing equivalents.

Flavodoxins are relatively small (M_r = 16000 - 25000) electron transfer proteins containing a single, noncovalently bound FMN prosthetic group (Mayhew and Ludwig, 1975). Flavodoxins have been extensively studied (Simondson and Tollin, 1980). High resolution crystal structures of a few of them have been reported (for a survey see Mayhew and Ludwig, 1975). The fluorescence of FMN in flavodoxin is strongly quenched. It amounts to about 1 - 5% of the fluorescence of free FMN. The fluorescence is, however, not absent and can be analyzed with modern fluorescence equipment.

We have undertaken a time-resolved fluorescence study of the flavo-doxin from _Desulfovibrio vulgaris_, from which the 3-dimensional structure is known (Watenpaugh et al., 1973). The fluorescence decay is complex and consists of at least two lifetime components. It is important to note that the lifetime of free FMN does not show up in the analysis (Visser, 1987), which indicates that the residual fluorescence does not arise from dissoci-ated FMN.

We have reanalyzed the fluorescence decay data using free FMN as a reference compound (Visser et al., 1984b). The results are shown in Figure 6A. The analysis shows that a very short lifetime component has the most predominant contribution. This short lifetime component (in the order of 40 ps) can be ascribed to isoalloxazine fluorescence that is strongly quenched by tyrosine and tryptophan residues, which are within van der Waals contact of the flavin (Watenpaugh et al., 1973). The longer component must be assigned to a flavin which is flexibly linked to the protein. The

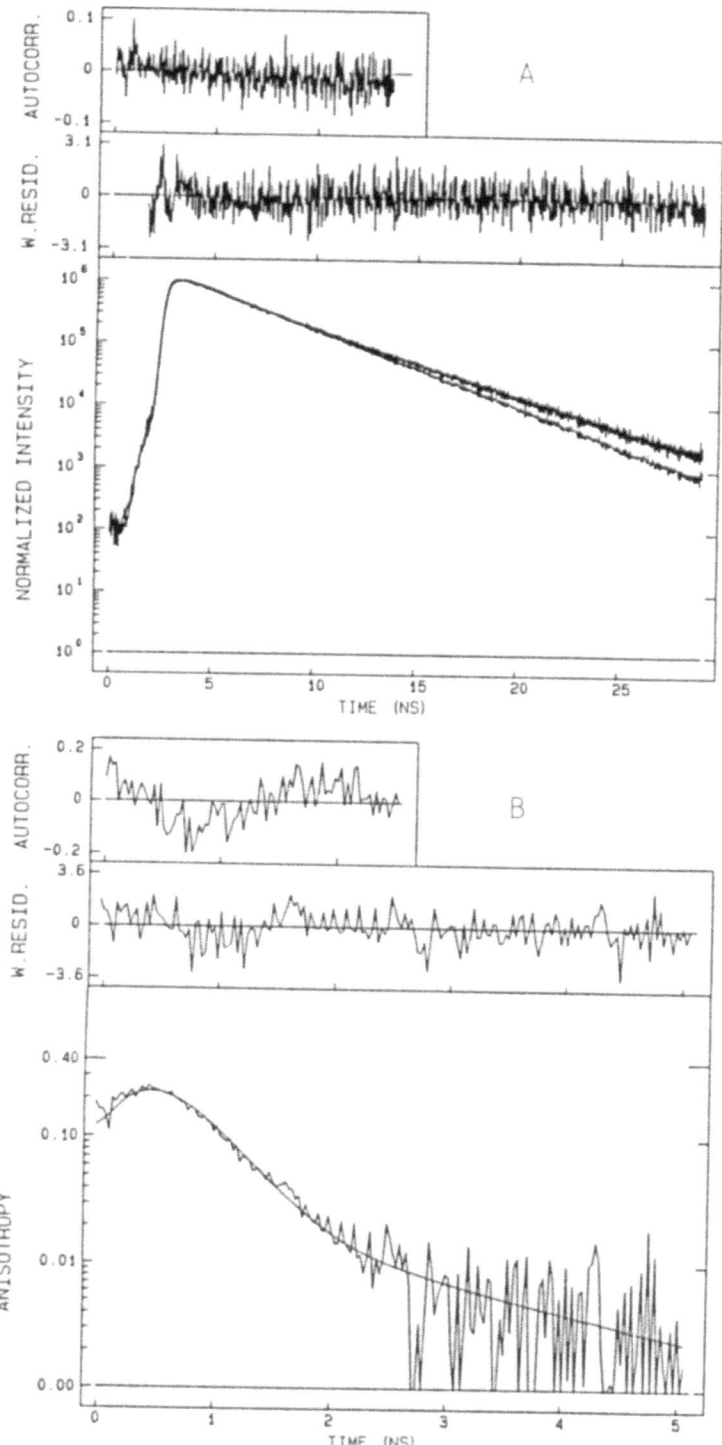

Figure 6. (Continued on next page)

latter conclusion is corroborated by fluorescence anisotropy decay (Figure 6B). NMR data provide strong support for tightly associated FMN in a compact conformation (Vervoort et al., 1986). The very short lifetime component must be associated with tightly bound FMN and it is at least 95% present. The relative intensity contributions can be estimated for the strongly quenched and moderately fluorescent components in flavodoxin by calculating the amplitude lifetime product (0.039 and 0.143, respectively). The latter data indicate that the steady state fluorescence arises mainly from the long living component.

The presence of a small amount of a flavodoxin conformation (about 3%) with appreciable fluorescence seems to be a general property of flavodoxins. Other flavodoxins show similar fluorescent behavior in which the fluorescence is depolarized (unpublished observations). The pressure-induced denaturation of flavodoxins (Visser et al., 1977b) may be explained with this equilibrium between two conformations. Upon increasing the pressure the equilibrium may be shifted to the less compact protein conformation after which the prosthetic group is released. A time-resolved fluorescence study of this process may provide more conclusive results to support this hypothesis.

ACKNOWLEDGEMENTS

My colleagues Arie van Hoek and Kees Vos are thanked for important contributions. Aart de Kok is acknowledged for providing results on glutathione reductase and lipoamide dehydrogenase. Discussions with Franz Müller are appreciated.

Figure 6. A. Fluorescence decay analysis of FMN in <u>Desulfovibrio vulgaris</u> flavodoxin at 20°C using FMN as reference compound $\tau_r=4.7$ ns). Shown are the fluorescence responses of FMN free and bound in flavodoxin and the calculated flavodoxin fluorescence based on a biexponential decay model with $\alpha_1=0.975$, $\tau_1=40$ ps, $\alpha_2=0.025$ and $\tau_2=5.7$ ns. See Figure 2 for details of analysis.

B. Anisotropy decay analysis showing rapid decay of FMN with a correlation time $\Phi=0.23$ ns ($\beta_1=0.38$) superimposed on a lower contribution ($\beta_2=0.02$) of a longer correlation time of about 2 ns.

REFERENCES

Ameloot, M., and Hendricks, H., 1982, Criteria for Model Evaluation in the Case of Deconvolution Calculations, J. Chem. Phys., 76:4419.

Barrio, J. R., Tolman, G. L., Leonard, N. J., Spencer, R. D., and Weber, G., 1973, Flavin 1,N^6-ethenoadenine Dinucleotide: Dynamic and Static Quenching of Fluorescence, Proc. Natl. Acad. Sci. USA, 70:941.

Bebelaar, D. , 1986, Time Response of Various Types of Photomultipliers and its Wavelength Dependence in Time-correlated Single Photon Counting with an Ultimate Resolution of 47 ps FWHM, Rev. Sci. Instrum., 57:1116.

Bootsma, J.P.C., Challa, G., Visser, A.J.W.G., and Müller, F., 1985, Polymer-bound Flavins: 5. Characterization by Static and Time-resolved Fluorescence Techniques, Polymer, 26:952.

Bosma, H., de Graaf-Hess, A.C., de Kok, A., Veeger, C., Visser, A.J.W.G., and Voordouw, G., 1982, Pyruvate Dehydrogenase Complex From Azotobacter Vinelandii: Structure, Function, and Inter-enzyme Catalysis, Ann. N.Y. Acad. Sci., 378:265.

Cross, A. J. and Fleming, G. R., 1984, Analysis of Time-resolved Fluorescence Anisotropy Decays, Biophys. J., 46:454.

Dale, R. E., Chen, L. A., and Brand, L., 1977, Rotational Relaxation of the "Microviscosity" Probe Diphenylhexatriene in Paraffin Oil and Egg Lecithin Vesicles, J. Biol. Chem., 252:7500.

de Kok, A., and Visser, A.J.W.G., 1984, Mobility of Lipoamide Dehydrogenase In and Out of the Pyruvate Dehydrogenase Complex From Azotobacter Vinelandii, in: Flavins and Flavoproteins, R. C. Bray, P. C. Engel, and S. G. Mayhew, eds., Walter De Gruyter, Berlin.

de Kok, A., and Visser, A.J.W.G., 1987, Flavin Binding Site Differences Between Lipoamide Dehydrogenase and Glutathione Reductase as Revealed by Static and Time-resolved Flavin Fluorescence (submitted).

Eweg, J. K., Müller, F., Bebelaar, D., and van Voorst, J.D.W., 1980, Spectral Properties of (Iso)alloxazines in the Vapour Phase, Photochem. Photobiol., 31:435.

Falk, M.C., and McCormick, D. B., 1976, Synthetic Flavinyl Peptides Related to the Active Site of Mitochondrial Monoamine Oxidase. II Fluorescence Properties, Biochemistry 15:646.

Fleming, G. R., 1986, Subpicosecond Spectroscopy, Ann. Rev. Phys. Chem., 37:81.

Ghisla, S., Massey, V., Lhoste, J-M., and Mayhew, S.G., 1974, Biochemistry Fluorescence and Optical Characteristics of Reduced Flavins and Flavoproteins, 13:589.

Gilbert, C. W., 1983, A Vector Method for the Non-linear Least Squares

Reconvolution-and- Fitting Analysis of Polarized Fluorescence Decay
Data, in: "Time-Resolved Fluorescence Spectroscopy", R. B. Cundall,
and R. E. Dale, eds., Plenum Press, New York.

Grande, H. J., Visser, A.J.W.G., and Veeger, C., 1980,Protein Mobility
Inside Pyruvate Dehydrogenase Complexes as Reflected by Laser Pulse
Fluorometry, Eur. J. Biochem. 106:361.

Karen, A., Ikeda, N., Mataga, N., and Tanaka, F., 1983, Picosecond Laser
Photolysis Studies of Fluorescence Quenching Mechanisms of Transfer
State Formation in Solutions and Flavoenzymes, Photochem. Photobiol.,
37:495.

Kulinski, T., Visser, A.J.W.G., O'Kane, D. J., and Lee, J., 1987,
Spectroscopic Investigations of the Single Tryptophan Residue and of
Riboflavin Bound to Lumazine Apoprotein from Photobacterium
Leiognathi, Biochemistry, 26:540.

Lakowicz, J. R., and Weber, G., 1973), Quenching of Fluorescence by Oxygen.
A Probe for Structural Fluctuations in Macromolecules, Biochemistry,
12:4161.

Lipari, G., and Szabo, A., 1980, Effect of Librational Motion on
Fluorescence Depolarization and Nuclear Magnetic Resonance Relaxation
in Macromolecules and Membranes, Biophys. J., 30:489.

Löfroth, J-E., 1985, Deconvolution of Single Photon Counting Data With a
Reference Method and Global Analysis, Eur. Biophys. J., 13:45.

Mayhew, S. G., and Ludwig, M. L., 1975, Flavodoxins and Electron-transferring
Flavoproteins, in: "The Enzymes", P. Boyer, ed., Vol. 12, Academic
Press, New York.

Müller, F., 1983, The Flavin Redox-system and its Biological Function, in:
"Topics in Current Chemistry", Vol. 108, F. L. Boschke, ed., Springer,
Berlin.

Nakashima, N., Yoshihara, K., Tanaka, F., and Yagi, K., 1980, Picosecond
Fluorescence Lifetime of the Coenzyme of D-amino Acid Oxidase, J.
Biol. Chem., 255:5261.

Nikolaus, B., and Grischkowsky, D., 1983, 90-fs Tunable Optical Pulses
Obtained by Two-Stage Pulse Compression, Appl. Phys. Lett., 43:3.

O'Connor, D. V., and Phillips, D., 1984, "Time-Correlated Single-Photon
Counting", Academic Press, London.

Perrin, F., 1929, La Fluorescence des Solutions. Induction Moléculaire. -
Polarisation et Durée d'émission. Photochimie. Ann. de Phys. 10e
série, t.XII:169.

Phillips, D., Drake, R. C., O'Connor, D. V., and Christensen, R. L., 1985,
Time-Correlated Single Photon Counting (TCSPC) Using Laser Excitation,
Anal. Instrum., 14:267.

Platenkamp, R. J., van Osnabrugge, H. D., and Visser, A.J.W.G., 1980, High-Resolution Fluorescence and Excitation Spectroscopy of N3-undecyllumi-flavin in n-decane, Chem. Phys. Lett., 72:104.

Platenkamp, R. J., Palmer, M. H., and Visser, A.J.W.G., 1987, Ab Initio Molecular Orbital Studies of Closed Shell Flavins, Eur. Biophys. J., in press.

Reed, L. J., and Cox, D. J., 1970, Multienzyme Complexes, in: "The Enzymes", P. Boyer, ed., Vol. 1, Academic Press, New York.

Rigler, R., Claesens, F., and Kristensen, O., 1985, Picosecond Fluorescence Spectroscopy in the Analysis of Structure and motion of Biopolymers, Anal. Instrum., 14:525.

Schierbeek, A. J., van der Laan, J. M., Groendijk, H., Wierenga, R. K., and Drenth, J., 1983, Crystallization and Preliminary X-ray Investigation of Lipoamide Dehydrogenase from Azotobacter Vinelandii, J. Mol. Biol., 165:563.

Simondson, R. P., and Tollin, G., 1980, Structure-function Relations in Flavodoxins, Mol. Cell. Biochem., 33:13.

Spencer, R. D., and Weber, G., 1969, Measurements of Subnanosecond Fluorescence Lifetimes with a Cross-correlation Phase Fluorometer, Ann. N.Y. Acad. Sci., 158:361.

Spencer, R. D., and Weber, G, 1972, Thermodynamic and Kinetics of the Intramolecular Complex in Flavin- Adenine Dinucleotide, in: "Structure and Function of Oxidation-Reduction Enzymes", Å. Åkeson, and A. Ehrenberg, eds., Pergamom, Oxford.

Thieme, R., Pai, E. F., Schirmer, R. H., and Schulz, G. E., 1981, Three-dimensional Structure of Glutathione Reductase at 2 Å Resolution, J. Mol. Biol., 152:763.

van Hoek, A., and Visser, A.J.W.G., 1981, Pulse Selection System With Electro-optic Modulators Applied to Mode-locked CW Lasers and Time-resolved Single Photon Counting, Rev. Sci. Instrum., 52:1199.

van Hoek, A., and Visser, A.J.W.G., 1985, Artefact and Distortion Sources in Time-correlated Single Photon Counting, Anal. Instrum., 14:359.

Veeger, C., Visser, A.J.W.G., Krul, J., Grande, H. J., de Abreu, R.A., and de Kok, A., 1976, Fluorescence Studies on Lipoamide Dehydrogenase, Pyruvate Dehydrogenase Complexes and Transhydrogenase, in: "Flavins and Flavoproteins", T. P. Singer, ed., Elsevier, Amsterdam.

Vervoort, J., Müller, F., Mayhew, S. G., van den Berg, W.A.M., Moonen, C.T.W. and Bacher, A., 1986, A Comparative Carbon-13, Nitrogen-15, and Phosphorous-31 Nuclear Magnetic Resonance Study on the Flavodoxins From Clostridium MP, Megasphaera elsdenii and Azotobacter vinelandii, Biochemistry, 25:6789.

Visser, A.J.W.G., Grande, H. J., Müller, F., and Veeger, C., 1974, Intrinsic Luminescence Studies on the Apoenzyme and Holoenzyme of Lipoamide Dehydrogenase, Eur. J. Biochem., 45:99.

Visser, A.J.W.G., Li, T. M., Drickamer, H. G. and Weber, G., 1977a, Volume Changes in the Formation of Internal Complexes of Flavinyl Tryptophan Peptides, Biochemistry, 16:4883.

Visser, A.J.W.G., Li, T. M., Drickamer, H. G. and Weber, G., 1977b, Effect of Pressure Upon the Fluorescence of Various Flavodoxins, Biochemistry, 16:4879.

Visser, A.J.W.G., and van Hoek, A., 1979, The Measurement of Subnanosecond Fluorescence Decay of Flavins Using Time-correlated Photon Counting and a Mode-locked Ar Ion Laser, J. Biochem. Biophys. Methods, 1:195.

Visser, A.J.W.G., Grande, H. J., and Veeger, C., 1980, Rapid Relaxation Processes in Pig Heart Lipoamide Dehydrogenase Revealed by Subnanosecond Resolved Fluorometry, Biophys. Chem., 12:35.

Visser, A.J.W.G., and van Hoek, A., 1981, The Fluorescence Decay of Reduced Nicotinamides in Aqueous Solution After Excitation With a UV Mode-locked Ar Ion Laser, Photochem. Photobiol., 33:35.

Visser, A.J.W.G., Scouten, W. H., and Lavalette, D., 1981, Rotational Diffusion of Eosin Labeled Pyruvate Dehydrogenase Complex of Escherichia coli, Eur. J. Biochem., 121:233.

Visser, A.J.W.G., Penners, N.H.G., and Müller, F., 1983a, Dynamic Aspects of Protein-Protein Association Revealed by Anisotropy Decay Measurements, in: "Mobility and Recognition in Cell Biology", H. Sund, and C. Veeger, eds., Walter De Gruyter, Berlin.

Visser, A.J.W.G., Santema, J. S., and van Hoek, A., 1983b, Energy Transfer in Biflavinyl Compounds as Studied with Fluorescence Depolarization, Photobiochem. Photobiophys., 6:47.

Visser, A.J.W.G., 1984, Kinetics of Stacking Interactions in Flavin Adenine Dinucleotide from Time-Resolved Flavin Fluorescence, Photochem. Photobiol., 40:703.

Visser, A.J.W.G., Penners, N.H.G., van Berkel, W.J.H., and Müller, F., 1984a, Rapid Relaxation Processes in P-hydroxybenzoate Hydroxylase From Pseudomonas Fluorescens revealed by Subnanosecond-Resolved Laser-Induced Fluorescence, Eur. J. Biochem., 143:189.

Visser, A.J.W.G., Santema, J. S., and van Hoek, A., 1984b, Spectroscopic and Dynamic Characterization of FMN in Reversed Micelles Entrapped Water Pools, Photochem. Photobiol., 39:11.

Visser, A.J.W.G., Ykema, T., van Hoek, A., O'Kane, D. J., and Lee, J., 1985, Determination of Rotational Correlation Times From Deconvoluted Fluorescence Anisotropy Decay Curves. Demonstration With 6,7-dimethyl-

8-ribityllumazine and lumazine protein from <u>Photobacterium leiognathi</u>
as Fluorescent Indicators, <u>Biochemistry</u>, 24:1489.

Visser, A.J.W.G, 1987, Excited States of Flavins, <u>in:</u> "Excited State Probes
in Biochemistry and Biology", A. G. Szabo, and L. Masotti, eds.,
Plenum, New York, in press.

Vos, K., van Hoek, A., and Visser, A.J.W.G., (1987), Application of a
Reference Convolution Method to Tryptophan Fluorescence in Proteins.
A Refined Description of Rotational Dynamics, <u>Eur. J. Biochem.</u>, in
press.

Wahl, P. H., Auchet, J-C., and Donzel, B., 1974, The Wavelength Dependence
of the Response of a Pulse Fluorometer Using the Single Photoelectron
Counting Method, <u>Rev. Sci. Instrum.</u>, 45:38.

Watenpaugh, K. D., Sieker, L. C., and Jensen, L. H., 1973, The Binding of
Riboflavin-5-Phosphate in a Flavoprotein: Flavodoxin at 2.0 Å Resolution,
<u>Proc. Natl. Acad. Sci. USA</u>, 70:3857.

Weber, G., 1948, The Quenching of Fluorescence in Liquids by Complex
Formation. Determination of the Mean Life of the Complex, <u>Trans.
Faraday Soc.</u>, 44:185.

Weber, G., 1950, Fluorescence of Riboflavin and Flavin-Adenine Dinucleotide,
<u>Biochem. J.</u>, 47:114.

Weber, G., 1952, Rotational Brownian Motion and Polarization of the
Fluorescence in Solutions, <u>Adv. Protein Chem.</u>, 8:415.

Weber. G., and Teale, F.W.J., 1957, Determination of the Absolute Quantum
Yield of Fluorescent Solution, <u>Trans. Faraday Soc.</u>, 53:646.

Weber, G., 1966, Intramolecular Complexes of Flavins, <u>in:</u> "Flavins and
Flavoproteins", E. C. Slater, ed., Elsevier, Amsterdam.

Weber, G., Tanaka, F., Okamoto, B. Y., and Drickamer, H. G., 1974, The
Effect of Pressure on the Molecular Complex of Isoalloxazine and
Adenine, <u>Proc. Natl.Acad. Sci. USA</u>, 71:1264.

Wyaendts van Resandt, R. W., Vogel, R. H., and Provencher, S. W., 1982,
Double Beam Fluorescence Lifetime Spectrometer with Subnanosecond
Resolution: Application to Aqueous Tryptophan, <u>Rev. Sci.Instrum.</u>,
53:1392.

Yagi, K., Tanaka, F., Nakashima, N., and Yoshihara, K., 1983, Picosecond
Laser Fluorometry of FAD of D-amino Acid Oxidase-Benzoate Complex, <u>J.
Biol. Chem.</u>, 258:3799.

Zuker, M., Szabo, A. G., Bramall, L., Krajcarski, D. T., and Selinger,
B., 1985, Delta Function Convolution Method (DFCM) for Fluorescence
Decay Experiments, <u>Rev. Sci. Instrum.</u>, 56:14.

FINAL WORDS AT BOCCA DI MAGRA

Gregorio Weber

Biochemistry Department
University of Illinois
1209 W. California St.
Urbana, IL   61801   USA

As I thought some time ago about what I was going to tell you, I
realized that I would be obliged to say about myself more than what I
should or what you may wish to hear. I must then ask you to forgive me if
I reminisce a little about my initiation in fluorescence which was probably
quite close to the initiation of the uses of fluorescence in Biochemistry.
I went to Cambridge from the Argentine in 1943 with a British Council
Fellowship and spent my first six months at the laboratory of Eric Riddeal
learning Surface Chemistry. I convinced myself that it had no future in
Biochemistry. You can see how wrong I was; I could have preceded relevant
work on membranes by a long period but I really missed my chance. And it
was not for lack of hints: At that time, Jim Danieli, in Cambridge itself,
and Dawson were telling everybody that cell membranes were essentially
lipid bilayers but nobody listened to them for quite a time. I was one of
the non-listeners. I then decided that I would be better off to work in
Biochemistry with Malcolm Dixon and went to talk to him about doing some
work applying Physical Chemistry to Biochemistry. In those days Physical
Chemistry was beginning to appear as important to Biochemistry, in relation
to a subject that is still very much alive: the Physical Chemistry of
Proteins. Since then we have made some progress in that area but as you
know, it has not been spectacular. Malcolm Dixon suggested that I investigate the fluorescence of flavins and flavoproteins. I had not heard much
about fluorescence and nothing about flavins but I soon learnt that there
were a number of low molecular weight flavin compounds that differed
greatly in fluorescence intensity, like riboflavin and FAD. A few flavoproteins had been "purified" but only one of them showed fluorescence

343

comparable to the free prosthetic group. I was given the task of sorting all that out. At Cambridge, the first thing that I discovered about fluorescence was that it had greatly interested, even fascinated, the physicists first and then the biologists; as to the chemists, they were quite innocent of it. However, they did make a great contribution: they synthesized fluorescein. In their investigations of the fluorescence of solutions the physicists utilized fluorescein and occasionally ventured into observations on solutions of one or two similar xanthydrols like rhodamine or pyronine. The reason for this limitation was that the only excitation sources available were the Tungsten filament - good for sub-stances with absorption in the visible region like fluorescein - and the carbon arc. If you go today to a military museum of World War II and you find one of the searchlights then used you will see that it had two carbon electrodes separated by a small gap. The current that jumped the gap volatilized the carbon producing a continuous spectrum with much visible and a little ultraviolet light. The Hydrogen arc and the Xenon arc came into use only after the war. The photoelectric measuring devices were primitive and just barely able to detect the strongest fluorescence emis-sions in the visible region. The eye was much better and one had to decide what the fluorescence was like by means of visual comparisons. You may judge this method as imprecise and primitive but it has its applications: I have seen often my students putting samples in the spectrofluorimeter and turning knobs and finding that nothing happens. Remembering my early training I would show them how a visual observation on simple excitation with a mercury arc was sufficient to decide that, to begin with, there was no fluorescence in the sample. Looking at fluorescence in this simple way, before you do some complicated measurement is something that I highly recommend. It is just as important as remembering to plug in the electri-cal measuring instrument, an equally common student failure. It is inter-esting to realize today that the basic rules for exciting, measuring and interpreting the data of fluorescence have existed since the 1920's and only the technology necessary for exact experimentation was missing. Wavilov had demonstrated that the quantum yield of fluorescence of fluorescein solutions could not be very different from unity. Gaviola, another native of Argentina, worked in Germany and made the first direct measurements of fluorescence lifetimes by the original phase fluorometer. In 1924-26 when he developed it he did not have at hand the famous equation $\tan \Phi = \omega\tau$, in other words, tangent of the phase shift angle equals circular modulation frequency of the excitation times lifetime of the excited state. That came ten years afterwards, I believe, with Duschinsky. What Gaviola did was to delay the exciting light through a longer reflection path until

344

it would match the time of arrival to the eye of the fluorescence that it had previously excited. From the length of the path and the velocity of light, he derived the delay and therefore the time between excitation and emission. He got 4.5 ns for the lifetime of fluorescein. We cannot do much better today. Evidently the fluorescence from fluorescein has remained the same that it was in 1924. In the thirties Jablonski and Szymanowski, another Polish physicist who died during World War II, greatly improved the design of the phase fluorometer, but still it was done visually. However, they were able to demonstrate that the components of the fluorescence polarized parallel and perpendicular to the exciting light had different lifetimes. Perhaps the work that most influenced mine, and certainly the one that I liked best, was that of Francis Perrin. I remember that Malcolm Dixon came to me one day, handed me a little piece of paper, and said that somebody at King's College - I wish I could remember his name - had said that there was a paper on fluorescence that I should read. The little piece of paper had written on it: F. Perrin, J. de Physique, 1926. So I went to the Cambridge Library, which I positively thought of as a temple of learning and looks indeed like one, and I read the famous paper of Perrin on depolarization of the fluorescence by Brownian rotations, not one but many times. Argentine secondary education in the first half of the century included French language and literature so that I could not only understand the scientific content, but also enjoy the literary quality of the writing. It was written in that transparent, terse style of XVIII century France, which I have tried, perhaps unsuccessfully, to imitate from then onwards. The clarity of Perrin's thought and his ability to do the right experiment were really remarkable. No doubt he had inherited some of that, as well as the interest in the subject, from his father, Jean Perrin, who, as you know, determined Avogadro's number from the study of the diffusional rotation and translation of the pollen grains. In 1936 Francis Perrin extended the theory of rotational depolarization to molecules of irregular shape, and this is still one of the mainstays of our work in fluorescence. Generally once a year, and sometimes twice, I am called to referee a theoretical paper by a young man who has rediscovered again the theory of anisotropic rotation. All these are welcome, because there are many ways of arriving to the same place and it is good to know about them, but already Perrin had arrived to most of the points that they make. A contribution of Perrin that has been almost lost, and of which I was reminded yesterday when Bernard Valeur spoke, was his extraordinary paper of 1932 on electronic energy transfer between atoms or molecules [*Editors' note: Perrin, F., 1932, Théorie Quantique des Transferts D'Activation Entre Molécules de Même Espèce. Cas des Solutions Fluorescentes, Ann. de Phys.*,

10e *série, t. XVII:20*]. You may remember that London provided the explanation of what we now call the London-van der Waals forces as due to energy
exchange by electromagnetic induction between atomic oscillators. He had
made a passing remark that if one of the oscillators was in an excited
state the energy exchanged between the oscillators would be the whole of
the excitation energy. In the late nineteen twenties Gaviola and Pringsheim
observed the depolarization of the fluorescence from concentrated solutions
of fluorescein and correctly attributed it to electronic energy transfer.
Jean Perrin himself had done a few experiments with colored quenchers in
fluid solutions that demanded a similar explanation. Francis Perrin was
the first to propose a detailed explanation of the phenomenon: He treated
first the case of identical, motionless atomic oscillators, and calculated
that transfer of energy could occur at distances as large as one quarter of
the emitted wavelength. When it came to extending the results to molecules
in solution he realized the two important differences with his model: The
coupling between the oscillators cannot be maintained for long because of
the continuous perturbation from the solvent molecules and unlike the
atomic case the bands of absorption and emission in molecules overlap only
partially. With good intuition he assessed that from these two features
one would lose approximately about an order of magnitude in the probability
of transfer so that the critical distance for transfer between molecules
would be less than 100 Å instead of the 1,000 Å for atoms in vacuum. His
paper enumerates all the variables that enter into the famous Förster's
equation, but he never computed explicitly the overlap integral. In 1941,
Förster introduced this last and most important refinement, everything came
into place and a quantitative, and practically useful, treatment of
electronic energy transfer became available. Very few people have read
this paper of Perrin, which I consider one of the most important ever
published on the subject of fluorescence, and I have wished to remind you
of its existence. It was from reading Perrin's papers that I conceived
three ideas on the use of polarization: determination of the change in
fluorescence lifetime as one quenches the fluorescence by addition of an
appropriate chemical, determination of the molecular volume of proteins by
fluorescent conjugates with known dyes and determination of the viscosity
of a medium through the polarization of the emission from a known fluorescent probe. I was thus able to determine the formation of non-fluorescent
complexes between fluorophore and quenchers: they do not change the
fluorescence lifetime of the uncomplexed molecules and therefore they do
not change the fluorescence polarization either. Collisional quenchers on
the other hand change lifetime and yield in similar proportions. The best
example of quenching by complex formation that I found was that of

riboflavin fluorescence by purines and pyrimidines. The physicists had seen cases like those that I studied but a physicist always looks for a general explanation, not one that depends on the specific properties of the system with which he deals. Molecular complexes did not appear to attract them and they preferred to invent spheres of action and things like that to explain the phenomenon. As a biochemist I was well schooled in molecular complexes, Michaelis and Menten and all that, so it was not difficult for me to tumble upon the correct explanation. Even today I think that the most direct evidence for the existence of short-lived molecular complexes comes from fluorescence observations: The concentrations in the experiments are thermodynamically ideal and we can identify conditions in which a "sticky collision" merges with a definite molecular complex. Another important concept that would be orphaned without fluorescence observations to support it is the collisional rate in fluid solvents: You must have used many times the figure 10^{10} liters per mol per second. It is difficult to find a direct experimental proof for this figure outside fluorescence observations.

It would be wrong for me to say that I was attracted to the study of fluorescence by all the possibilities that I have enumerated and many more that became reality in the following years. More than anything else I was attracted by the counterpoint of the esthetic and scientific aspects: that a change in fluorescence color or intensity could be immediately related to a molecular event appeared to me then, as now, as really extraordinary. Now that people put their fluorescent solution into instruments and only look at the numbers that come out I wonder if they can develop a similar esthetic rapport with the subject. Most investigators marry their research subjects at the time of the Ph.D., and I have noticed that, just like in life, they remain attached to it in spite of occasional infidelities. I did the same.

This will do for my relations with fluorescence in the past and I will try now to make one or two comments about the present and future of fluorescence as it relates to its use in Chemistry and Biology. There are many levels of explanation in science and I was being made particularly aware of it a moment ago when I was listening to Dr. Califano. The aim of science is to reduce the complex phenomenon that you are observing to something simpler, but reductionism is just like an elevator that goes only one floor at a time. If you wish now to go to a floor higher or lower than that you must take another elevator and this functions on a completely different principle from the elevator that you have first taken. Even if you know well how to manipulate the first, you can hardly apply your knowledge to make the second one work. It follows that when one thinks of unresolved problems one is always pitching them at some, perhaps arbitrary, level. An important unresolved problem in molecular fluorescence is the relation of absorption and emission. The changes in fluorescence that follow changes in the molecular environment are of a considerable magnitude and we can propose for them reasonable explanations, at least in most cases. This arises because of the increased polarizability and often large charge separation characteristic of excited states. On the other hand, the changes in absorption are usually much smaller and arise from combinations of small influences none of which is truly dominant, the opposite of what happens with the fluorescence changes. A better knowledge of the effects of environment upon absorption would permit us to provide an explanation for the usually small changes that we see in proteins upon binding the substrates, changing temperature or pressure, etc. Another area that has hardly been touched is that of radiationless deactivation of the excited state. We know that there is a triplet state and that there is something called internal conversion that depends primarily upon some anharmonicity that permits the large electronic energy to be divided into smaller energy packets. But we simply cannot make a decent calculation of the quantum yield. The literature is full of examples of truly spectacular changes of fluorescence yield with solvent changes or with a very modest substitution in the fluorophore, and we are pretty well devoid of explanations. Even reasoning by analogy and a posteriori rationalizations are hard to come by in these cases. Passing on to proteins we are beginning to be able to correlate lifetimes, spectral distribution and rotational rates of the tryptophan residues. These studies will provide us with good estimates of the very short term dynamics - picoseconds to nanoseconds - of proteins. It is something of a paradox that we can do so well at these very short times. Who would have thought that we can determine what happens in a natural system in times of picoseconds, while being much less aware of the origin of molecular events in the microsecond and longer time ranges. And it is in these longer time ranges, milliseconds, in which the biologically important molecular changes connected with muscle contraction, brain communication, sense organ transduction take place. Now, what about thinking of devising new ways of using fluorescence in the investigation of long-term dynamics? I know that it will certainly take you more than five minutes during which you will think that in any subject the "pioneers" are those that solve the easy problems leaving the harder ones for future workers. To end on a more cheerful note, I will say that we now have all the necessary equipment to attack many complex biological problems by employing as probes the properties of the excited state that reveal themselves through their light emission. The consequence will be that in the years to come, the focus of fluorescence studies, as they apply in Biology, will be on the properties of systems as such rather than in the molecular interpretation of the observed fluorescence changes. It will be undesirable if these two aspects were to part company completely and it will be much better if those applying fluorescence to purely biological problems keep an eye open towards the elementary physical properties that can help in the description, and in the finer interpretation of the molecular events in the system.

It only remains for me to express how much I have enjoyed this unique occasion. I have learned a lot from your communications; you have done me much honor, and I am duly humble. Thank you.

ABSTRACTS FROM POSTER CONTRIBUTIONS

A NON-LINEAR LEAST-SQUARES APPROACH TO THE ISOTHERMAL PERRIN-WEBER PLOT IN TERMS OF ROTATIONAL CORRELATION TIMES

A. U. Acuña and Mª. P. Lillo

Instituto de Quimica Fisica, C.S.I.C.
119 Serrano
28006 Madrid, Spain

The study of the stationary fluorescence anisotropy, $\bar{r}$, of chromophores attached to biopolymers, as originally developed by Weber (1953), can be used to detect rotational and segmental motions in these molecules. However, when several motions appear simultaneously it is often impossible to quantify each of them, because the value of $\bar{r}$ at a given viscosity and temperature contains contributions from all the orientational processes active in these conditions (Weber, 1953). On the other hand, since the values of $\bar{r}$ can be measured easily to a very high accuracy with present day instrumentation, it would be helpful to have methods to interpret those measurements on the frequent occasions where several reorientation processes are likely to occur. Here we present a semi-quantitative approach to the isothermal Perrin-Weber plots that could yield at least an order of magnitude estimation of the rotational correlation times present in a particular system.

It is postulated first that all the fluorescence depolarizing motions in a labeled protein can be reduced to combinations of two archetypes: restricted and free rotations. Moreover, it is assumed that these motions are mutually independent and that their combined effect on the total time-dependent anisotropy can be expressed by the appropriate Soleillet products (Soleillet, 1929; Steiner and McAlister, 1957; Dale and Eisinger, 1975). Thus, for example, in the case of a free rotating protein (Figure 1) with a chromophore with two local restricted motions one can build a model by writing first a Soleillet product:

$$r(t) = r_o\left(\frac{3}{2} <\cos^2\theta_1, t> -\frac{1}{2}\right)\cdot\left(\frac{3}{2} <\cos^2\theta_2, t> -\frac{1}{2}\right)\left(\frac{3}{2} <\cos^2\theta_p, t> -\frac{1}{2}\right) \qquad [1]$$

where r_o is the anisotropy in the absence of any motion. The restricted motions are approximated by the wobbling in a cone expression (Kinosita, et al. 1977) and the free rotation by a single exponential term:

Restricted: $\left(\frac{3}{2} <\cos^2\Theta_i,t> -\frac{1}{2}\right) = A_{i_1} \exp(-t/\Phi_i) + A_{i_2}$ [2]

Free: $\left(\frac{3}{2} <\cos^2\Theta_p,t> -\frac{1}{2}\right) = \exp(-t/\Phi_p)$ [3]

where Φ's are rotational correlation times.

Each factor in Eq. [1] can be replaced by its value from Eqs. [2] and [3]. To obtain the final expression one should consider that these orientational motions must take place in very distant time ranges, to be really independent. In that case:

$$\frac{1}{\Phi_1} + \frac{1}{\Phi_p} \simeq \frac{1}{\Phi_1}; \quad \frac{1}{\Phi_2} + \frac{1}{\Phi_p} \simeq \frac{1}{\Phi_2}; \quad \frac{1}{\Phi_1} + \frac{1}{\Phi_2} + \frac{1}{\Phi_p} \simeq \frac{1}{\Phi_1}$$ [4]

Then,

$$r(t) = r_o [a_1\exp(-t/\Phi_1) + a_2\exp(-t/\Phi_2) + a_3\exp(-t/\Phi_p)]$$ [5]

where $\Sigma a_i = 1$.

The preexponentials a_i give an idea of the fraction of the initial anisotropy which decays through each motion. They might be replaced by residual anisotropies, cone amplitudes, etc., but this probably would be physically meaningless, considering the approximations involved.

The steady-state anisotropy is finally obtained by time-averaging r(t) for a chromophore with a single fluorescence lifetime τ_F

$$\bar{r} = \frac{\int_o^\infty r(t)\cdot\exp(-t/\tau_F)}{\int_o^\infty \exp(-t/\tau_F)}$$ [6]

Therefore,

$$\bar{r} = r_o \left[\frac{a_1}{1+\tau_F/\Phi_1} + \frac{a_2}{1+\tau_F/\Phi_2} + \frac{1-(a_1+a_2)}{1+\tau_F/\Phi_p}\right]$$ [7]

Within the hydrodynamical approximation the rotational correlation times can be expressed by $\Phi_i = f_i (\eta/T)$, where f_i is a geometrical factor

354

(size and shape) independent of viscosity (η) and temperature (T). Then:

$$\bar{r} = r_o \left[\frac{a_1}{1+(\tau_F/f_1)L} + \frac{a_2}{1+(\tau_F/f_2)L} + \frac{1-(a_1+a_2)}{1+(\tau_F/f_p)L} \right] \qquad [8]$$

where $L = \dfrac{T}{\eta}$.

In proteins f_i might not be independent of temperature and, in fact, Weber already mentioned (1953) the possibility of "thermally activated motions". These thermal effects would foil the simple interpretation presented here, therefore we applied the present methods only to the isothermal fluorescence depolarization.

Eq. [8], or any variant of it, is easily fitted to an experimental set of $\bar{r}$ values vs L by standard least-squares techniques. In Figure 2 we show an example taken from dansylated fibrinogen (Acuña, Gonzalez and Lillo, to be published), a protein that was immobile during the lifetime of the fluorophore. In that case, two restricted motions assigned to the dansyl label were used to interpret the experimental depolarization.

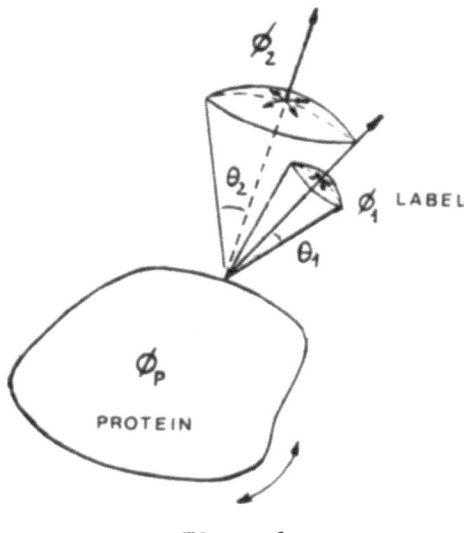

Figure 1

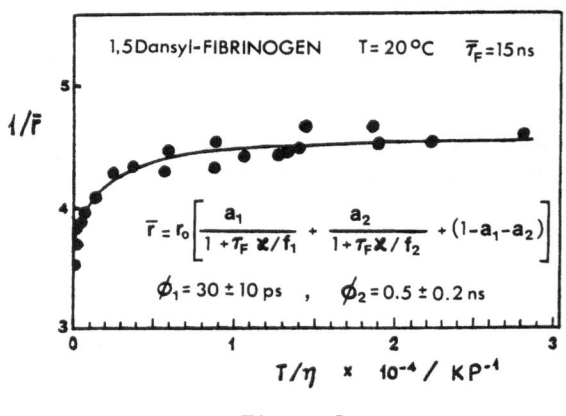

Figure 2

ACKNOWLEDGMENTS

This work was supported by Grant No. 1197/81 from the Comision Asesora de Investigacio Cientifica y Tecnica.

REFERENCES

Dale, R. E, and Eisinger, J., 1975, in: "Biochemical Fluorescence", R. F. Chen and H. Edelhoch, eds, M. Dekker, NY.

Kinosita, K., Kawato, S., Ikegami, A., 1977, Biophys. J., 20:289.

Soleillet, P., 1929, Ann. Phys., 12:23.

Steiner, R. F., and McAlister, A., 1957, J. Polym. Sci., 24:105.

Weber, G., 1953, Adv. Prot. Chem., 8:415.

FLUORESCENCE STUDIES ON MEMBRANES FROM HUMAN SKIN FIBROBLASTS DEFICIENT IN PLASMALOGENS

A. Hermetter, B. Rainer, G. Schwabe, A. Roscher and F. Paltauf

Department of Biochemistry
Technical University of Graz, Austria

Plasmalogens (1-O-alkenyl-2-acyl-glycerophospholipids) are important constituents of most animal cell membranes. In artificial membranes, they differ from the "common" diacyl glycerophospholipids with regard to physical properties and interaction with sterols and proteins. Nothing is known, however, about the impact of plasmalogens on the properties of biological membranes. In order to address this problem, we started a comparative fluorescence study on the membrane properties of cultured fibroblasts deficient in plasmalogens (from patients with the cerebro-hepato-renal "Zellweger-Syndrome") and cells with a normal plasmalogen level (controls).

The effect of plasmalogen deficiency on membrane fluidity was monitored by fluorescence anisotropy. We used TMA-DPH, a label probing exclusively the cell surface membrane, as well as DPH, this probe being distributed among plasma membranes and organelle membranes. Fibroblasts deficient

Table 1. Fluorescence anisotropy of TMA-DPH in Zellweger (HA, GM) and control (MO, SA) fibroblasts.

CELL STRAIN	FLUORESCENCE ANISOTROPY OF TMA-DPH	PLASMALOGEN PHOSPHOLIPID	CHOLESTEROL PHOSPHOLIPID	PROTEIN PHOSPHOLIPID
HA	$0.307 \pm 5\%$	2.5%	0.49	10.8
GM	$0.285 \pm 8\%$	3.1%	0.32	12.3
MO	$0.346 \pm 5\%$	11.7%	0.32	11.7
SA	$0.331 \pm 10\%$	13.9%	0.30	12.8

Fibroblasts grown on glass cover slips were labeled in the presence of 5 μM TMA-DPH at 37°C. After washing with Hank's solution at 37°C, cells were transferred for measurement into a cuvette containing Hank's solution at 37°C. Fluorescence anisotropies r were determined from the fluorescence intensities measured at parallel ($I_{\shortparallel}$) and vertical orientations ($I_{\perp}$) of the emission polarizer relative to the vertically oriented excitation polarizer. $r = (I_{\shortparallel} - G \cdot I_{\perp})/(I_{\shortparallel} + 2G \cdot I_{\perp})$

in plasmalogens consistently exhibit lower fluorescence anisotropies and, therefore, higher membrane fluidity as compared to controls (Table 1). Restoring normal plasmalogen levels in the cell membranes after supplementation with suitable biosynthetic precursors (alkylglycerols), in turn, reduces membrane fluidity. Cell monolayers exhibit higher fluorescence anisotropies as compared to cells in suspension irrespective of their plasmalogen content, but differences due to plasmalogen content were observed in both systems.

For further characterization of the plasmalogen effect on membranes, we synthesized fluorescently labeled 1-0-alkenyl-2-DPH-propionyl-sn-glycero-3-phosphocholines and 1-0-alkenyl-2-DPH-propionyl-sn-glycero-3-phospho-ethanolamines as well as their 1-acyl analogs. When incorporated in choline plasmalogen vesicles, both of these DPH-phospholipids exhibit higher fluorescence anisotropies as compared with diacyl lecithin vesicles. The fluorescence anisotropy of the DPH-labeled 1-0-alkenyl-phospholipids is higher than the anisotropy of the 1-acyl analogs when both are in the same phospholipid matrix. A linear correlation was found between fluorescence anisotropy and temperature for the DPH-phospholipids in fluid plasmalogen and palmitoyloleoyllecithin bilayers (20° - 40°C) at a lipid to label ratio of 1000/1.

When DPH-phospholipids are taken up by fibroblasts from sonicated lipid dispersions, the emission maximum of the label shifts from 480 nm to 430 nm. In addition, the fluorescence intensity at 430 nm increases several fold. This allows a time-dependent measurement of the lipid transfer to cells. Cells deficient in plasmalogens take up choline and ethanolamine plasmalogens at a higher rate as compared to control cells. Restoring normal plasmalogen levels in the cell membranes after supplementation with suitable biosynthetic precursors (alkylglycerols) significantly reduces the rate of plasmalogen uptake by the cells (Figure 1). Alkenyl-acyl glycerophospholipids are transferred more rapidly into the fibroblast membranes than are the diacyl derivatives. Experiments with choline phospholipid vesicles as acceptor membranes again showed a much faster uptake of alkenylacyl as compared to diacyl DPH-glycerophospholipids. The results obtained for cells and vesicles clearly show that the chemistry of the membrane-water interface (1-alkenylether or 1-acylester bonds in the glycerophospholipid molecules) is very important for interface-mediated phenomena such as intermembrane lipid transfer.

In conclusion, it can be said that plasmalogens have an impact on the

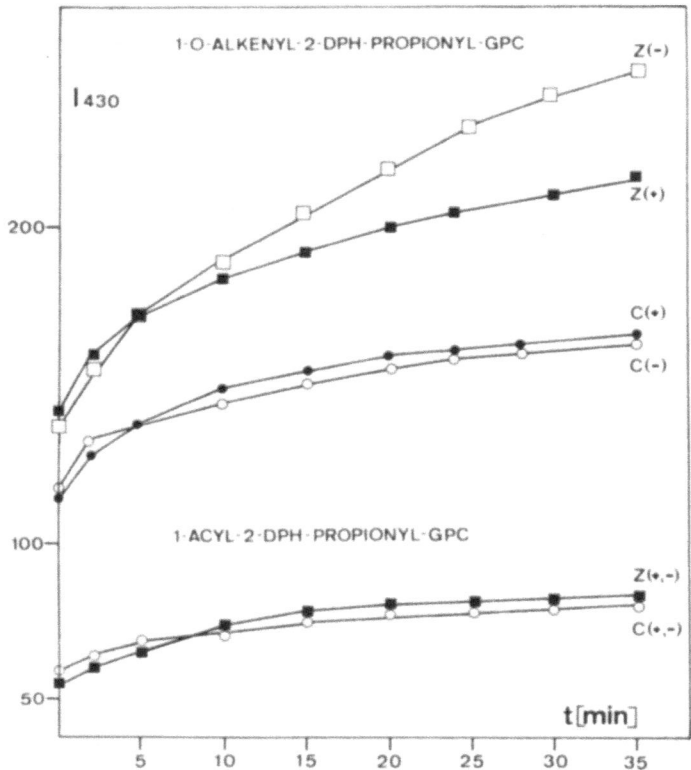

Figure 1: Uptake of DPH-propionyl-GPCs by Zellweger (Z) and control (C) Fibroblasts supplemented (+) and unsupplemented (-) with alkylglycerol. Cell monolayers (~25 nmol phospholipid) were incubated with sonicated aqueous dispersions of 1-O-alkenyl or 1-acyl-2-DPH-propionyl-glycero-phosphocholines (5 μM in Hank's solution) in a fluorescence cuvette at 37°C. Uptake of the fluorescent phospholipid by the acceptor membranes was determined from the dequenching of the DPH-fluorescence at 430 nm.

fluidity state of biological and artificial membranes as well as on membrane-membrane interaction. They exert a membrane-rigidifying effect and may represent a highly exchangeable subclass of membrane glycerophospholipids as compared to diacyllipid compounds.

NANOSECOND TIME-RESOLVED FLUORESCENCE STUDIES OF THIOREDOXIN

M. Han, M. VandeVen, D. Walbridge, J. Knutson, R. Lessick, C. Anfinsen, and L. Brand

The Johns Hopkins University, Department of Virology, Baltimore, MD 21218 USA
The Laboratory of Technical Development, NHLBI, NIH, Bethesda, MD 20205 USA

Thioredoxin from E. coli contains two tryptophan residues, trp 28 and trp 31, close to two cysteine residues, cys 32 and cys 35, which form a disulfide bond. Time-resolved fluorescence decay of tryptophan was monitored at emission wavelength of 340 nm with excitation wavelength of 295 nm in 100 mM potassium phosphate buffer, pH 5.7, and 1 mM EDTA at 5°C. Oxidized thioredoxin exhibits a triple fluorescence decay with lifetimes of 0.3, 1.4, and 5.9 nsec. Upon reduction of the disulfide bond by dithiothreitol (DTT), the steady-state fluorescence intensity increased and decay times of 0.4, 2.4, and 5 ns were found. The kinetics of reduction were also studied with a pulse fluorometer by collecting thirty consecutive decay curves at a rate of 2 min per curve for one hour under computer control. Simultaneous global analysis of the data, with lifetimes linked, indicates that the amplitude associated with the longest lifetime increases significantly, while the amplitude associated with the short lifetime decreases. Rate constants for the amplitude changes with time were estimated by a non-linear least squares method. Enzyme denatured by guanidine-HCl shows increased fluorescence. The fluorescence decay exhibits a triple exponential decay with lifetimes of 0.4, 2.0, and 4.4 nsec. Addition of DTT to the denatured enzyme does not increase the fluorescence intensity nor change the lifetimes. The time-resolved emission anisotropy of both reduced and oxidized enzyme was also studied. A mean rotational correlation time of 4.7 and 5.1 nsec was obtained at 22.8°C for reduced and oxidized enzyme, respectively. This indicates a minor change in the global structure upon reduction. Therefore, the changes of the tryptophan lifetimes due to reduction may reflect local conformational changes. Supported by NIH grant No. GM 11632.

ANALYZING THE DISTRIBUTION OF DECAY CONSTANTS IN PULSE FLUORIMETRY USING THE MAXIMUM ENTROPY METHOD

J. C. Brochon and A. K. Livesey

L.U.R.E. Laboratoire C.N.R.S.-C.E.A.-M.E.N.
Bat. 209D, Université Paris-Sud
F 91405 Orsay Cedex - France

Fluorescence kinetics are usually very complicated arising from the heterogeneity of the types of fluorophores and from their varied physical or/and chemical environments. In the most general case the fluorescence kinetics obeys an exponential law and the central problem in data analyses is the recovery of the distribution $\alpha(\tau)$ of exponentials describing the decay of fluorescence (i.e. inverting the Laplace transform).

It was previously demonstrated that the use of Shannon-Jaynes entropy function is justified (Jaynes, 1983):

$$S = \int p(\tau) \, \log \frac{p(\tau)}{m(\tau)} \, d\tau$$

where $p(\tau) = \alpha(\tau)/\int \alpha(\tau) \, d\tau$ are the normalized amplitudes and $m(\tau)$ is the model which encodes our prior knowledge about the system. If one has no knowledge of the values of τ expected one must choose $m_i = \tau_i^{-1}$ (Jaynes, 1983, p 114). Working with equally spaced points in log τ space the model m_τ becomes constant. The MEM solution displays useful properties:
1) The spectrum $\alpha(\tau)$ is smooth.
2) It only shows features (in particular the resolution of close peaks) if demanded by the data.
3) Provided the digitization is fine enough to show all the relevant details, the shape of the spectrum $\alpha(\tau)$ is independent of the number points used to display it.
4) It is "robust" to noise.
5) For data which are linear functions of the spectrum (this includes the Laplace transform and convolutions) the MEM solution is unique.

In general the MEM solution does not have a statistical interpretation.

Maximum entropy method is shown to give high quality results from both computer-generated "noisy" data and experimental data from chemical and biological molecules; an examples is given in figure 1. The MEM results

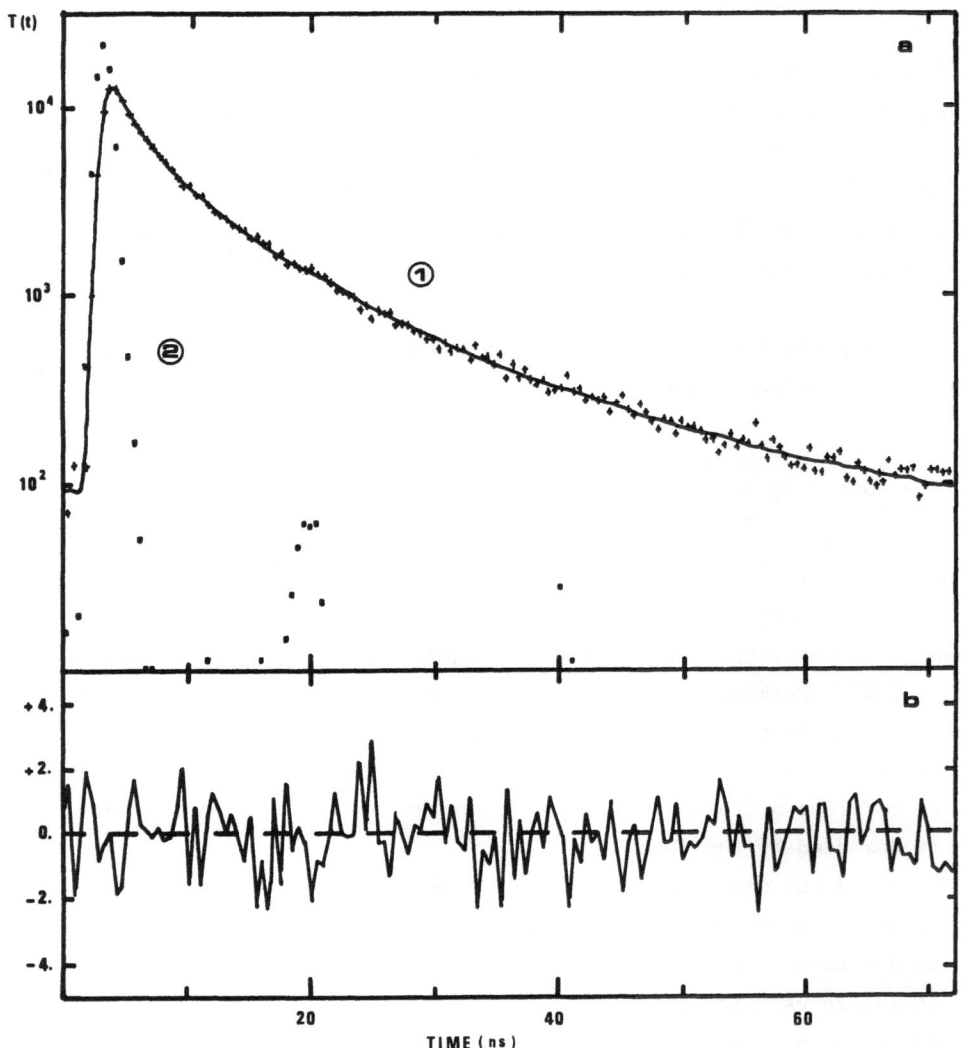

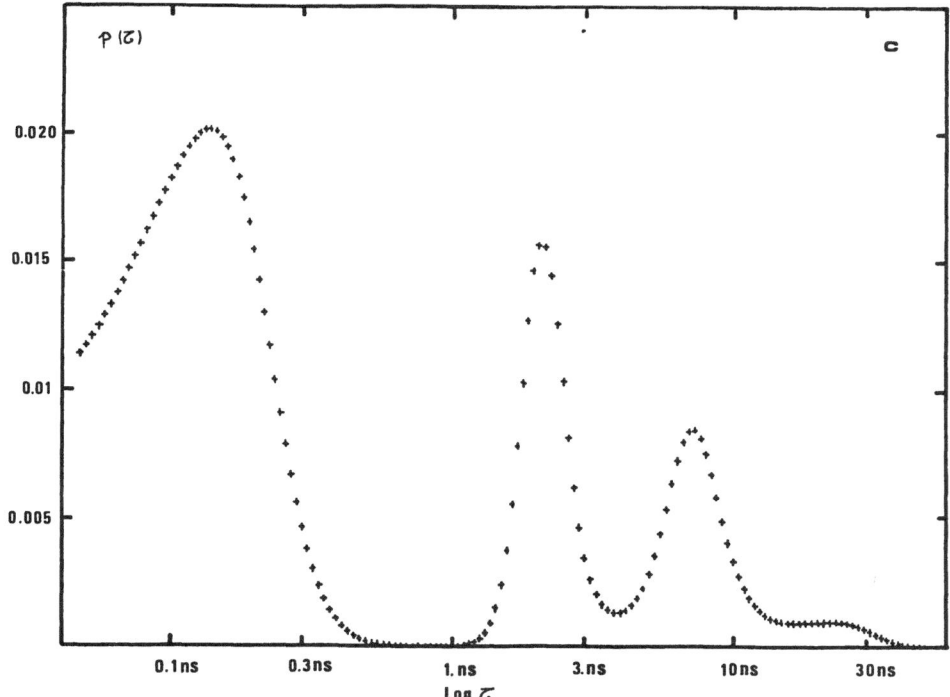

Figure 1. a) Fluorescence decay of labeled AEDANS ATCase. The excitation and emission wavelengths were respectively 350 nm ($\Delta\lambda$ = 16 nm) and 448 nm ($\Delta\lambda$ = 4.5 nm) and temperature was stabilized at 20°C ± 0.10°C. (1) (+++ experimental curve, ⸺ calculated curve) 2) is the experimental flash profile.
b) Plot of the weighted residuals.
c) MEM reconstructed spectrum of $\alpha(\tau)$ on 150 points equally spaced in Log τ between 0.05 ns and 50 ns. There is strong evidence for the three major peaks and a fourth minor contribution at long decay times.

are much easier to interpret than the results from the current data analysis techniques and help to overcome some experimental limitations. The fluorescence decay of the IAEDANS labeled ATCase enzyme demonstrates that MEM is a powerful tool to analyze the heterogeneity of fluorescent emission which could be correlated with their conformational dynamics in solution.

REFERENCES

Jaynes, E. T., 1983, Papers on the Probability, Statistics and
    Statistical Physics, ed. R. D. Rosenkrantz, D. Reidel, Dordrecht.

THERMODYNAMIC PROPERTIES OF ANTIFLUORESCYL ANTIBODIES

J. N. Herron[*], D. M. Jameson[†], D. M. Kranz[Θ] and E. W. Voss, Jr.[§]

[*]Dept. of Engineering, University of Utah, Salt Lake City, UT
84112 USA
[†]Dept. of Pharmacology, University of Texas Southwestern Medical
Center, Dallas, TX 75235 USA
[Θ]Dept. of Biochemistry and [§]Dept. of Microbiology, University of
Illnois at Urbana-Champaign, Urbana, IL 61801 USA

The thermodynamic properties of three monoclonal antifluorescyl antibodies (4-4-20,20-19-1,20-20-3) were determined over a temperature range of
2°-70°C, using fluorescence methodology. Curvilinear van't Hoff plots (ln
K_a vs. T^{-1} were observed for all three antibodies indicating that their
standard enthalpy changes ($\Delta H°$) were temperature dependent. This phenomenon
was further investigated by plotting the changes in unitary free energy
(ΔG_u), standard enthalpy ($\Delta H°$), and unitary entropy (ΔS_u) versus temperature.
At low temperatures (4°C), entropy played a major role in the binding of
fluorescein by all three antibodies, while enthalpy dominated at higher
temperatures. Negative heat capacity values ($\Delta C_p°$) were observed for all
three antibodies, which indicated that the hydrophobic effect was instrumen-
tal in the binding of fluorescein. However, ΔS_u values were lower than
expected for hydrophobic binding alone, suggesting that other forces were
acting to mitigate the hydrophobic effect. One possibility was that the
binding of fluorescein acted to restrain vibrational fluctuations in the
active site region, producing negative changes in both heat capacity and
entropy. This hypothesis was confirmed by determining the hydrophobic and
vibrational contributions to binding, using the method of Sturtevant (1977).
The phenomenon was most pronounced with the 20-20-3 protein, and the least
with the 4-4-20 protein. Considering that 4-4-20 exhibited the highest
affinity of the three proteins, it is possible that one mechanism of gener-
ating high affinity active sites may involve structural changes which reduce
the size of the unfavorable vibrational condition. By criteria established
by Torgerson et al. (1979), antibody active sites should exhibit negative
standard volume changes (soft sites) because they are formed from six
disjointed regions of polypeptide chain. However, a negative $\Delta V°$ value was
observed for only one of the antifluorescyl antibodies (20-19-1). We
believe that this discrepancy is due to the hydrophobic effect, which makes
a positive contribution to $\Delta V°$.

REFERENCES

Torgerson, P. M., Drickamer, H. G., and Weber, G., 1979, Biochemistry,
18:3079.

APPLICATIONS OF FLUORESCENT PROBES FOR MONITORING ENERGY-LINKED PHENOMENA IN BIOMEMBRANES AND PROTEOLIPOSOMES

R. Kraayenhof, G. J. Sterk, H. S. van Walraven, F. A. de Wolf, K. Krab and H. W. Wong Fong Sang

Department of Molecular and Cellular Biology
Vrije Universiteit
De Boelelaan 1087, 1081 HV Amsterdam
The Netherlands

Our main research is focussed on the resolution of bioenergy-transducing mechanisms in photosynthetic and respiratory membranes and simplified or reconstituted model membranes derived therefrom. Within this framework we are using several types of commercially available or home-synthesized fluorescent probes for monitoring dynamic structural and functional phenomena related to energy-transducing reactions. We will discuss the following examples of such probes:

(1) PROBES FOR MONITORING pH CHANGES INSIDE CELLS, ORGANELLES OR PROTEOLIPOSOMES

(a) In reconstituted ATPase proteoliposomes, prepared from a thermophilic cyanobacterium with extremely good membrane integrity, we have measured internal and external pH changes due to ATP synthesis or hydrolysis and were able to quantify the established pH gradients and proton translocation kinetics, using pH-sensitive probes (van Walraven et al., 1984) .

(b) For measuring the internal pH in intact cells we have synthesized some families of pH-sensitive fluorescent compounds with a dihydropyridine moiety. Being rather lipophilic, these (non-fluorescent) compounds can easily pass a cell membrane after which they can be converted by an intracellular NAD-dependent dehydrogenase or molecular oxygen to a very hydrophilic (and fluorescent) pyridinium analogue. The pyridinium is prereduced by borohydride to the dihydropyridine, which is then, after permeation, reoxidized to the impermeant pyridinium. We have observed a strong accumulation of some of these compounds inside hepatoma cells, whereas very slow back-leakage occurs. This was tested with quantitative fluorescence microscopy of immobilized cells as well as fluorescence measurements in suspensions (Sterk et al., 1986).

(c) pH measurements inside organelles (chloroplasts, mitochondria) proved to be very difficult. We did not succeed to trap fluorescent probes inside the organelles by fusion with probe-loaded liposomes without damaging the membranes or loosing the probe. For trapping externally added probes inside, the required internal probe-converting enzymes are not always present (e.g. chloroplasts). Therefore, we have also developed some lipophilic compounds with pH-dependent fluorescence, which, upon reaction with oxygen, or by slow hydrolysis in water, are converted into very hydrophilic species.

(2) PROBES FOR MONITORING INTRAVESCIULAR OXYGEN CONCENTRATIONS

Using the same principle as in 1c we have used in chloroplasts boro-hydride-prereduced acryflavin, which easily permeates into the thylakoid lumen and, upon illumination (inducing oxygen production) is reconverted to the impermeant quaternary compound, trapping it inside. Thus, the internal light-induced oxygen production could be easily monitored in chloroplasts and intact cyanobacteria with 2-3 times higher sensitivity and much faster response than the oxygen electrode. Moreover, this probe would offer an interesting tool to study fast kinetics of oxygen evolution in single-turn-over flashes.

(3) PROBES FOR ESTIMATION OF MEMBRANE SURFACE CHARGE DENSITY AND SURFACE pH

In the past we have observed a large fluorescence quenching of amino-acridine derivatives, occurring upon energization of photosynthetic or respiratory membranes, and found these probes useful to monitor the energy state under different conditions (R. Kraayenhof, 1970). This quenching was correlated with increased electrostatic binding of the probes due to increased external negative surface charge density. The pKa of these compounds was shown to increase significantly, allowing estimates of the pH in the direct vicinity of the membrane surface (Kraayenhof and Arents, 1977; de Wolf et al., 1985).

There is little information about the size and dynamics of the inter-facial pH (and surface potential) profile between membrane surface and bulk solution. In chloroplasts the electrokinetic (zeta) potentials at the plane of shear are definitely different from the `local' potentials regis-tered by membrane-associated cationic probes and this difference is changed by energization (light or ATP hydrolysis). Therefore, it would be of interest to calibrate' the interfacial pH profile by using pH-sensitive fluorophores, anchored at varying distance from the surface, which may allow extrapolation to a more realistic surface pH' value. We have synthesized a series of compounds consisting of a hydrophobic part for anchoring in the lipid bilayer, a very hydrophilic and rigid spacer group of varying length that extends into the aqueous phase, and a pH-sensitive fluorophore. It appears that the thickness of the layer between bulk solution and membrane surface, in which the pH changes, may not exceed the length of one butylene chain. Obviously, the presence of mono- and multivalent cations, buffers and energy-linked structural changes will affect this local pH profile and these aspects are subjects of our present investigations.

(4) PROBES FOR MEMBRANE POTENTIAL CHANGES

(a) In photosynthetic membranes the absorbance of native carotenoids can be used to monitor electric field changes upon (flash) illumination, ATP hydrolysis or relaxation of ion diffusion potentials. However, the carotenoid signal not only monitors trans-membrane potentials but intramembrane fields (close to the photosynthetic reaction centers) as well. In this respect the extrinsic probe oxonol VI is a much better tool for membrane potential measurements (Peters et al., 1983).

(b) In reconstituted ATPase proteoliposomes both of these probes are useful to some extent. While the carotenoid response is independent of the polarity of the electric field (generated by diffusion potentials), as expected, oxonol VI only responds to a diffusion potential that is negative at the side of the membrane where the oxonol was added. On the other hand, oxonol VI is better suited for calibration purposes (van Walraven et al., 1985).

(5) COVALENT FLUOROPHORE AS CONFORMATIONAL PROBE FOR H^+-ATPase

We have attempted to resolve the kinetics of the conformational change of the membrane-bound ATPase in chloroplasts when these were energized by light. This conformational change is possibly related to the latent→active transition, required for ATP synthesis and affinity changes for adenine nucleotides. The fluorogenic label fluorescamine was highly suitable for this purpose. Under carefully chosen conditions the ATPase was labeled in situ, then isolated and rapidly reconstituted with ATPase-depleted unlabeled membranes. All energy-linked functions are almost fully restored in the reconstituted system and the kinetics of the transition and light-induced spectral changes were characterized (Kraayenhof and Slater, 1975). Interestingly, these kinetics are very similar to those of the membrane surface charge rearrangements, as observed with membrane-bound cationic fluorescent probes.

(6) FLUORESCENCE POLARIZATION PROBE FOR DETERMINING LIPID PHASE BEHAVIOR IN RECONSTITUTED CYANOBACTERIAL MEMBRANES

We have used the classical probe 1,6-diphenyl-1,3,5-hexatriene for determining the phase behavior of the major lipids of thermophilic cyano-bacteria, in order to characterize the specific role of non-bilayer-forming lipids (e.g. MGDG) in determining the membrane lipid order and insertion of membrane proteins.

REFERENCES

de Wolf, F. A., Groen, B. H., van Houte, L.P.A., Peters, F.A.L.J., Krab, K., and Kraayenhof, R., 1985, Biochim. Biophys. Acta, 809:204.

Kraayenhof, R., 1970, FEBS Letters., 6:161.

Kraayenhof, R., and Slater, E. C., 1975, in: Proc. 3rd. Int. Congr. on Photosynthesis, Israel, M. Avron, ed., Elsevier, Amsterdam.

Kraayenhof, R., and Arents, J. C., 1977, in: Electrical Phenomena at the Biological Membrane Level, E. Roux, ed., Elsevier, Amsterdam.

Peters, F.A.L.J., Van Spanning, R., and Kraayenhof, R., 1983, Biochim. Biophys. Acta, 724:159.

Sterk, G. J., et al., 1986, in: Short Rep. 4th Eur. Bioenergetics Conf., Prague.

van Walraven, H. S., Koppenaal, E., Marvin, H.J.P., Hagendoorn, M.J.M., and Kraayenhof, R., 1984, Eur. J. Biochem., 144:563.

van Walraven, H. S., Krab, K., Hagendoorn, M.J.M., and Kraayenhof, R., 1985, FEBS Letters., 184:96.

FLUORESCENCE QUENCHING STUDIES ON ENZYME I OF THE BACTERIAL
PHOSPHOTRANSFERASE SYSTEM

Paolo Neyroz, Bonnie Bassler, Ludwig Brand and Saul Roseman

Department of Biology
The Johns Hopkins University
Baltimore, MD 21218 USA

INTRODUCTION

Enzyme I is the first protein involved in the series of phosphoryl
transfer steps catalyzed by the glycose:pyruvate phosphotransferase system
(PTS) (Meadow et al., 1984). The PTS is responsible for the phosphorylation and concomitant transport of its sugar substrates across the bacterial
membrane, and in this process phosphoenolpyruvate (PEP) serves as the
phosphoryl donor. It has been previously observed (Kukuruzinska et al.,
1982) that the subunits of Enzyme I associate in a temperature-dependent
manner to form dimers. The enzyme contains two tryptophans per subunit and
their intrinsic emission have been characterized at 2°C for the monomer and
23°C for the dimer. The decay of the fluorescence intensity has been
resolved into a double exponential decay for both the conformations of the
protein (Neyroz et al., 1984). At 2°C a short lifetime of 3 ns and a long
one of 7.5 ns were found that account for the 17% and 83% of the total
emitted light, respectively. At 23°C the recovered decay constants were
3.7 ns and 7.5 ns with an increase of the fractional contribution of the
short lifetime to 43% of the total intensity. Two different decay associated spectra (DAS) (Knutson et al., 1982) were obtained for each decay
component. A spectrum with its maximum at 330 nm was found for the short
lifetime and a second spectrum centered at 345 nm was resolved for the
longer lifetime. Changes in the proportions of the two spectra obtained at
2°C and 23°C well reproduced the difference of amplitudes reported above
when the protein was excited at 295 nm and the fluorescence observed at 340
nm.

In view of these results, it would be tempting to associate each
tryptophan residue (or each class of tryptophan residues in the dimer) with
a single lifetime, as in the case of the two tryptophans of horse liver
alcohol dehydrogenase (HLADH). In fact, by means of fluorescence quenching
techniques Ross et al. (1981) showed that the short lifetime of HLADH,
which was not quenched by KI, was associated with the buried trp-314, while
the long lifetime accessible to the quencher was associated with the
exposed trp-15.

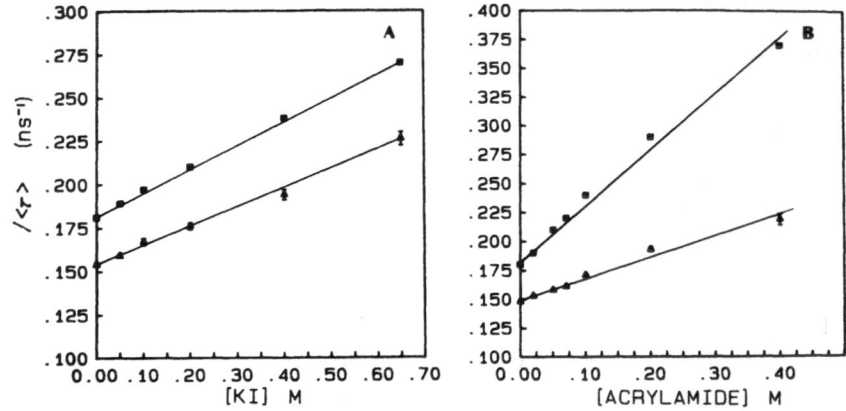

Figure 1: Time-resolved Stern-Volmer plots of the quenching of Enzyme I by KI (A) and acrylamide (B). The mean lifetimes obtained in the absence and the presence of the quencher were plotted in accord to eq. 1. The square symbols refer to the experiment performed at 23°C and the triangles to the experiments done at 2°C. The samples protein concentration was 0.5 mg/ml and the buffer composition was, 75 mM potassium phosphate pH 6.5, 1 mM EDTA, 0.5 mM DTT and 5 mM MgCl$_2$.

In this report, we therefore present the results obtained for the quenching of the intrinsic fluorescence of Enzyme I using two different quenchers, KI and acrylamide.

MATERIALS AND METHODS

The fluorescence decay of Enzyme I has been studied at two temperatures, 2°C (monomer) and 23°C (dimer) at increasing concentrations of the quenchers (0-0.65 M). Experimental curves were collected at five emission wavelengths (330, 340, 350, 360 and 380 nm) exciting the samples at 295 nm. The data was then analyzed by the global method (Knutson et al., 1983). The protein was dissolved in 75 mM potassium phosphate buffer pH 6.5, containing 1 mM EDTA, 0.5 mM DTT and 5 mM MgCl$_2$. Samples containing iodide were prepared at constant ionic strength.

Calculations: The bimolecular quenching constants Kq, were obtained by Stern-Volmer graphical analysis plotting the fluorescence lifetimes according to the following relationship:

$$1/\tau = 1/\tau_0 + Kq[Q] \tag{1}$$

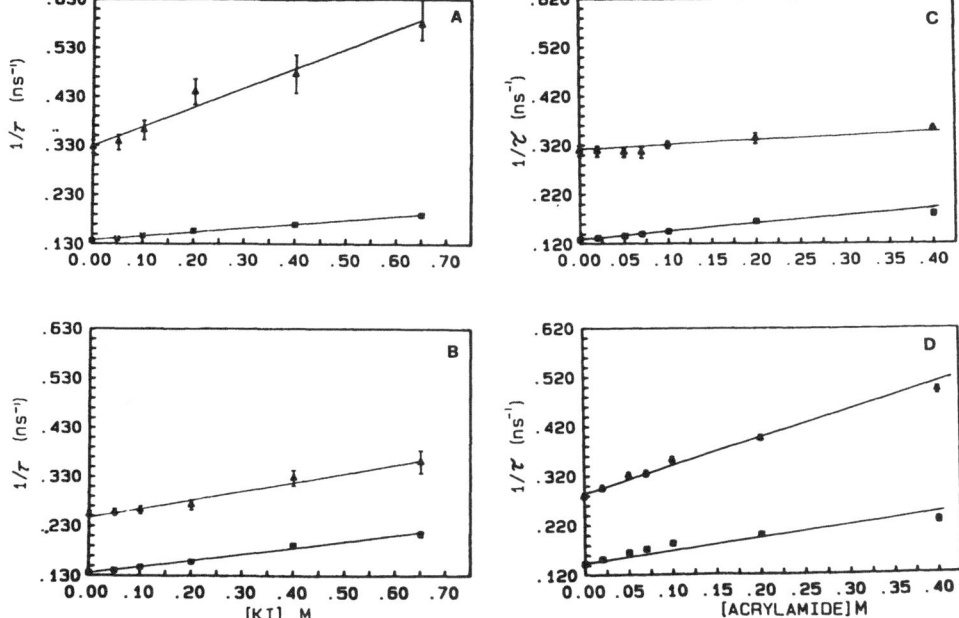

Figure 2: Stern-Volmer plots resolved for each lifetime of the fluorescence decay of Enzyme I. The square symbols refer to the long lifetime recovered in the presence and absence of the quencher and the triangular symbols refer to the short lifetime under the same conditions. The bars refer to the error we found in recovering the decay constants. A and B, results obtained for the quenching of Enzyme I by KI at 2°C and 23°C, respectively. C and D, results obtained by acrylamide at 2°C and 23°C, respectively.

where τ_0 is the fluorescence lifetime in the absence of the quencher. The steady-state fluorescence data was treated according to the classical Stern-Volmer equation:

$$F_0/F = 1 + K_{sv}[Q] \tag{2}$$

where K_{sv}, the Stern-Volmer quenching constant is equal to $Kq \cdot \tau_0$, F_0 and F are the fluorescence intensities in the absence and presence of the quencher. The static component of the quenching was evaluated by a modified form of the Stern-Volmer equation (Birks, 1970):

$$F_0/F e^{V[Q]} = 1 + K_{sv}[Q] = 1 + Kq\tau_0[Q] \tag{3}$$

RESULTS AND DISCUSSION

In Figure 1 are reported the Stern-Volmer plots for the quenching of
the mean fluorescence lifetimes by KI (Figure 1A) and acrylamide (Figure
1B). The recovered Kq were the following: KI-Kq at 2°C = $0.10 \times 10^9 M^{-1} sec^{-1}$,
KI-Kq at 23°C = $0.14 \times 10^9 M^{-1} sec^{-1}$, acrylamide-Kq at 2°C = $0.18 \times 10^9 M^{-1} sec^{-1}$,
acrylamide-Kq at 23°C = $0.48 \times 10^9 M^{-1} sec^{-1}$. These quenching rate constants
were about ten-fold lower than what one would expect for a fully exposed
tryptophyl residues (Eftink and Ghiron, 1976). The quenching constants
(K_{sv} and Kq) that we calculated for KI by fluorescence intensity measurements were in excellent agreement with those obtained by time-resolved
experiments (data not shown). This result and the linear plots we
obtained, imply that the quenching mechanism was prevalently collisional.
Fluorescence intensity data obtained with acrylamide showed, instead, an
upward curvature of the Stern-Volmer plot relative to the monomer. According to equation 3 a static quenching constant $V = 0.7 M^{-1}$ was calculated.

The Stern-Volmer plots resolved for each decay component are shown in
Figure 2 (A-D) and the calculated Kq are reported below in Table I. At
23°C (dimer) the short lifetime was more quenched than the longer one and
the results obtained with the different quenchers were in agreement. At
2°C (monomer) we found that increasing concentrations of acrylamide had
very low effect on the short lifetime, while the negatively charged I^-
ions, again, quenched this short component to an extent higher than the
longer one.

On the basis of these results we conclude that Enzyme I does not
follow the classical example of HLADH in which the two tryptophan residues
were distinctly separated in "accessible" and "non-accessible" to the
quencher. In addition, we have found that for the dimer form of Enzyme I
both the quenchers used are more effective on the short than on the long
lifetime. The pattern of the quenching parameters found for the dimer, was
also found for the quenching of the monomer by iodide, while acrylamide
showed a very low Kq for the shorter lifetime. This discrepancy could be
explained by the fact that the quenching by acrylamide of the monomer has a
static component, and this might result in an underestimation of the
parameters obtained by time-resolved measurements.

In order to provide a complete interpretation in terms of preferential
accessibility of the tryptophan residues of Enzyme I, experiments are now
in progress using CsCl.

372

Table 1: Bimolecular quenching constants resolved for each fluorescence decay time. K_{q1} and K_{q2} refer to the results obtained for the short and the long decay times respectively.

Temp. (°C)	KI		Acrylamide	
	K_{q1}	K_{q2}	K_{q1}	K_{q2}
2°	0.40	0.08	0.13	0.18
23°	0.19	0.14	0.52	0.21

Bimolecular quenching constants resolved for each fluorescence lifetime Kq_1 and Kq_2 refer to the results obtained for the short and the long lifetime, respectively. The units are $Kq \times 10^{-9} M^{-1} sec^{-1}$.

ACKNOWLEDGEMENTS

This work was supported by National Institutes of Health Grants, Nos. GM11632 and CA21901.

REFERENCES

Birks, J. B., 1970, Photophysics of Aromatic Molecules, New York, N.Y., Wiley-Interscience, pp. 433-447.

Eftink, M. R., and Ghiron, C. A., 1976, Biochemistry, 15:672.

Knutson, J. R., Beechem, J. M., and Brand, L., 1983, Chem. Phys. Lett., 102:501.

Knutson, J. R., Walbridge, D., and Brand, L., 1982, Biochemistry, 21:4671.

Kukuruzinska, M. A., Harrington, W. G., and Roseman, S., 1982, J. Biol. Chem., 257:14470.

Meadow, N. D., Kukuruzinska, M. A., and Roseman, S., 1984, in: "Enzymes of Biological Membranes", A. Martonosi, ed., Plenum Press, New York, 3:523.

Neyroz, P., Beechem, J. M., Roseman, S., and Brand, L., 1984, Photochem. and Photobiol., 39:41s.

Ross, J.B.A., Schmidt, C. J., and Brand, L., 1981, Biochemistry, 20:4369.

SPONTANEOUS TRANSFER AND MOLECULAR ORGANIZATION OF DEHYDROERGOSTEROL IN LIPID BILAYERS

Parkson Lee-Gau Chong, L. Bar, Y. Barenholz and T. E. Thompson

Department of Biochemistry
University of Virginia
Charlottesville, VA 22908 USA

Dehydroergosterol (DHE), a fluorescent cholesterol analogue, was used as a cholesterol mimic to study both the transfer kinetics between phospholipid bilayers and the molecular organization of this sterol in the bilayer vesicles during transfer. The fluorescence intensity was used to monitor the transfer rate of DHE between 1-palmitoyl-2-oleoyl-phosphatidylcholine small unilamellar vesicles at different temperatures. The equation that best fits the data is a single exponential decay plus a base value due to a non-exchangeable pool, $A \exp(-kt)+B$. The fitted halftimes ($t_{\frac{1}{2}}=\ln 2/k$) and base values (B) are: 61 min and 45.4 at 16°C, 27 min and 25.1 at 24°C, 18 min and 20.0 at 37°C. These values are comparable to those obtained using ^{3}H-labeled cholesterol (Bar et al., 1986), suggesting that DHE resembles cholesterol and that a non-exchangeable pool exists for DHE as well as cholesterol. In order to examine the changes in DHE organization during the transfer, acrylamide quenching was performed on the donor vesicles both at the beginning of transfer, designed as t_o, and at the end of transfer, t_∞. Our results show that the quenching rate constant at t_o is greater than that at t_∞. However, three days after the transfer, no apparent difference in the quenching rate constant was observed between t_o and t_∞. This indicates that a reorganization of DHE occurs within this time period. The fluorescence lifetime is 0.88 ns and 0.76 ns for t_o and t_∞, respectively. Rotational correlation times at t_o and t_∞ also exhibit a small difference. In all probability these differences can be attributed to changes in the DHE organization. A tentative working model is proposed to interpret these results. (Supported by NIH grants GM-14628 and HL-17576.)

REFERENCES

Bar, L. K., Barenholz, Y., and Thompson, T. E., (1986) Biochemistry, 26:6701.

STANDARDIZATION OF FLUOROPHORES USING PHASE FLUOROMETRY

Richard B. Thompson and Enrico Gratton*

Bio/Molecular Engineering Branch
Code 6190, Naval Research Laboratory
Washington, DC, 20375-5000, USA
*Dept. of Physics and Laboratory for Fluorescence Dynamics
University of Illinois
Urbana, IL, 61801, USA

Measurements of fluorescence lifetimes, like any other analytical procedure, require the use of standards for instrument calibration and comparison. Scattering materials have been widely used in time-resolved fluorometry as standards because, on the nanosecond time scale of fluorescence, their femtosecond "lifetime" is accurately approximated as zero. As is well known (Wahl et al., 1974; Rayner et al., 1976), the use of scattering compounds can introduce artifacts into lifetime determination because the wavelength of emission is necessarily different from that of excitation; e.g., "color effects" in photomultiplier tubes, and other, more subtle optical artifacts (R.B.T. and E.G., unpublished results; Bernard Valeur and David Jameson, personal communication). To minimize these difficulties "standard fluorophores" have long been used, since they permit matching the standard's emission wavelength with that of the sample (Lakowicz et al., 1981; F. Castelli, 1985; Kolber and Barkley, 1986). Unfortunately, there is no wide agreement (to better than 5%) on the lifetimes of the standards now in use, even though the precision of current multifrequency phase and pulse instrumentation is perhaps 0.5%. This may be due to differences in the samples themselves, and particularly the concentration of the important quencher, oxygen. Thus a transferable standard might be difficult to specify. Moreover, the dozen or so compounds widely used as standards cover only half the spectral region of interest (250-1000 nm). Evidently a large array of compounds would be necessary to cover this entire region of the spectrum. Rather than specify a number of compounds as standards which might be poorly transferable, we propose a phase fluorometric method for accurately determining the lifetime of standard fluorophores in one's own laboratory, without assuming any reference value.

In our method, the phase differences and demodulation ratios between two different standard fluorophores (both assumed to be monoexponential) are measured over a range of modulation frequencies; the data are easily obtained by misinforming the data acquisition computer that the shorter lifetime is zero. The data obtained for two monoexponential decays is

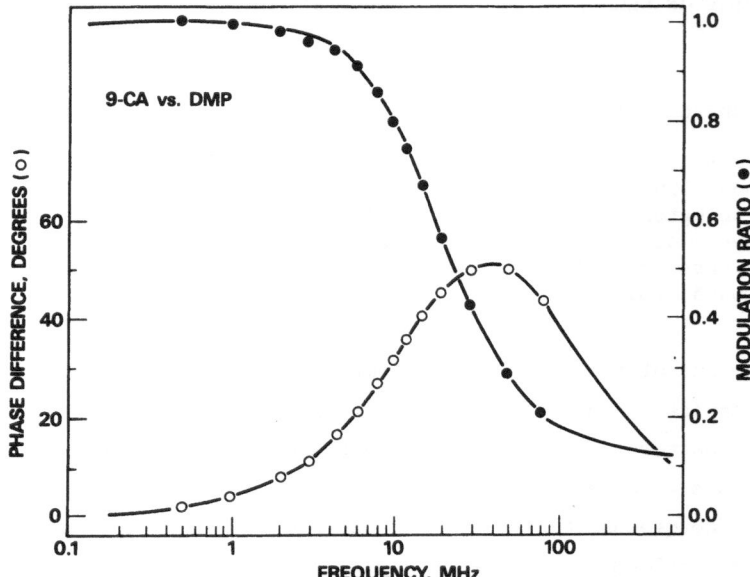

Frequency-dependent phase differences (o) and demodulation ratios (•) between 9-cyanoanthracene and dimethyl POPOP, both in ethanol at 25 C. The lines indicate the best fit to the data, yielding values of 11.94 and 1.49 nsec for 9-CA and dimethyl POPOP, respectively. Excitation was at 325 nm, emission was observed through a Corning 0-52 filter to eliminate Rayleigh and Raman scattered excitation, and no attempt was made to exclude oxygen.

always of the form depicted in the Figure. However, the maximum phase difference and demodulation ratio, and their positions in frequency space uniquely define the values of the two monoexponential decays. The data are fit without using any prior assumption of the lifetime values, only that they are monoexponential; the veracity of that assumption is tested by the usual criteria of goodness-of-fit.

We measured the lifetimes of several fluorophores known or likely to have monoexponential decays using the above method, and using a scatterer. The compounds were of the highest available purity, filtration was chosen to avoid Rayleigh and Raman scattering, and the solvents were shown to have negligible background fluorescence. Multi-frequency phase fluorometric measurements were performed on an ISS Greg-200 instrument (Industria Strumentazioni Scientifiche, La Spezia, Italy) essentially as previously described (Lakowicz et al., 1984; Gratton et al., 1984). The values from the new method were typically close to the best single exponential fits of the data obtained using a scatterer, and to published values (Wahl et al.,

1976; Lakowicz et al., 1981; F. Castelli, 1985; Kolber and Barkley, 1986).

A more stringent test of the accuracy of the new method was performed using a fluorophore-quencher system which is known to exhibit a linear Stern-Volmer plot; in this case the correlation coefficient of the best fit straight line to the derived lifetime values is a good indication of the accuracy of the lifetime measurements (Kolber and Barkley, 1986). We measured the intensities and lifetimes of N-(3-sulfopropyl)acridinium quenched by micromolar concentrations of potassium bromide using the new method (Wolfbeis and Urbano, 1983), and also using a scatterer. The correlation coefficients obtained using intensity data, scatterer lifetime data, and new method lifetime data were .9945, .9974, and .9941, respectively. Thus the new method provides results at least comparable to those obtained using a scatterer, and without assuming any reference lifetime.

ACKNOWLEDGEMENTS

The authors thank NIH, NSF, and DARPA for support, and Drs. Jay Knutson, Bernard Valeur, and David Jameson for helpful discussion. R.B.T. is grateful for an NRC Associateship, during which some of this work was performed.

REFERENCES

Castelli, F., 1985, Rev. Sci. Instrum., 56:538.

Gratton, E., Limkeman, M., Lakowicz, J. R., Maliwal, B. P., Cherek, H., and Laczko, G., 1984, Biophys. J., 46:479.

Kolber, Z. S., and Barkley, M. D., 1986, Anal. Bioch., 152:6.

Lakowicz, J. R., Cherek, H., and Balter, A., 1981, J. Biochem. Biophys. Meth., 5:131.

Lakowicz, J. R., Laczko, G., Cherek, H., Gratton, E., and Limkeman, M., 1984, Biophys. J., 46:463.

Rayner, D. M., MacKinnon, A. E., Szabo, A. G., and Hackett, P. A., 1976, Can. J. Chem., 54:3246.

Wahl, Ph., Auchet, J. C., and Donzel, B., 1974, Rev. Sci. Instrum., 45:28.

Wolfbeis, O. S., and Urbano, E., 1983, Fres. Z. Anal. Chem., 314:577.

PHASEFLUOROMETRY USING THE HARMONIC CONTENT OF A SYNC-PUMPED MODE-LOCKED CAVITY-DUMPED DYE LASER

K. Clays, Y. Engelborghs and A. Persoons

Laboratory for Chemical and Biological Dynamics
University of Leuven
Celestijnenlaan 200 D
B-3030 Leuven, Belgium

As proposed by E. Gratton and R. Lopez-Delgado (1980), impressive time resolution in fluorescence lifetime measurements can be achieved by coupling repetitive pulsed excitation with phase shift and amplitude modulation techniques.

We have previously built a phase fluorimeter using a Pockels cell modulator (Ide et al., 1983). In this work, however, we present some results of fluorescence lifetime determination using the harmonic content of a train of light pulses. A schematic description of our instrument is given in figure 1. The repetitive light source is a sync-pumped, injection-locked, cavity-dumped dye laser, pumped by a mode-locked Ar$^+$-ion laser. With a pulse width of about 10 psec, the sample may be considered as simultaneously excited with a set of modulation frequencies up to 100 GHz. The frequency spacing is determined by the pulse repetition rate, which is normally set at 4 MHz.

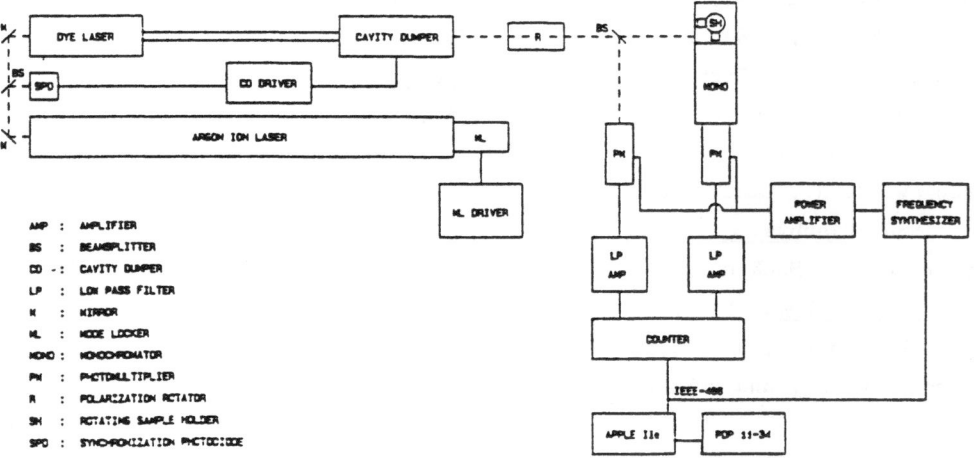

AMP : AMPLIFIER
BS   : BEAMSPLITTER
CD - : CAVITY DUMPER
LP   : LOW PASS FILTER
M   : MIRROR
ML   : MODE LOCKER
MONO : MONOCHROMATOR
PM   : PHOTOMULTIPLIER
R   : POLARIZATION ROTATOR
SH   : ROTATING SAMPLE HOLDER
SPD : SYNCHRONIZATION PHOTODIODE

Figure 1.

To measure the phase shift at a single harmonic component of the pulse classical cross-correlation techniques are applied (Spencer and Weber, 1969). The sensitivity of the method is increased compared with direct measurements, because the phase shift is measured in the low frequency region. The photomultiplier risetime of 2 nsec implies a limiting band-width of about 300 MHz, which is a considerable extension compared with electro-optic modulation techniques. Using an integration time of 5 seconds, the standard deviation of the phase shift is typically 0.1 degree, even at the higher frequencies. This results in a lifetime resolution of about 1 picosecond.

Figure 2 shows the results of a phase measurement and the data analysis (non-linear least square, Marquard algorithm) for a solution of 10^{-6} M DODCI (3,3'-diethyloxadicarbocyanine iodide) and 6.10^{-6} M malachite green in ethanol. The very good precision of our instrument is not only revealed by the very low standard deviation from the fitting procedure (0.8 psec), but also by the reproducibility of the lifetime determination. A 10^{-6} M solution of DODCI in ethanol was measured on several days. The standard deviation was 1.9 psec. The accuracy of our instrument can be estimated by comparing our lifetime with that obtained from single photon counting. The values in the literature are in good agreement with our value of 1099 ± 2 psec. To demonstrate the potential of our instrument, the lifetime of

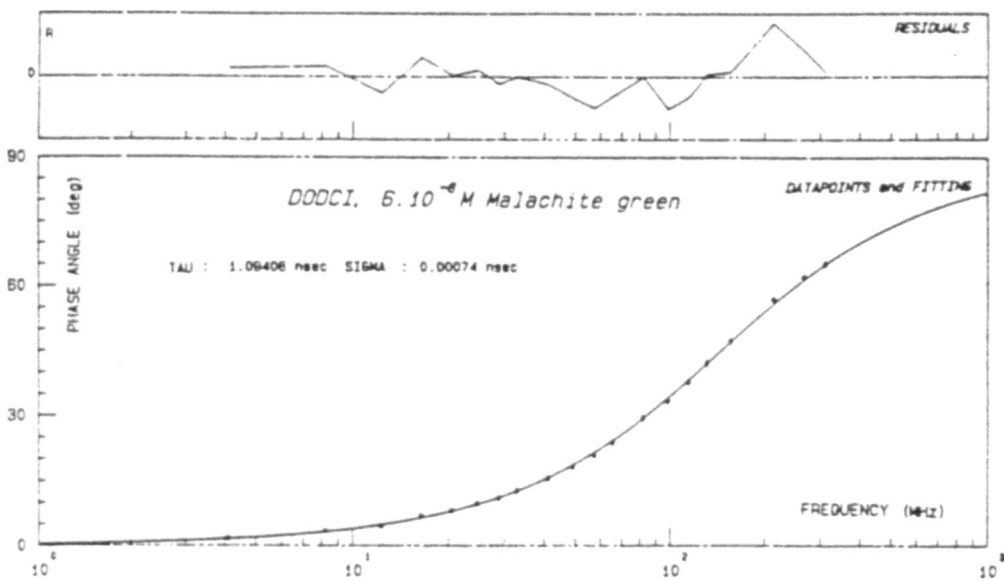

Figure 2.

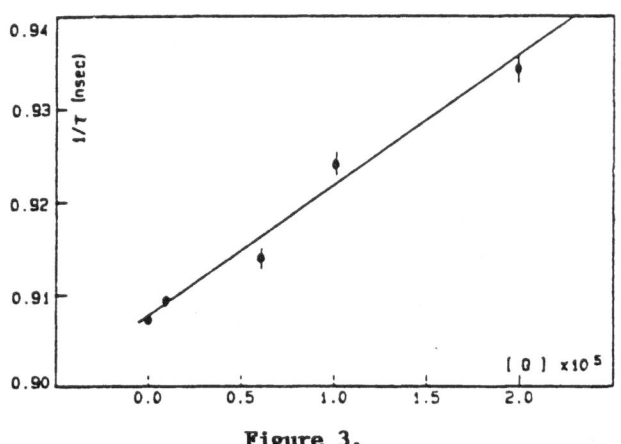

Figure 3.

[Q]	τ(ns)	σ(ns)
0.	1.1022	0.0006
1.10^{-6}	1.0997	0.0005
6.10^{-6}	1.0941	0.0007
1.10^{-5}	1.0822	0.0006
2.10^{-5}	1.0703	0.0008

Table 1

DODCI in a series of solutions with increasing malachite green concentration in ethanol was measured. The results of these measurements are listed in table 1. Figure 3 is a Stern-Volmer plot for the quenching of DODCI by malachite green from fluorescence lifetimes. It's clearly obvious that very small differences in fluorescence lifteimes can be measured very accurately.

ACKNOWLEDGEMENTS

We are deeply indebted to Drs. B. Van Wonterghem for valuable discussions and practical help and to R. Strobbe for the electronics. We would also like to thank Spectra Physics for providing the cavity dumper.

NOTES AND REFERENCES

K. Clays is a research assistant and Y. Engelborghs is a senior research associate of the Belgian National Fund for Scientific Research.

Gratton, E., and Lopez-Delgado, R., 1980, Il Nuovo Cimento, 56B:110.
Ide, G., Engelborghs, Y., and Persoons, A., 1983, Rev. Sci. Instrum., 54:841.
Spencer, R. D., and Weber, G., 1969, Ann. N. Y. Acad. Sci., 158:361.

THE STUDY OF TUBULIN-COLCHICINE INTERACTIONS USING FLUORESCENCE TECHNIQUES

Yves Engelborghs

Laboratory of Chemical and Biological Dynamics
Katholieke Universiteit Leuven
Celestijnenlaan 200 D, B-3030 Heverlee BELGIUM

Colchicine is a very potent antimitotic drug, through its direct interaction with the protein tubulin. Tubulin is the building block of microtubules. These hollow fibers form the spindle figure, responsible for the transport of chromosomes during cell division. The tubulin-colchicine complex is unable to form the correct structures. Colchicine is a tricyclic component, containing trimethoxybenzene, and two seven rings, the external one of which is tropolone. It is extremely weakly fluorescent in aqueous or organic solutions. Upon binding to tubulin fluorescence appears with a maximum around 400 nm (Bhattacharyya and Wolff, 1974). Fluorescence is also induced when the molecule is immobilized by freezing (Leterrier and Rieger, 1975) or in very viscous solutions at room temperature (Bhattacharyya and Wolff, 1984). We also observed fluorescence when colchicine was dried on a glass substrate. The low quantum yield is therefore probably due to the possibilities of radiationless deactivation coupled to intramolecular flexibility. A fluorescence lifetime of 1.14 (± 0.02) ns was determined with phase fluorimetry for the tubulin-colchicine complex (Ide and Engelborghs, 1981). The binding is almost irreversible, suggesting the burial of the molecule into a deep cleft on the protein. Iodide was therefore used to estimate the exposure of the molecule. It was found to expel colchicine from its complex. Other chaotropic anions behaved in the similar way (Ide and Engelborghs, 1981). Quenching, therefore, had to be studied by extrapolation of the kinetic curves to time zero. The quenching curve obtained in this way showed downward curvature, either indicating an inaccessible fraction, or a screening effect of bound anions. KSCN was found to inhibit iodide quenching suggesting the second hypothesis. The situation is thus analogous to the quenching of the single tryptophan in human serum albumin (Lehrer, 1976). Using the technique of fluorescence stopped flow, with front surface detection, the kinetics of binding could be studied, despite severe inner filter effects. Binding occurs in two steps: a fast preequilibrium and a very slow conformational change (Garland, 1978; Lambeir and Engelborghs, 1981). Fluorescence increase occurs largely during the second step. From this we conclude that the conformational freedom of the molecule is only lost in the final stage. The fast preequilibrium is endothermic (Lambeir and Engelborghs, 1981) and the thermodynamic parameters correspond nicely with the data for tropolone binding (Andreu and Timasheff, 1982) and colchicine dimerization (Engelborghs, 1981). It was therefore suggested that initial binding occurs through the topolone ring (Andreu and Timasheff, 1982). A comparison is made with the bicyclic analog AC (Engelborghs and Fitzgerald, 1986). This molecule has a much higher flexibility and binds much faster, but also dissociates faster. Its initial binding is, however, exothermic. It is, therefore, probable that already in the initial binding of AC, the trimethoxybenzene ring contributes. In colchicine itself, substituents at the middle ring also prove to have an influence on the binding kinetics and on the fluorescence increase upon binding (Bhattacharyya et al., 1986).

ACKNOWLEDGEMENTS

The author is senior research associate of the National Fund for Scientific Research of Belgium, and wishes to thank this fund and the research fund of the university for financial support.

REFERENCES

Andreu, J. M., and Timasheff, S. N., 1982, Biochemistry, 21:6465.

Bhattacharyya, B., and Wolff, J., 1974, Proc. Natl. Acad. Sci. U.S.A., 71:2627.

Bhattacharyya, B., and Wolff, J., 1984, J. Biol. Chem., 259:11836.

Bhattacharyya, B., Howard, R., Maity, S. N., Brossi, A., Sharma, P. N., and Wolff, J., 1986, Proc. Natl. Acad. Sci. U.S.A., 83:2052.

Engelborghs, Y., 1981, J. Biol. Chem., 256:3276.

Engelborghs, Y., and Fitzgerald, T. J., 1986, Ann. N.Y. Acad. Sci.,

Garland, D. L., 1978, Biochemistry, 17:4266.

Ide, G., and Engelborghs, Y., 1981, J. Biol. Chem., 256:11684.

Lambeir, A., and Engelborghs, Y., 1981, J. Biol. Chem., 256:3279.

Lehrer, S., 1976, in: "Biochemical Fluorescence Concepts II", R. F. Chen and H. Edelhoch, ed., M. Dekker, Inc., N.Y., pp. 515.

Leterrier, F., and Rieger, 1975, C.R.S.S.A. Trav. Scientif., 5:224.

STRUCTURAL DYNAMICS OF MYELIN BASIC PROTEIN IN THE ASSEMBLY OF A MODEL MEMBRANE

P. Cavatorta°, L. Masotti*, A. G. Szabo[$], P. Riccio[§] and E. Quagliariello[§]

°Dept. of Biphysics-GNCB and *Institute of Biological Chemistry, University of Parma, 43100 Parma, Italy, [$]Division of Biological Sciences, National Research Council of Canada, Ottawa, Canada, K1A OR6, [§]Inst. of Biological Chemistry, University of Bari, 70124 Bari, Italy

INTRODUCTION

The encephalitogenic myelin basic protein (MBP), with a MW of 18.4 KDa protein, is the smallest of the CNS myelin proteins, and accounts for about 30% of their total content (Boggs et al., 1982). MBP, when extracted with organic solvents, is soluble in water and has been extensively studied because of its ability to induce inflammatory demyelination (Kies et al., 1965). Recently Riccio et al. (1986) reported EM evidence that MBP, extracted with endogenous lipids employing the detergent CTAB, was able to induce lysolecithin to undergo a structural transition from micelles into multilamellar vesicles. On the other hand lipid-free MBP interacted with lysolecithin but was not able to induce such a transition. The aim of this work is to study the ability of lipids to modulate protein conformation and of a protein such as MBP to influence lipid supramolecular organization. Static and time resolved fluorescence of the single tryptophan residue of MBP and CD measurements are presented that illustrate interesting details of the molecular dynamics involved in lipid protein interactions.

MATERIALS AND METHODS

Water-soluble MBP (Deibler-MBP) was isolated and purified from bovine brains according to the procedure of Deibler et al. (1972). Lipid-bound MBP (CTAB-MBP) was isolated from bovine brains as described earlier (Riccio et al., 1986). All other chemicals used were of analytical grade purity. The instrumentations used in this paper are described elsewhere in this volume (Masotti et al., this volume).

Deibler-MBP-lysolecithin complexes were prepared in 10 mM Tris buffer, pH 7, by adding lysolecithin micelles to Deibler-MBP to give a L/P = 500. The preparation of CTAB-MBP-lysolecithin vesicles has been described previously (Riccio et al., 1986).

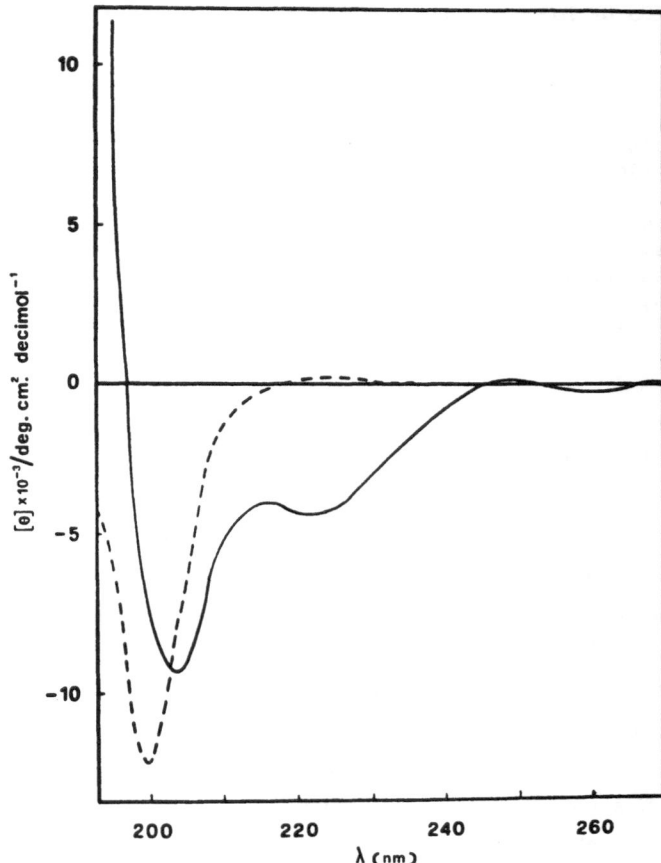

Figure 1. Circular dichroism spectra of DEIBLER-MBP (dashed line) and CTAB-MBP (solid line). In both cases the protein concentration is 0.3 mg/ml and the optical pathlength of the cell is 0.2 mm.

RESULTS AND DISCUSSION

The far-UV CD spectra of Deibler-MBP and CTAB-MBP in 10 mM Tris Buffer, pH 7, are shown in Figure 1. The water-soluble Deibler protein showed a single negative maximum centered at 200 nm, typical of a peptide in a random coil conformation. In the lipid-bound form, however, the pattern of the spectrum changed dramatically. The spectrum had two negative bands, one at 220-222 nm and one at 203 nm, and one positive peak around 190 nm. These results clearly indicate that the CTAB-MBP assumed a degree of helical conformation due to the interaction with endogenous lipids and detergent.

Table 1. Fluorescence Parameters of MBP

	Q (20°C)	λ_{max}(nm)	τ (20°C) (ns)	k_q (20°C) ($10^{-9}M^{-1}sec^{-1}$)
DEIBLER	.090±.005	350 ± 1	4.10 1.90 .30	3.3
DEUBKER+Lysolecithin	.180±.005	335 ± 1	5.40 1.90 .10	---
CTAB+MBP	.210±.005	345 ± 1	7.00 2.10 .30	1.8
CTAB+MBP+Lysolecithin	.095±.005	340 ± 1	7.40 2.00 .15	---

Q = quantum yield; λ_{max} = emission maximum wavelength; τ = experimental lifetimes; k_q = Acrylamide quenching rate constant.

Several fluorescence parameters of the single tryptophan of the protein in the two forms without and in the presence of lysolecithin are listed in Table 1. The fluorescence results in the absence of lysolecithin confirmed that the two forms of proteins were quite different in structure. The quantum yield of CTAB-MBP was higher than that of Deibler-MBP; the fluorescence maximum of the former was positioned at 345 nm while that of the latter was red shifted at 350 nm. Furthermore, the long lifetime was higher in CTAB-MBP in accord with the quantum yield behavior and the acrylamide quenching rate constant was smaller. These data indicated that the tryptophan in CTAB-MBP was located in a more hydrophobic environment with respect to the Deibler form, probably as a consequence of a conformational change due to the presence of the endogenous lipids and of the detergent.

The addition of lysolecithin caused, in both cases, a blue shift of the fluorescence emission maximum indicative of a shielding of the tryptophan from the aqueous solvent. The fluorescence quantum yield, however, increased in the case of Deibler but decreased in the case of CTAB-MBP-lysolecithin complexes. The fluorescence anisotropy dependence on temperature, shown in Figure 2, proved to be quite interesting. The anisotropy decreased slowly with temperature in all cases except for

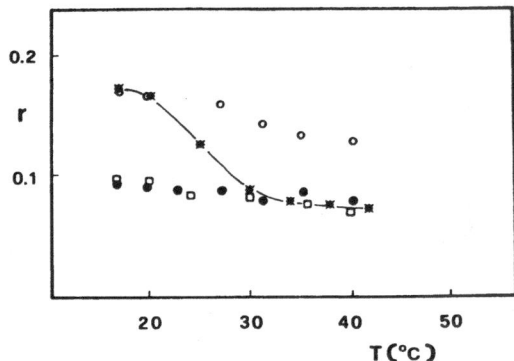

Figure 2. Fluorescence anisotropy dependence on temperature. DEIBLER-MBP (•); CTAB-MBP (□); DEIBLER-MBP+lysolecithin (O); CTAB-MBP+lysolecithin (*).

CTAB-MBP + lysolecithin complex, where a temperature transition near 25°C was evident. This temperature transition was irreversible: in fact lowering the temperature stepwise back to 17°C, the anisotropy remained practically constant at a value near 0.1.

Since lysolecithin alone did not show a phase transition in this temperature range, the result clearly indicated that CTAB-MBP was able to organize the lipid in a bilayer-type structure.

In Figure 3 the time-resolved emission spectra of MBP in the presence and in the absence of lysolecithin are shown. In all the samples the protein exhibited three distinct lifetime associated spectra. The spectra have three maxima positioned at different wavelengths and different relative intensities. These results suggest that the protein exists in three different average conformations. An alternative interpretation of a conformational distribution might be considered. However, the former seems to be preferable because the spectra have different maxima that behave in a different manner as a consequence of the interaction of the protein with lipids.

ACKNOWLEDGEMENTS

This work was supported by grants from the National Science and Engineering Research Council of Canada and MPI (40%) Rome and travel grants from NATO.

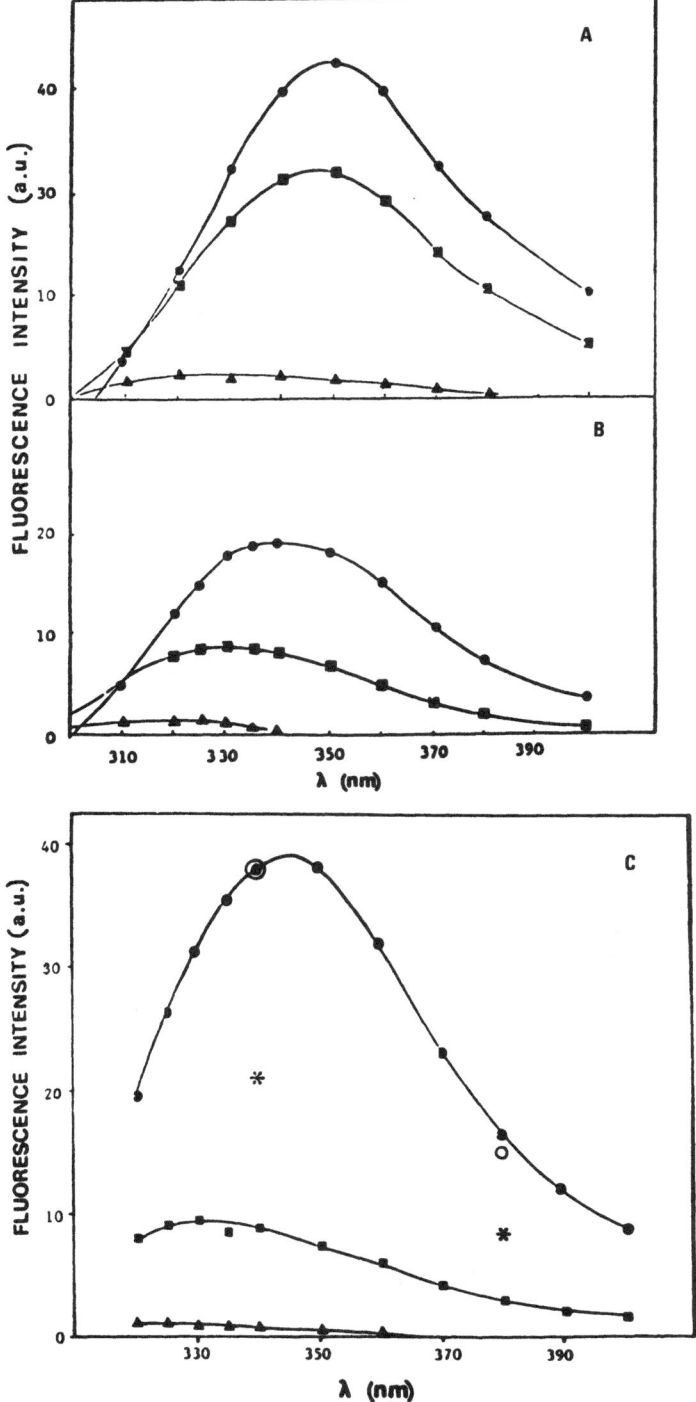

Figure 3. (Continued on next page)

Figure 3. Time resolved fluorescence spectra of MBP.

A) DEIBLER-MBP

pH 7 $\tau_1 = 4.1$ ns ($\bullet$) $\tau_2 = 1.9$ ($\blacksquare$) $\tau_3 = 0.3$ ns ($\blacktriangle$)

B) DIEBLER-MBP+lysolecithin

pH 7 $\tau_1 = 5.4$ ns ($\bullet$) $\tau_2 = 1.9$ ($\blacksquare$) $\tau_3 = 0.1$ ns ($\blacktriangle$)

C) CTAB-MBP

pH 8.5 $\tau_1 = 7.0$ ns ($\bullet$) $\tau_2 = 2.1$ ($\blacksquare$) $\tau_3 = 0.3$ ns ($\blacktriangle$)

CTAB-MBP+lysolecithin

pH 7 $\tau_1 = 7.4$ ns (o) $\tau_2 = 2.0$ (*)

REFERENCES

Boggs, J. M., Moscarello, M. A., and Papahadjopoulos, D., 1982, in: Lipid-Protein Interaction, Vol. 2, P. C. Jost and O. H. Griffith, eds., page 1-51, John Wiley, New York.

Deibler, G. E., Martenson, R. E., and Kies, M. W., 1972, Prep. Biochem., 2:139.

Kies, M. W., Alvord, E. C., Mortenson, R. E., and LeBaron, F. N., 1965, Science, 151:821.

Masotti, L., Cavatorta, P., Szabo, A. G., Farruggia, G., Sartor, G., this volume.

Riccio, P., Masotti, L., Cavatorta, P., DeSantis, A., Juretic, D., Bobba, A., Pasquali-Ronchetti, I., and Quagliariello, E., 1986, Biochem. Biophys. Res. Comm., 134:313.

ANALYSIS OF COMPLEX FLUORESCENCE DECAY DATA

Peter Bayley and Stephen Martin

Division of Physical Biochemistry
National Institute for Medical Research
Mill Hill, London NW7 1AA

There are a number of methods of representing time-resolved fluorescence decay data currently in use, e.g. i) a discrete number of exponential decays with usually n < 3 or 4. ii) a continuous distribution of exponential decays, using sums of either symmetrical or unsymmetrical functions, such as Gaussians or Lorentzians (Alcala et al., 1987); and iii) one or more exponential decays including a diffusional (sqrt t) term to represent a diffusional mechanism (Nemzek and Ware, 1978).

We have therefore examined the behavior of a typical non-linear least squares analysis of simulated decay data representing either simple Gaussian distribution functions or diffusional characteristics. We examine the ability of these complex decays to be fitted within statistical criteria by the sum of a small number of discrete exponentials:

$$I(t) = \Sigma a_i \exp[-t/tau_i] + BL, \ (i=1 \text{ to } n), \text{ cf. (James and Ware, 1985).}$$

The continuous distribution data sets are generated from a Gaussian distribution of amplitudes with Gaussian width, σ, from 0 to 1.5 ns for lifetimes distributed evenly over 41 values from a mean value of tau = 5 ns +/- 3σ. Each lifetime decay curve is convoluted with a 'pump' profile (FWHM = 0.2 ns), and the resultant curves are summed. The peak counts are normalized to a fixed value (e.g. 25000, 250000, etc) and statistical noise added.

The diffusional data is generated from the equation:

$$I(t) = I(0) \cdot \exp [-a \cdot t - 2 \cdot b \cdot sqrt(t)]$$

where a = 1/tau and b is proportional to sqrt(D), D being a diffusion coefficient. For a given value of b, a (Non-convoluted) curve is generated with tau = 5 ns, and normalized with noise added as above.

An individual simulated decay curve is analyzed by a non-linear least squares program based on the Marquardt algorithm, with the statistical criteria of a) residuals; b) χ^2; c) the autocorrelation function; d) the Durbin-Watson parameter (DW) (O'Connor and Phillips, 1984). Acceptable values are DW > 1.65 or DW > 1.75 for single or double exponentials respectively.

RESULTS

(1) Continuous Distributions: At $\sigma = 0.5$, and 25 K peak counts, a single exponential analysis is still acceptable (DW = 1.925); (two components gives DW = 2.168). AT $\sigma = 0.5$, and 250 K peak counts, the single exponential is unacceptable (DW = 0.765); the two component fit is acceptable (DW = 2.008).

As σ increases, the single component analysis breaks down at a point determined by the peak counts: e.g., at 2.5 K peak counts, $\sigma < 1.0$ can be tolerated; at 250 K counts $\sigma > 0.3$ appears as a two component system. DW provides a more sensitive criterion than χ^2.

Even at 250 K peak counts, the two component analysis is valid for up to 1.5 ns. (Note: Only a negligible baseline term (BL) is involved in the two component fit, even at the largest σ; a small baseline term improves the single component fit; this increases with σ, but is no greater than 0.06% at $\sigma = 1.5$ ns). The values of tau1 and tau2 diverge progressively from the central value of 5 ns, independently of the number of peak counts, with typical values of tau1 = 3.30 and tau2 = 6.17 at = 1.5 ns.

DISCUSSION

The results with the continuous distribution of lifetimes show that, on the basis of the data alone, it is not possible to assign a model either of discrete exponentials or a distribution of lifetime parameters with respect to normal statistical criteria.

The range of tolerance of the distribution is somewhat surprising; at a Gaussian width of 10% of the mean value (5 ns) (i.e. $\sigma = 0.5$), the data deviates from a single exponential only if the peak counts are taken to a value in excess of 25 K. This suggests that the analysis is rather insensitive to the 'wings' of the distribution and could probably equally well

Table 1. Results of Multi-Exponential Analysis.

	σ	K	a1	tau1	a2	tau2	BL	χ^2	DW
1.1	0.5	25	0.199	5.02			12.6	1.023	1.92
			0.124	5.40	0.076	4.36	-2.1	0.918	2.17
1.2	0.5	250	0.199	5.03			48.0	2.687	0.77
			0.124	5.40	0.076	4.35	-0.5	0.924	2.01
1.3	0.3	25	0.123	5.33	0.077	4.48	-6.1	0.918	2.01
	0.6	25	0.123	5.50	0.076	4.20	-3.5	1.005	1.94
	0.9	25	0.122	5.71	0.078	3.86	-2.5	1.035	2.06
	1.2	25	0.120	5.93	0.080	3.60	0.8	1.039	1.82
	1.5	25	0.119	6.17	0.080	3.30	-4.3	1.017	1.75
1.4	0.2	250	0.125	5.18	0.075	4.70	-3.3	0.953	2.11
	0.4	250	0.125	5.32	0.076	4.47	-1.8	1.014	1.91
	0.6	250	0.126	5.46	0.074	4.42	1.8	1.001	1.87
	0.8	250	0.125	5.60	0.075	3.99	5.6	1.006	1.87
	1.0	250	0.127	5.74	0.073	3.75	0.7	1.029	1.83

	b	K	a1	tau1	a2	tau2	BL	χ^2	DW
2.1	0.06	10	0.182	4.320			12.3	2.396	0.597
			0.165	4.573	0.029	1.315	0.2	1.132	1.951
2.2	0.08	10	0.159	4.400	0.034	1.060	0.5	1.159	1.630
2.3	0.20	10	0.119	3.640	0.066	0.845	3.4	1.820	1.150
2.4	0.20	10	0.097	3.920					
			0.066	1.587					
			0.036	0.154			-0.6	1.143	1.762

be fitted with alternative functions, e.g. Lorentzians or even 'top hat', with only minor changes in the derived parameters.

The general conclusion is that the analysis of decay curves is necessarily highly degenerate and ultimately model-dependent. With optimal data (i.e. purely random noise) it is not possible to distinguish by the normal statistical criteria whether an experimental curve indicates multiple discrete lifetimes, or a continuous distribution. Similar conclusions are given in James and Ware (1985).

Similar degeneracy is found in the analysis of the diffusional data including sqrt(t); multiple exponentials can readily describe the decay. Highly accurate data at short times is essential for evaluating this case.

In both cases, additional information is essential to justify the
choice of physical model. In the case of an externally added quencher
contributing a diffusional term, the dependence of parameters a and b on
[Q] supports the diffusional mechanism. For intramolecular diffusion (e.g.
movement of an adjacent chemical group) this criterion cannot be applied.
In the case of distributional analysis, the additional information might
derive from data at different emission/excitation wavelengths and different
physical conditions, subjected to global analysis (Beechem et al., 1985).
The need for statistical significance is emphasized; one possible approach
is described by Brochon elsewhere in this volume (Brochon and Livesey).

REFERENCES

Alcala, J. R., Gratton, E., and Prendergast, F. G., 1987, Biophys. J.,
    51:587.
Beechem, J. M., Ameloot, M., and Brand, L., 1985, Analytical Instrum.,
    14:379.
Brochon, J. C., and Livesey, A. K. (This volume).
James, D. R., and Ware, W. R., 1985, Chem. Phys. Lett., 120:455.
Nemzek, J. L., Ware, W. R., 1978, J. Chem. Phys., 62:477.
O'Connor, D. V., and Phillips, D., 1984, 'Time-correlated Photon
    Counting'; Academic Press, London. pp 187-189.

TIME RESOLVED FLUORESCENCE DEPOLARIZATION OF DPH IN THE NEMATIC LIQUID CRYSTAL ZLI-1167

A. Arcioni, F. Bertinelli, R. Tarroni and C. Zannoni

Instituto de Chimica Fisica, Universita,
Viale Risorgimento 4,
40136 Bologna, Italy

A time dependent fluorescence depolarization experiment is reported and analyzed for the first time for a probe dissolved in an aligned thermotropic liquid crystal. We have investigated the popular dye all trans 1,6 diphenyl hexatriene (DPH) in the transparent mesophase ZLI-1167 (a ternary mixture of 4'-propyl, 4'-pentyl and 4'-heptyl, 4-cyano-bicyclohexyls) at a series of temperatures within the nematic and isotropic range. The probe order parameter $\langle P_2 \rangle$ and its perpendicular rotational diffusion coefficient $D_\perp$ have been determined, together with the fluorescence lifetime τ_F in the liquid crystal and isotropic phase. This information is extracted applying a previously proposed theory (Zannoni, 1979) for rotational fluorescence depolarization in a mesophase and introducing a global (Knutson et al., 1983) variant of the target analysis deconvolution procedure (Arcioni and Zannoni, 1984). Our results demonstrate the possibility of employing fluorescence depolarization as a useful technique for investigating aligned mesophases.

Main assumptions made

-DPH probe is rigid and with effective cylindrical symmetry.
-DPH transition moments are parallel to the long axis.
-The reorientation dynamics of the probe is diffusional in a second rank potential.

Global Target Analysis

All measurements in the nematic or in the isotropic phase are analyzed simultaneously and the global reduced chi-square χ_r^2 is minimized in terms of

    i) probe order parameter $\langle P_2 \rangle$

    ii) rotational diffusion coefficients.

Here we assume an Arrhenius type law for the component of the rotational diffusion tensor $\underset{\sim}{D}$ perpendicular to the long axis

$$D_\perp(t) = D_\perp^o \exp(-E_a/RT)$$

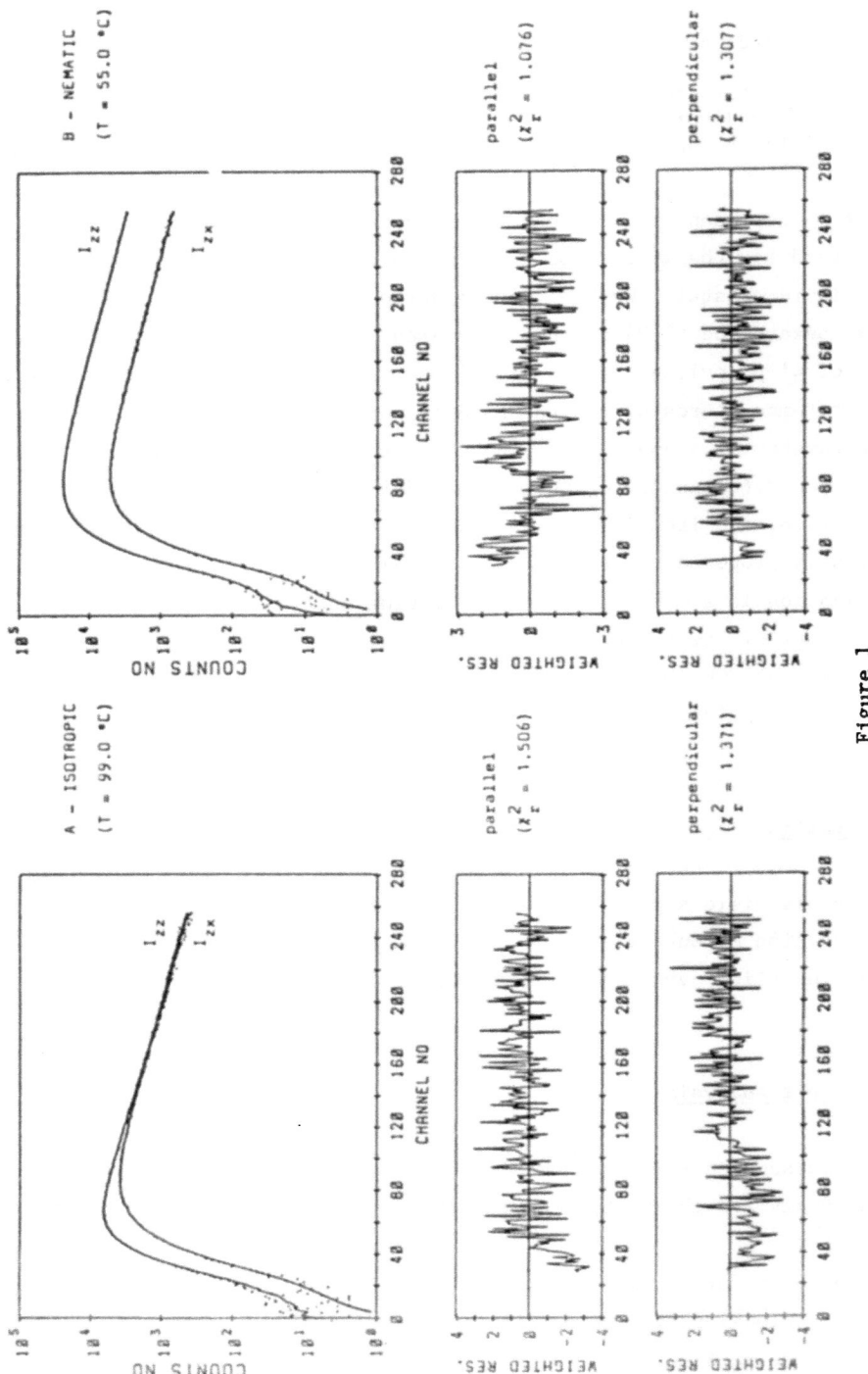

Figure 1

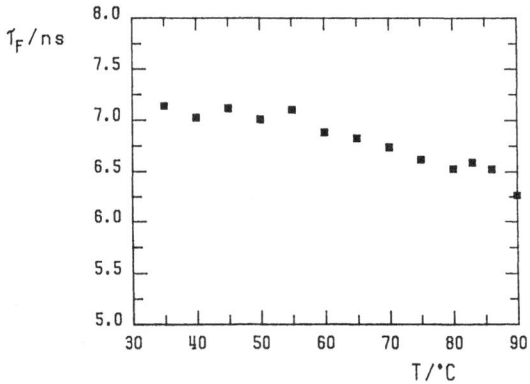

Figure 2

and optimize $D_\perp^\circ$ and the activation energy E_a. The fluorescence decay time of DPH is measured at various temperatures in a separate set of experiments using magic angle analyzer.

RESULTS

Raw polarized intensities $I_{zz}(t)$ and $I_{z\chi}(t)$ for DPH in ZLI-1167 for a temperature in the isotropic (A) and one in the nematic (B) phase are shown

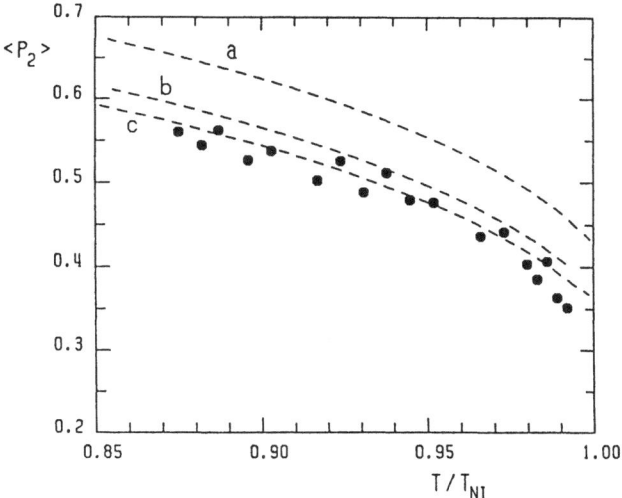

Figure 3

in figure 1. They illustrate clearly the difference in the long time behavior expected for an ordered and a disordered system.

The results for τ_F of DPH in ZLI-1167, obtained from a separate magic angle experiment, are shown in figure 2 as a function of temperature. Using this τ_F we find that all the polarization measurements in the nematic can be fitted simultaneously with a global reduced $\chi_r^2 = 1.65$. The order parameters $<P_2>$ obtained ($\bullet$) are plotted in figure 3 as a function of reduced temperature T/T_{NI}. Here T_{NI} is the nematic-isotropic transition temperature (355.2K). We also find $D_\perp$ in the range $0.2 - 0.5$ ns^{-1}.

The curves in figure 3 are those expected for a Humphries-James-Luckhurst effective potential $U(\beta) = \kappa Tu_2[<0_2P_2(\cos \beta) + \lambda P_4>P_4(\cos \beta)]$ with $\lambda = 0.0$ (a), -0.2 (b), -0.3 (c).

REFERENCES

Arcioni, A., and Zannoni, C., 1984, Chem. Phys., 88:113.

Knutson, J. R., Beechem, J. M., and Brand, L., 1983, Chem. Phys. Letts., 201:501.

Zannoni, C., 1979, Molec. Phys., 38:1813.

MOLECULAR DYNAMICS AND STRUCTURE OF PEPTIDE HORMONES AT MEMBRANE-WATER INTERFACES

Jacques Gallay*, Michel Vincent*, Claude Nicot° and Marcel Waks°

*Laboratoire pour l'Utilisation du Rayonnement Electromagnetique
CNRS-CEA-MEN. Bâtiment 209 D, Universite Paris Sud
91405 ORSAY Cedex France
°E.R. 64 CNRS 45 rue des Saints-Peres
75270 PARIS Cedex France

Peptide hormones of low molecular weight are among the most flexible polypeptides in living organisms (Blundell and Wood, 1982). In order to express their specific functions they must acquire a more restricted bioactive conformation conformation (Kaiser and Kezdy, 1983; 1984). As recently pointed out by Braun et al. (1983), it appears rather unlikely that the initial interaction of peptide hormones with the target cells will result in the immediate formation of a specific complex with the membrane-

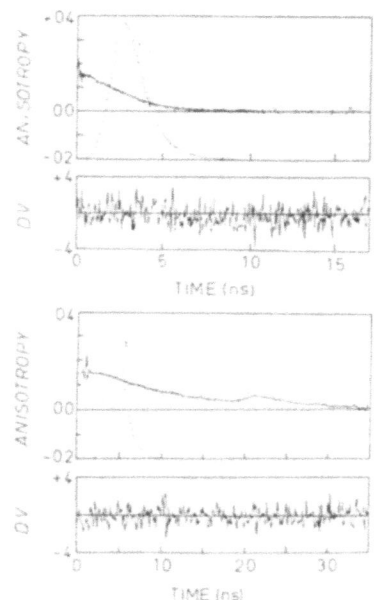

Figure 1. <u>Fluorescence anisotropy decay of Trp-9 in ACTH(1-24).</u>
t: p-terphenyl total intensity decay; dotted curve: experimental anisotropy decay; full curve: calculated decay; DV: deviation function. Upper panel: ACTH(1-24) in phosphate buffer pH 7.6. Biexponential model: $\Phi_1 = 0.10 \pm 0.04$ ns, $\Phi_2 = 1.59 \pm 0.06$ ns, $\beta_1 = 0.106 \pm 0.019$, $\beta_2 = 0.081 \pm 0.004$, $\chi^2 = 1.07$. Time calibration: 0.047 ns/channel. Lower panel: ACTH(1-24) in reverse micelles (water surfactant molar ratio = 5.6. Biexponential model: $\Phi_1 = 0.58 \pm 0.16$ ns, $\Phi_2 = 11.30 \pm 0.38$ ns, $\beta_1 = 0.030 \pm 0.003$, $\beta_2 = 0.128 \pm 0.002$, $\chi^2 = 1.01$. Time calibration: 0.090 ns/channel.

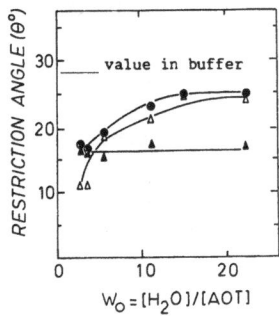

Figure 2. <u>Variation of the restricted angle of rotation</u> (θ) of the Trp residue in the peptides as a function of the water content of the micelles was calculated according to the expression $\beta_2/r_o = [(3\cos^2\theta-1)/2]^2$. (▲) ACTH(1-24, (o) glucagon, (Δ) ACTH(5-10).

bound receptors at the plasma membrane level. Thermodynamic as well as kinetic considerations indeed suggest that the lipid matrix of the cell plasma membrane may have functional implications in the transmembrane signaling process by playing as an "antenna" for the capture of the peptide (Sargent and Schwyzer, 1986). Moreover, the interaction of the peptide with the membrane lipids may select a bioactive conformational state among the many existing in aqueous solutions (Behnam and Deber, 1984; Deber and Behnam, 1985; Schwyzer, 1986). The interfacial characteristics of the lipid matrix (charge density, head-group packing, availability of hydrogen-bond forming groups, mobility of the interfacial water molecules and local proton activity) are likely to be involved in the selection of a dynamic and conformational state of the peptide.

Experimentally, the conformational adaptability of peptide hormones has been investigated in homogeneous solvents and after interaction with surfactant micelles and lipid vesicles in aqueous solutions (Blundell and Wood, 1982; Braun et al., 1983; Gremlich et al., 1983). In the present report, reverse micelles of surfactant/water in organic solvent were chosen to mimic the membrane-water interface since the encased water in such system is thought to be in a physical state comparable in many respects to the interfacial water in biomolecules and more specially in biomembranes (Fendler, 1984). In this study, we have incorporated the peptide hormones ACTH(1-24) and Glucagon in reverse micelles of Sodium bis(2-ethylhexyl) sulfosuccinate (AOT) and water in isooctane. The rotational dynamics of their single tryptophan residue and their secondary structure were monitored in reverse micelles as compared to bulk water, by time-resolved

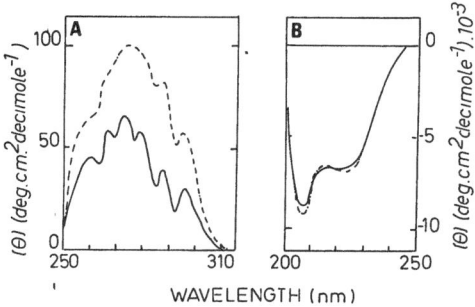

Figure 3. <u>Near (A) and far (B) ultraviolet Circular Dichroism spectra</u> of glucagon in micellar solutions at water/surfactant molar ratios of 3.5 (——) and 22.4 (—·—).

fluorescence anisotropy measurements using the synchrotron radiation as pulsed excitation light source, and Circular Dichroism, respectively. The results evidence:

1) a severe slowing down of the internal motion of the Trp residue for both peptides as shown on figure 1, where the time-resolved fluorescence anisotropy decays in bulk water and in reverse micelles of ACTH(1-24) are compared. The average angle of rotation becomes also more restricted in reverse micelles as compared to water as shown on figure 2. This effect is more pronounced for the basic peptide ACTH(1-24), in closer contact with the negatively charged groups of the surfactant than glucagon.

2) a specific induction of α-helical structure for glucagon and of β-pleated sheet far ACTH(1-24) as evidenced in figures 3 and 4.

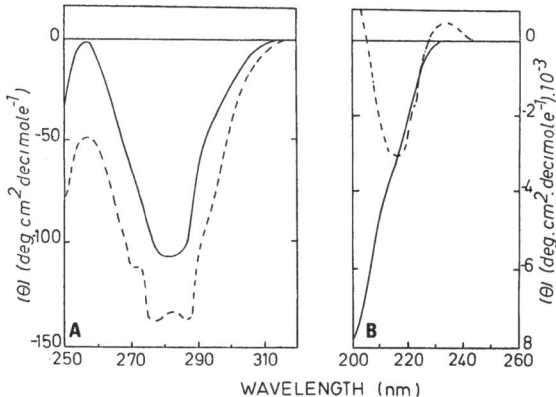

Figure 4. Near (A) and far (B) ultraviolet Circular Dichroism spectra of ACTH(1-24) in micellar solutions at water/surfactant molar ratios of 3.5 (——) and 22.4 (—·—).

These results suggest that the specific amino-acid sequence of each peptide, as well as the physico-chemical properties of the sequestered water, play a prominent role in the mechanism of dynamic and structural adaptability of these peptides at a membrane-water interface.

REFERENCES

Behnam, B. A., and Deber, C. M., 1984, J. Biol. Chem., 259:14935.

Blundell, T., and Wood, S., 1982, Ann. Rev. Biochem., 51:123.

Braun, W., Wider, G., Lee, K. H., and Wuthrich, K., 1983, J. Mol. Biol., 169:921.

Deber, C. M., and Behnam, B. A., 1985, Biopolymers, 24:105.

Fendler, J. H., 1984, Ann. Rev. Phys. Chem., 35:137.

Gremlich, H. U., Fringeli, P., and Schwyzer, R., 1983, Biochemistry, 22:4257.

Kaiser, E. T. and Kezdy, J. F., 1983, Proc. Natl. Acad. Sci. U.S.A., 83:5774.

Kaiser, E. T., and Kezdy, J. F., 1984, Science, 223:249.

Sargent, D. F., and Schwyzer, R., 1986, Proc. Natl. Acad. Sci. U.S.A., 83:5774.

Schwyzer, R., 1986, Biochemistry, 25:4281.

TRYPTOPHAN FLUORESCENCE AS A STRUCTURAL PROBE OF INTERACTION OF AMPHIPHILIC PEPTIDES WITH CALMODULIN

Gautam Sanyal[*] and Franklyn G. Prendergast[+]

[*]Department of Chemistry, Hamilton College, Clinton, NY 13323 USA
[+]Department of Biochemistry and Molecular Biology, Mayo Foundation, Rochester, MN 55905 USA

Several peptides exhibit calcium-dependent high affinity binding (dissociation constant $^{-}10^{-9}$M) to calmodulin (CaM), the ubiquitous calcium binding protein (Malencik and Anderson, 1983; Malencik & Anderson, 1984; Cox et al., 1985). These peptides, such as the three mastoparans, competitively displace enzymes (e.g. myosin light chain kinase) off the 1:1 CaM-enzyme complex to form the CaM-peptide complex (Malencik and Anderson, 1983). Consequently they are considered to be viable structural models for the CaM-interacting surfaces on the CaM-dependent enzymes. It was proposed independently by us and by others, that the ability to form a basic amphiphilic helix is probably the common driving force for these peptides to bind to CaM (Cox et al., 1985; McDowell et al., 1985). The purpose of this study was to probe the environment of the CaM-interacting region of some amphiphilic peptides by using the fluorescence signal from their single tryptophan (Trp) residues (CaM does not contain Trp). The single letter amino acid sequences of these peptides are shown in Table 1. With the exception of melittin the Trp in each peptide forms a part of the hydrophobic face of the putative helical wheel (not shown, see McDowell et al., 1985). They all have large helical hydrophobic moments (Eisenberg et al., 1982) and difference circular dichroic (CD) spectra suggest that a high degree of helicity is induced in the peptides upon CaM binding (Maulet and Cox, 1983; McDowell et al., 1985).

Table 1.

Melittin: G-I-G-A-V-L-K-V-L-T-T-G-L-P-A-L-I-S-W-J-K-R-K-R-Z-Z

Polistes mastoparan: V-D-W-K-K-L-G-Z-H-I-L-S-V-L

Mastoparan X: I-N-W-K-G-I-A-A-M-A-K-K-L-L

CBP1: L-K-W-K-K-L-L-K-L-L-K-K-L-L-K-L-G

CBP2: K-L-W-K-K-L-L-K-L-L-K-K-L-L-K-L-G

Table 2. Tryptophan Fluorescence Maxima (λmax), Steady-State Anisotropies (r_{ss}) and Limiting Anisotropies [$r(o)$ and $r(o)'$] of CaM-Bound Peptides

Sample	λmax[a] (nm)	r_{ss}[b] (at 25°)	$r(o)$[c]	$r(o)'$[d] (extrapolated from temp-perrin plot)
Melittin - CaM	331	0.115	0.186	0.162
Polistes mastoparan-CaM	325	0.149	0.230	0.196
Mastroparan X-CaM	325	0.181	0.230	0.215
CBP1 - CaM	321	0.177	0.218	0.232
CBP2 - CaM	321	0.165	0.231	0.202

[a]For free peptides the λmax values are between 348 and 350 mm.
[b]The r_{ss} values (25°) are between 0.021 and 0.032 for free peptides.
[c]Obtained by extrapolation of $1/r_{ss}$ versus τ plots to $\tau = 0$.
[d]Obtained by extrapolation of $1/r_{ss}$ versus T/η plots to $T/\eta = 0$.

The peptides containing Trp on the hydrophobic face of the helical wheel exhibit a marked blueshift (by 23-28 nm) of Trp fluorescence spectrum upon CaM binding (Table 2). The peptides that contain Trp on the hydrophilic face show a less marked blueshift (e.g. 17 nm for melittin) of Trp fluorescence upon binding to CaM. This is suggestive of a direct involvement of the hydrophobic surface of the peptide in forming the interface with CaM. We have made quantum yield (Φ) measurements for the single Trp residue in the free and CaM-bound peptides, using tryptophan (Φ = 0.14) as the standard. The changes in the Φ values, upon CaM binding, have not shown any apparent correlation with the position of Trp in the peptide. For example, for CBP1 the Φ is increased from 0.10 in the free peptide to 0.17 in the CaM-bound peptide. However, for mastroparan X, which also has a Trp 3 on the apolar face of the putative amphiphilic helix (presumably formed on the CaM surface), Φ is decreased from 0.09 to 0.04 as a result of binding to CaM. These results suggest that local interactions of the bound peptide with residues on the interacting CaM surface may be the major determinant of Φ for the Trp of the CaM-peptide complex. The steady-state Trp fluorescence anisotropy (r_{ss}) values were increased from ~0.025 for the free peptide to $\geq$015 (at 25°) for the CaM-bound peptide for peptides containing Trp on the apolar face (Table 2). Lifetime (τ)

resolved anisotropy measurements using the technique of oxygen quenching (Lakowicz and Weber, 1980) yielded limiting anisotropy values [r(o)] at extrapolated zero τ which are listed in Table 3. These values, when compared with the r_o value obtained for motionally "frozen" Trp of several proteins in vitrified solutions at a very low temperature (Lakowicz and Weber, 1980; Lakowicz et al., 1983; Sanyal et al., 1987), indicate a limited degree of local Trp motion in the CaM-bound peptides. For melittin, where the Trp is presumably not an integral part of the hydrophobic CaM-peptide interface, the extent of local Trp mobility is greater compared to the other CaM-bound peptides. Similar conclusions can be drawn from determinations of temperature (T) dependence of r_{ss} which yielded, independently, limiting anisotropies [r(o)'] by extrapolation to the limit of T/η approaching zero (η = viscosity). These data are also shown in Table 2. The modified perrin plots ($1/r_{ss}$ vs τ and $1/r_{ss}$ vs T/η) were monophasic and the rotational correlation times determined from these plots reflected the slower overall motion of the CaM-peptide complex. This suggests that the correlation time for the local Trp motion was faster than the shortest τ (~1 nsec) that was accessible in our measurements.

The accessibility of the Trp residue to acrylamide was significantly reduced (6-13 fold) upon CaM binding for peptides containing Trp on the apolar face, again suggesting involvement of Trp in the hydrophobic peptide-CaM interface. The marked decrease in accessibility of the peptide Trp to iodine upon CaM binding is possibly due to the presence of negative charges in the peptide binding region of CaM. The quenching data are given in Table 3 in terms of the bimolecular quenching (k_q) obtained from Stern-Volmer plots assuming collisional quenching.

Ultraviolet circular dichroic difference spectra in the aromatic region have shown that Trp ellipticity is induced (a negative maximum at 291 nm and a smaller negative maximum at 284 nm) by CaM in CBP1, CBP2 and mastroparan X but not in melittin. This is consistent with the Trp being frozen in an asymmetric environment at the CaM-peptide interface for the former peptides.

It is important to point out that for the analysis of the steady-state fluorescence data an average τ value for each peptide Trp residue was used. These were determined from the weighted maxima of curves describing a continuous distribution of τ (Gratton and Barbieri, 1986) measured by means of multiple frequency phase fluorometry using modulation frequencies from 10 to 250 MHz (Gratton et al., 1984).

Table 3. Bimolecular Rate Constants for Collisional Quenching of Tryptophan Fluorescence in Free and CaM-Bound Peptides by Acrylamide and Iodide

$$kq \times 10^{-9}(M^{-1}s^{-1}) \text{ at } 25°C^{a}$$

Peptide	acrylamide		iodide	
	Free	CaM-Bound	Free	CaM-Bound
Melittin	5.2	1.2	7.1	0.11
Polistes mastoparan	3.5	0.60	3.2	0.16
Mastoparan X	3.9	0.39	6.0	<0.10
CBP1	3.2	0.24	6.2	0.15

^{a} $kq = \dfrac{K_{sv}}{\tau}$ where K_{sv} is the slope of the Stern-Volmer plot according to the equation $F_{o}/F = 1 + K_{sv}[Q]$, F_{o} being the unquenched fluorescence intensity and F being the intensity at a quencher concentration of $[Q]$.

In conclusion, our results indicate that the CaM-peptide interface is formed by a hydrophobic surface of CaM and the apolar face of the amphiphilic peptide helix. One would predict that the environment sensitivity of Trp fluorescence spectrum could be used to determine the periodicity of the CaM-bound peptide structure by using similar amphiphilic peptides with Trp placed in different locations of the sequence. This provides an excellent model for studying structures of peptides strongly bound to protein surfaces once the nature of the protein surface is known.

ACKNOWLEDGMENTS

We are grateful to Dr. William F. Degrado for kindly providing us with CBP1, CBP2 and other amphiphilic peptides. We thank Dr. Enrico Gratton for helping us with measurements and analysis of lifetime data. This work was supported in part by American Chemical Society PRF 18076-GB4 and Research Corporation Grants (G.S.), and by National Institutes of Health Grant GM 34847 (F.G.P.)

REFERENCES

Cox, J. A., Comte, M., Fitton, J. E., and Degrado, W. F., 1985, J. Biol. Chem., 260:2527

Eisenberg, D., Weiss, R. M., and Terwilliger, T. C., 1982, Nature (London), 299:371.

Gratton, E., and Barbieri, B., 1986, Spectroscopy, 1:28.

Gratton, E., Jameson, D. M., and Hall, R. D., 1984, Ann. Rev. Biophys. Bioeng., 13:105.

Lakowicz, J. R., and Weber, G., 1980, Biophys. J., 32:591.

Lakowicz, J. R., Maliwal, B. P., Cherek, H., and Balter, A., 1983, Biochemistry, 22:1741.

Malencik, D. A., and Anderson, S. R., 1983, Biochem. Biophys. Res. Commun., 114:50.

Malencik, D. A., and Anderson, S. R., 1984, Biochemistry, 23:2420.

Maulet, Y., and Cox, J. A., 1983, Biochemistry, 22:5680.

McDowell, L., Sanyal., G., and Prendergrast, F. G., 1985, Biochemistry, 24:144.

Sanyal, G., Charlesworth, M. C., Ryan, R. J., and Prendergast, F. G., 1987, Biochemistry, 26:1860.

FLUORESCENCE DECAY MEASUREMENTS OF 1-METHYLINDOLE IN ALCOHOL SOLUTIONS: DIRECT OBSERVATION OF SOLVENT REORIENTATION KINETICS AT AMBIENT TEMPERATURES

A. G. Szabo and S. Yamashita

Division of Biological Sciences
National Research Council
Ottowa, Canada, K1A OR6

The understanding of the photophysical processes of indoles is crucial to the application of fluorescence spectroscopy of the indole amino acid, tryptophan in studies of protein structure and dynamics. The elucidation of the origin of the large Stokes shift of indole fluorescence spectra when in polar environments continues to be an important issue. It is well established that there is a large change in the dipole moment of indoles when they are excited into lowest singlet state. Earlier we had shown that in dilute alcoholic solutions of indoles, 1-methylindole (1-MeI) formed weak ground state complexes with alcohols (Skalski et al., 1980). In this report we present time resolved fluorescence data on 1-MeI in alcohols which provides direct evidence for the solvent reorganization in these complexes and show that it is viscosity dependent.

MATERIALS AND METHODS

All solvents were of the highest grade spectral purity available. Octanol was purified by distillation from KOH. 1-MeI was purified prior to use by sublimation under reduced pressure. Time resolved fluorescence measurements were made on instrumentation described previously (Zuker et al., 1985) with the only change being the use of a Hamamatsu 1564U microchannel plate photomultiplier tube as the detector for the emission. The instrument response function was recorded at the excitation wavelength. The channel width was 21.6 ps and the data was collected in 1024 channels. The data was analyzed using a non-linear least squares iterative convolution technique. The results presented herein were those which satisfied the criteria for the adequacy of fit which have been discussed earlier (MacKinnon et al., 1977).

RESULTS AND DISCUSSION

The fluorescence maxima of 1-MeI in methanol, ethanol, propanol,

Table 1. Fluorescence Decay Times of 1-MeI in Alcohols.

Alcohol	$1/k_E$ (ns)	$1/(k_A + k_{EA})$ (ns)
propanol	5.12	0.072
butanol	5.51	0.102
octanol	5.38	0.345

Table 2. Effect of Temperature on the Fluorescence Decay Times of 1-MeI in Octanol.

Temperature (°C)	$1/_{kE}$ (ns)	$1/(k_A + k_{EA})$ (ns)	k_{EA} ($10^9 s^{-1}$)
15	5.35	0.39	2.37
20	5.32	0.34	2.94
30	5.26	0.24	4.08
40	5.24	0.17	5.68
50	5.18	0.11	8.96

butanol, and octanol were located at 331 nm, 325 nm, 324 nm, 322 nm, and 316 nm, respectively. While the spectrum of 1-MeI in octanol had the lowest spectral maximum the spectrum rose sharply on the high energy side and then had a very broad tail, suggesting a mixture of spectral components. Measurement of the fluorescence decay of 1-MeI in these alcohols provided an interesting set of results. In the case of methanol and ethanol single exponential decay kinetics were observed at all wavelengths (MeOH, 5.25 ns; EtOH, 6.05 ns). In the other alcohols, however, the decay profiles could only be adequately fit by double exponential decay parameters at most wavelengths. At 325 nm for octanol, 330 nm for butanol, and 335 nm for propanol the fluorescence decay could be satisfactorily fit to single exponential decay kinetics, the decay time being the same as the longest (ca. 5.2 ns) decay time found for the double exponential decay parameters. When double exponential decay kinetics were observed the same two decay times were observed over the entire wavelength range for the respective alcohols and the data could be analyzed using global analysis procedures (Knutson et al., 1982). Below the wavelengths listed above, the pre-exponential terms were always positive while above these wavelengths the pre-exponential term of the short decay component was negative. The decay times are summarized in Table 1.

The data can best be rationalized in terms of the following scheme where A* is the initially excited 1-MeI/alcohol ground state complex and E* is the equilibrated complex which results after solvent reorganization occurring because of the change in dipole moment when the 1-MeI is excited.

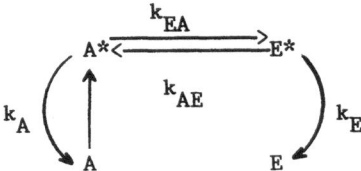

The kinetic expression which describes this behavior may be written as:

$$(I_A + I_E)(\lambda,t) = C_E(\lambda) \cdot A_o^* \cdot k_{FE} \cdot k_{EA} \cdot \exp^{-\gamma 1 \cdot t} / (k_A + K_{EA} - k_E)$$

$$+ (C_A(\lambda) \cdot A_o^* \cdot k_{FA} - C_E(\lambda) \cdot A_o^* \cdot k_{FE} \cdot k_{EA} / (k_A + k_{EA} - k_E)) \cdot \exp^{-\gamma 2 \cdot t}$$

where $\gamma 1 = k_e$ and $\gamma 2 = k_A + k_{EA}$ and it has reasonably been assumed that $k_{AE} \ll k_{EA}$ or even negligible. $C_A(\lambda)$ and $C_E(\lambda)$ are the spectral distributions of each component. Single exponential decay kinetics would then be observed when the pre-exponential of the second term equals zero.

The fluorescence decay of 1-MeI in the different alcohols was measured at several temperatures. In interests of brevity only the octanol results at 310 nm are presented in Table 2. Very similar values were also obtained for octanol at 360 nm. The long decay time is hardly affected over the temperature range studied. The relatively constant value of the long decay time component with temperature and for all the different alcohols allows some simplifying assumptions. It may be reasonable as a first approximation that the main deactivation pathway for A* and E* is radiative so that $k_{FE} = k_{FA}$ and so $k_E = k_A$. Then k_{EA} may be estimated at each temperature. Also this allows the resolution of the spectra of the two components since it can be shown that:

$$C_A(\lambda)/C_E(\lambda) = a_1/a_2 + 1$$

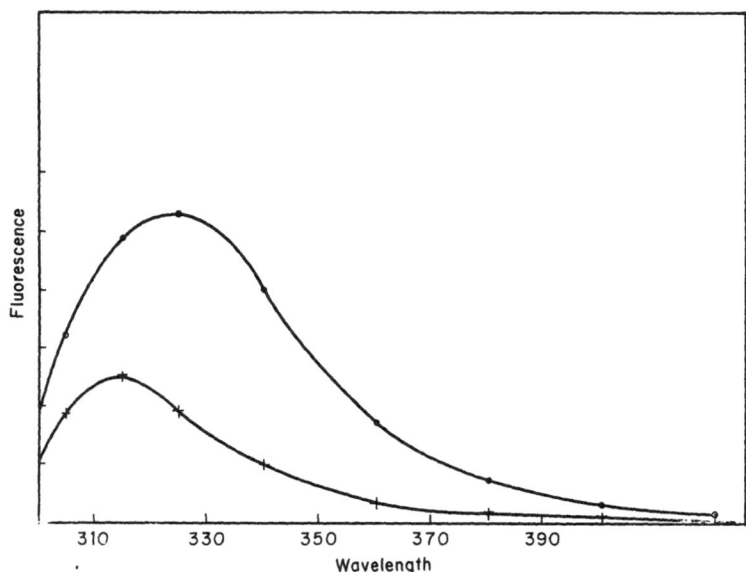

Figure 1. Fluorescence spectra of the two decay components in octanol assuming that $k_A = k_E$. a) 5.32 ns; b) 0.34 ns.

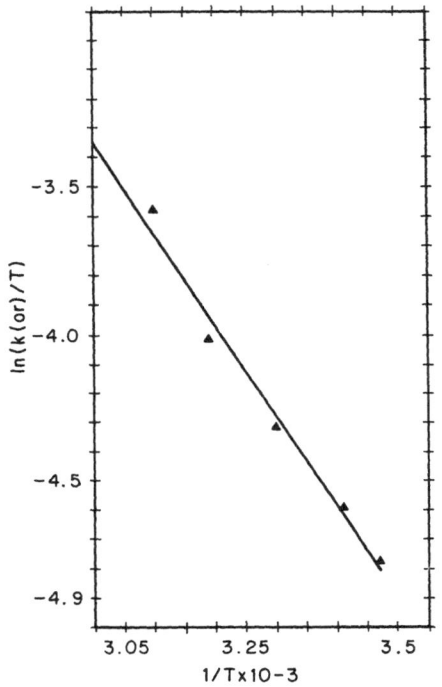

Figure 2. Plot of $\ln (k_{AE}/T)$ vs $1/T$.

The spectra of the two components are shown in Figure 1 and the spectral maximum of E* (325 nm) is shifted ca 10 nm from that of A* (315 nm).

A plot of $\ln(k_{EA}/T)$ vs $1/T$ (5) for the data in Table 2 is linear and provides an estimate of ΔE - 4.3 kcal mol^{-1}. This value is consistent with the proposal that the dielectric solvent reorganization around the 1-MeI excited singlet state is viscosity dependent. We conclude that the solvent molecules on the initially excited 1-MeI/alcohol complex undergoes a reorganization to form a complex with a different solvent structure in which their dipoles are realigned with the excited state dipole moment of 1-MeI.

REFERENCES

Knutson, J. R., Wallbridge, D. G., and Brand, L., 1982, Biochemistry, 21:4671.

MacKinnon, A. E., Szabo, A. G., and Miller, D. R., 1977, J. Phys. Chem., 81:1564.

Skalski, B., Rayner, D. M., and Szabo, A. G., 1980, Chem. Phys. Lett., 70:587.

Zuker, M., Szabo, A. G., Bramall, L., Krajcarski, D. T., and Selinger, B., 1985, Rev. Sci. Instrum., 56:14.

TIME-RESOLVED FLUORESCENCE STUDIES ON HORSE SPLEEN FERRITIN

N. Rosato[*], A. Finazzi Agro[+], E. Gratton[⊖], S. Stefanini[*] and
E. Chiancone[*]

*CNR Centre of Molecular Biology and Department of Biochemical
Sciences, University of Rome "La Sapienza" Rome, Italy
[+]Department of Experimental Medicine and Biochemical Sciences,
University of Rome "Tor Vergata" Rome, Italy
[⊖]Department of Physics, University of Illinois, Urbana, IL 61801
USA

The horse spleen apoferritin polymer (M=450,000) is characterized by
an intrinsic fluorescence emission at 315 nm when excited at 280 nm. The
emission maximum is shifted at 325 nm upon dissociation of the polymer into
subunits. We have studied the decay of fluorescence of the apoferritin
polymer and its subunits.

Table 1 shows a set of data relative to the polymer and the subunits
at pH 3.5 measured with the phase-shift technique. The experiments have
been performed at this pH to avoid extensive polymerization of subunits,
which under these conditions are essentially dimers. No significant
differences between the fluorescence data of the polymer at pH 3.5 and at
pH 7 were observed. The table shows that in the case of the polymer
fitting with a continuous distribution of lifetimes is substantially better
than that using two discrete values of lifetimes. The same is true when
three lifetimes were used. In contrast, no dramatic difference was observed
with the subunits using the two lifetimes analysis. Furthermore, two
gaussians were required to fit the data of the polymer while two lorentzians more adequately fitted the fluorescence decay of subunits. It must
be stressed that in the limit of width of 0.2 ns the fitting is not significantly improved using continuous distribution rather than a single value
of lifetime.

The increasing value of the width value going from apoferritin subunits
to polymer (Table 1) might be explained as a heterogeneity brought about by
the polymerization process. In fact, it should be recalled that the
apoferritin might show two kinds of heterogeneity - one due to the presence
of a variable amount of heavy and light subunits, the other related to the
assembly of the protein shell according to a 4:3:2 symmetry. These two
factors may contribute to the formation of slightly different environments
around the single tryptophan present in each subunit.

Table 1. Fluorescence decay parameters (phase shift technique)

Sample	Two exponential analysis				Two Gaussian continuous distribution analysis						
	τ_1	τ_2	f_1	χ^2	C_1	W_1	C_2	W_2	f_1	χ^2	F 2G/2τ
Apoferritin polymer pH 7	0.4	3.3	.29	11	0.8	0.4	4.3	1.7	.45	1.8	0.01
Apoferritin polymer pH 3.5	0.5	3.7	.46	12	0.7	0.3	3.9	2.5	.48	3.8	0.05
Subunits, pH 3.5	1.6	6.4	.58	6	1.8	0.4	6.8	0.4	.65	6	0.5
Subunits, pH 1.8	1.9	7.2	.61	5	1.8	0.1	7.5	0.2	.62	3.6	0.5

τ_1, τ_2 = lifetimes in ns, ($\Delta\tau_1 \sim \Delta\tau_2 \sim 0.2$ ns)

f_1 = fraction of the total fluorescence intensity relative to τ_1 (or C), ($\Delta f_1 \sim 0.05$)

C_1, C_2 = centers of Gaussian distributions in ns ($\Delta C_1 \sim \Delta C_2 \sim 0.2$ ns)

W_1, W_2 = widths of Gaussian distributions in ns ($\Delta W_1 \sim \Delta W_2 \sim 0.2$ ns)

χ^2 = reduced chi-square

F2G/2τ = significance level of F-test between two Gaussian distribution analysis and two exponential analysis

The addition of iron fully quenches the fluorescence of apoferritin when more than 200 irons per mole of polymer are bound. We have determined how the binding of the first 50 and 100 irons per mole influences the lifetime of fluorescence. Table 2 shows that the average lifetime is decreased upon binding of iron but not proportionally to the decrease in quantum yield. From this observation it is inferred that iron quenches the fluorescence both by static and dynamic mechanisms. Therefore, upon binding of iron, a perturbation of tryptophan not observable by CD becomes operating either by short range (contact quenching) or by long range (conformational changes) interactions.

Table 2. Effect of FE(III) on Apoferritin Fluorescence Lifetime

Sample	C1	W1	C2	W2	f1	$\langle\tau\rangle$	q
Apoferritin	1.6	2.7	8.5	8.2	0.48	1.0	48
+50 Fe (III)	0.1	35.0	6.0	10.0	0.10	0.9	38
+100 Fe (III)	0.1	25.0	6.0	10.0	0.15	0.8	24

$\langle\tau\rangle$ = average lifetime
q = quantum yield
    other symbols as for Table 1

METAL SUBSTITUTION AND FLUORESCENCE MODIFICATIONS IN ASCORBATE-ALKALINE PHOSPHATASE INTERACTION

G. E. Martorana, E. Meucci, G.A.D. Miggiano, A. Mordente, S. A. Santini and A. Castelli

Institute of Biological Chemistry
Catholic University
Rome, Italy

Alkaline phosphatase (EC 3.1.3.1), a zinc metallo-enzyme, is irreversibly inhibited by ascorbate, probably through a modification of active site environment (Martorana et al., 1986; Miggiano et al., 1984). A metal substituted form was prepared from bovine kidney alkaline phosphatase purchased from Calbiochem (La Jolla, CA, U.S.A.) and further purified as described (Martorana et al., 1986). Cobalt can replace zinc yielding a partially active phosphohydrolase, which lacks phosphotransferase activity (Ensinger et al., 1978) and presents a relevant visible absorption (Simpson and Vallee, 1968).

Steady state fluorescence spectroscopy was performed in an LS5 Luminometer (Perkin-Elmer, Beaconsfield, Bucks, U.K.) and absorbance was measured with an HP 8450A UV-Vis diodes-array spectrophotometer (Hewlett-Packard, Palo Alto, CA, U.S.A.), as previously described (Martorana et al., 1986). Excitation wavelength was set at 300 nm, where ascorbate does not present any significant absorption (A<0.05). Measurements were done at 25 (±0.1)°C with all the experiments carried out at least in duplicate.

The apoenzyme was prepared from Zn^{2+}-alkaline phosphatase by treatment with Chelex 100 and exhaustive dialysis (10 days) against large volumes of metal-free water (Csopak et al., 1972). The residual activity of the apoenzyme, which contains less than 1% of that of the native protein during the first 5 minutes of enzymatic assay, performed according to Miggiano et al. (1984). Enzyme forms were prepared by incubation (6 h) with spectrally pure metal ions (Puratronic, Johnson Matthey Chemicals, Royston, U.K.; 0.01 M cobalt sulfate in 0.003 M HCl or 0.01 M zinc chloride in water) and further dialyzed to remove unbound ions.

Zinc removal and replacement do not significantly modify the protein fluorescence. Only a slight increase in the emission can be observed in the Zn-compared to the Apo-enzyme (Figure 1A), similarly to what observed in the interaction of free tryptophan with zinc, where the metal ion

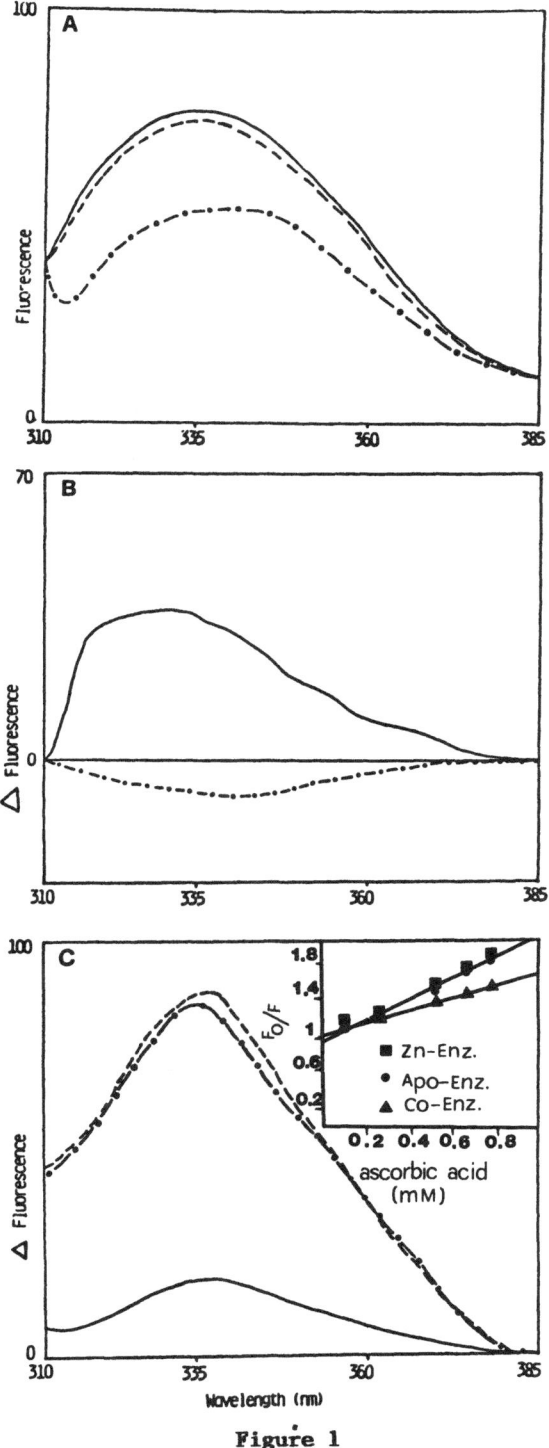

Figure 1
(Continued on next page)

displaces a quenching proton (Chen, 1971). Cobalt, instead, provokes a decrease of the emission. With excitation at 300 nm, the difference spectrum reflects the perturbation of internal tryptophans (Figure 1B). As already reported for the native enzyme (Martorana et al., 1986), ascorbate induces a marked quenching of the three protein forms (Figure 1C), among which Cobalt-enzyme seems the least affected (fluorescence is decreased in the order Zn-~Apo->Co- enzyme) (Figure 1C, inset). The shapes of the emission spectra are not greatly modified by ascorbate addition, particularly in the case of Cobalt-enzyme.

Difference absorption spectra in the far U.V. before and after ascorbate addition present a maximum at 240-245 nm preceded by a valley at about 212 nm (Figure 2A). These features are quite evident for the Apo- and Zn-enzyme and document a red shift, suggesting a change in the ionization of imidazole and thiol groups, beside that of tyrosines, contributing to the 242 nm difference peak (Donovan, 1969). Cobalt-enzyme again displays less pronounced modifications. The visible spectrum of the Cobalt-enzyme preparation indicates a form, similar to E. coli alkaline phosphatase, not absolutely free from phosphate (Chappelet-Tordo et al., 1974) and zinc (Simpson and Vallee, 1968). Absorbance is, however, decreased upon ascorbic acid addition (Figure 2B). Cobalt-enzyme, which retains only about 25% of the original activity, is also less inactivated than Zinc-enzyme by the same effector amount (with 0.28 mM ascorbate, 20% versus 40%, respectively). Ascorbate varying effect on the spectral characteristics of the metal substituted forms may then reflect their different kinetic and structural properties.

Even if spectroscopy is not conclusive regarding whether or not there is binding to the metal, a chelating action by ascorbate (Martorana et al.,

Figure 1: **A)** Fluorescence of 0.36 mg/ml Apo-ennzyme (- - - -); Zinc-enzyme (————) and Cobalt-enzyme (-•-•-•-). **B)** Difference fluorescence spectra: Apo-enzyme minus Zinc-enzyme (-•-•-•-) and minus Cobalt-enzyme (- - - -). Spectra multiplied threefold. **C)** Difference fluorescence spectra: untreated minus treated (0.5 mM ascorbate) Apo-enzyme (-•-•-•-), Zinc-enzyme (- - - -) and Cobalt-enzyme (————). Spectra multiplied fourfold. **Inset:** Stern-Volmer plots of ascorbate quenching of alkaline phosphatase. Slits: 5 nm; scan speed 240 nm/min, excitation 300 nm. Intensity in arbitrary units. Other experimental condiditons as in Martorana et al. (1986).

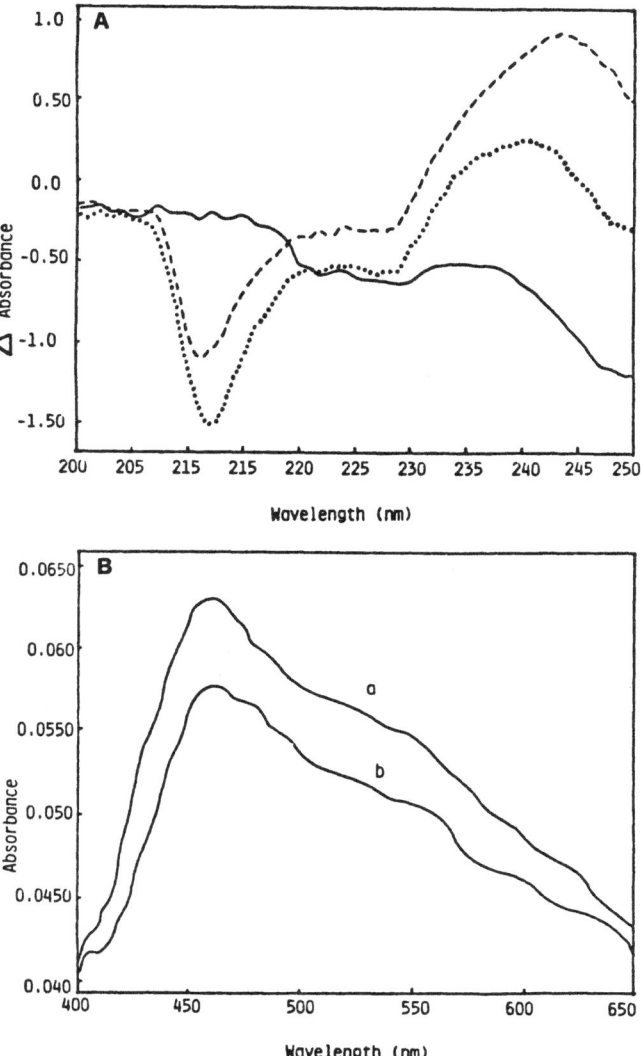

Figure 2. **A)** Difference absorption spectra between treated (0.5 mM ascorbate) and untreated phosphatase: (• • • • •) Apo-enzyme; (- - - -) Zinc-enzyme and (————) Cobalt-enzyme. **B)** Absorption spectra of Cobalt-enzyme (a) and the same treated with 0.5 mM ascorbate (b). Experimental conditions as in Martorana et al. (1986).

1986) can be excluded, since the activity cannot be restored, even partially, after ascorbate treatment, dialysis and addition of zinc and magnesium ions in molar excess.

Beside the structural changes, which could modify the conformational flexibility necessary for alkaline phosphatase catalysis, a charge transfer mechanism can, in our opinion, explain the spectral and kinetic data. Ascorbate inhibition could be provoked by damage at the topographical level (i.e., active site and/or metal-subsite), involving the effector binding at some molecular domain and migration of the electron through a rather low efficiency pathway (Myer et al., 1980). The effect produced is transfered to the metal catalyst, as shown by the modification of the visible spectrum of the Cobalt-enzyme.

REFERENCES

Chappelet-Tordo, D., Iwatsubo, M., and Lazdunski, M., 1974, Biochemistry, 13:3754.

Chen, R. F., 1971, Arch. Biochem. Biophys., 142:552.

Csopak, H., Falk, K. E., and Szajn, H., 1972, Biochim. Biophys. Acta, 258:466.

Donovan, J. W., 1969, in: "Physical Principles and Techniques of Protein Chemistry", S. J. Leach, ed., Academic Press, New York.

Ensinger, H. A., Pauly, H. E., Pfleiderer, G., and Stiefel, T., 1978, Biochim. Biophys. Acta, 527:532.

Martorana, G. E., Meucci, E., Ursitti, A., Miggiano, G.A.D., Mordente, A., and Castelli, A., 1986, Biochem. J., 240, in press.

Miggiano, G.A.D., Mordente, A., Martorana, G.E., Meucci, E., and Castelli, A., 1984, Biochim. Biophys. Acta, 789:343.

Myer, Y. P., Thallum, K. K., and Pande, A., 1980, J. Biol. Chem., 255:9666.

Simpson, R. T., and Vallee, B. L., 1968, Biochemistry, 7:4343.

Jorge E. Churchich and John A. Cidlowski

Department of Biochemistry, University of Tennessee, Knoxville,
TN 37999 USA
Department of Physiology, Biochemistry and Nutrition, University
of North Carolina, Chapel Hill, NC 27514 USA

INTRODUCTION

The interaction between gaba·transaminase and succinic semildehyde
dehydrogenase, two mitochondrial enzymes, have been studied using affinity
chromatography, polarization of fluorescence and gel filtration techniques
(Hearl and Churchich, 1984). In this report, we have investigated the
interaction between gaba·transaminase and monoclonal anti·p·pxy antibodies.

A variety of monoclonal antibodies to vitamin B_6 have been discovered
by Viceps-Madore et al. (1983). One of these antibodies (E_6[2]2) recog-
nizes native and reduced gaba·transaminase as demonstrated by elisa,

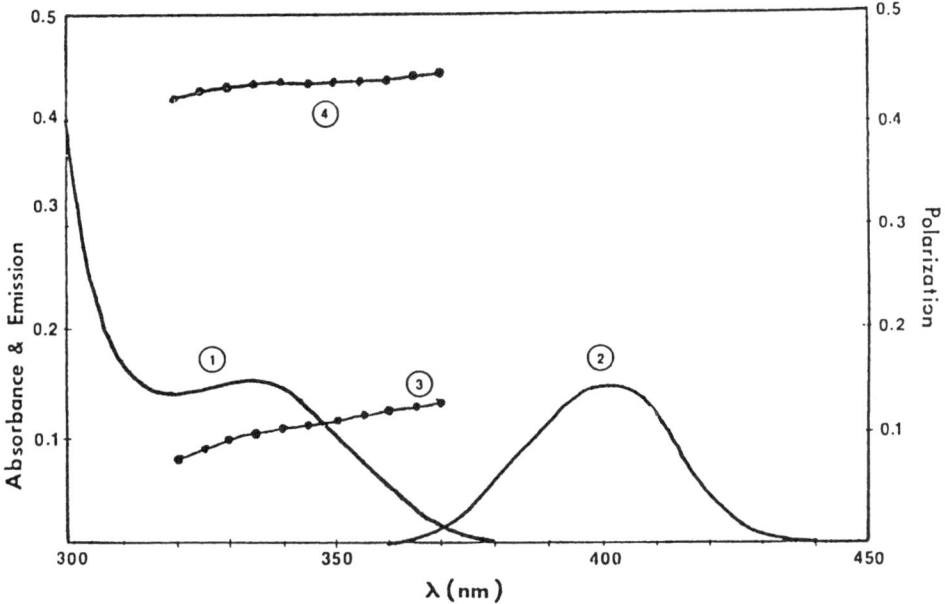

Figure 1. Absorption (1), emission (2) and polarization excitation spectra
(3) of PMP (5 μM) bound to monoclonal antibody in 0.1 M ethanolamine HCl
(pH 7). The polarization excitation spectrum of PMP in glycerol (4°C) is
included in the figure.

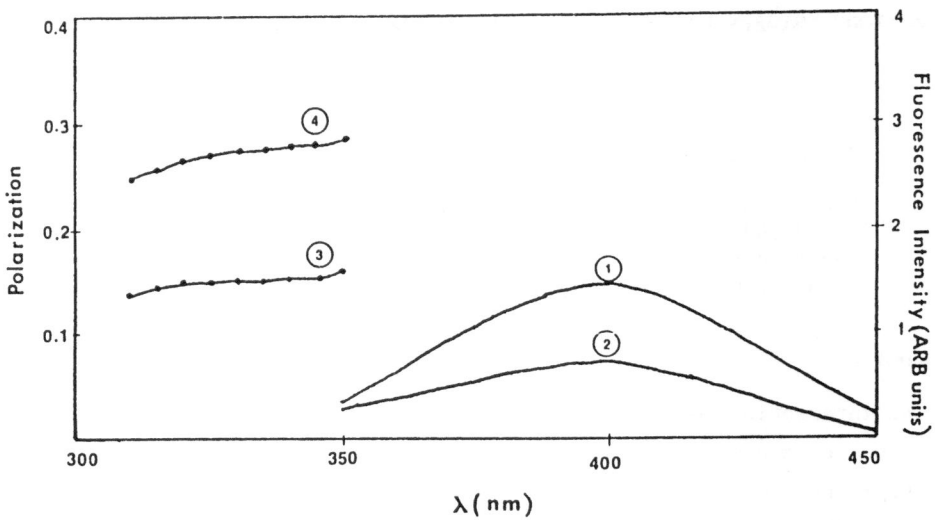

Figure 2. Fluorescence changes induced by binding of P-Pyr-GABA Transaminase (3 μM) to monoclonal antibody (20 μM) at pH 7 in 0.1 M triethanolamine-HCl buffer (pH 7). Emission spectra of P-Pyr-GABA Transaminase in the absence (1) and presence (2) of antibody. Polarization excitation spectra of P-Pyr-GABA Transaminase in the absence (3) and presence (4) of antibody.

immunoblotting and affinity chromatography techniques. Since pyridox-amine-5-p and reduced gaba-transaminase (p-pxy-enzyme) are fluorescent, it is possible to detect the binding of either the vitamin or the derivatized enzyme to specific monoclonal antibodies by means of fluorescence methodology.

RESULTS AND DISCUSSION

The binding of pyridoxamine-5-p to the monoclonal antibody (E$_6$[2]2) was monitored by polarization of fluorescence and by saturation analysis using the procedure developed by Viceps-Madore et al. (1983).

Purified monoclonal antibody, prepared by chromatography on QAE·sephadex G-50, binds pyridoxamine-5-p with a dissociation constant K$_D$=2x10^{-7}M. A similar K$_D$ value was obtained when a fixed concentration of antibody (1x10^{-7}M) was titrated with increasing concentrations of pyridoxamine-5-p and the binding process monitored by steady state polarization of fluores-cence.

As shown in figure 1, binding of pyridoxamine-5-p to the monoclonal immunoglobulin has no effect on the quantum yield of the ligand which exhibits an emission maximum at 400 nm and excitation maximum at 330 nm.

The polarization excitation spectrum of bound pyridoxamine-5-p is characterized by polarization values ranging from 0.1 to 0.15 over the excitation region coinciding with the longest wavelength absorption band of the ligand. Although an increase in the polarization values of pyridoxamine-5-p complexed to the immunoglobulin is easily detected over a wide range of excitation, it should be noted that pyridoxamine-5-p immersed in a viscous solvent (glycerol/water, 98/2) exhibits polarization values expected for a chromophore whose transition moments of absorption and emission are parallels (Figure 1).

Thus, the steady-state polarization values obtained for pyridoxamine-5-p in the presence of antibody indicate that the vitamin is not completely immobilized by the antibody binding site.

Additional experiments were therefore carried out to gain an understanding of the dynamics process. Nanoseconds excitation pulses and a single photon counting apparatus (Massey and Churchich, 1979; O'Connor and Phillips, 1984) were used to measure the emission kinetics of pyridoxamine-5-p bound to the monoclonal antibody. The mean lifetime of bound pyridoxamine-5-p bound to the antibody is 2.2 nanoseconds. In addition, the fluorescence anisotropy decays are not single exponentials; they include a short component ($\Phi=2$ nanoseconds), associated with a high amplitude, and a long component ($\Phi=80$ nanoseconds) of low amplitude.

The initial anisotropy decays very fast and it is lower (A=0.2) than the expected fundamental anisotropy ($A_o=0.32$); suggesting significant subnanosecond fluctuations of bound pyridoxamine-5-p.

Figure 2 shows the steady-state polarization of fluorescence values of reduced gaba·transaminase in the presence and absence of monoclonal antibody.

Reduced gaba·transaminase, carrying 2 moles of p-pxy residues for dimer, exhibits low polarization values; whereas the reduced enzyme in the presence of antibody displays higher polarization values (Figure 2).

As shown in figure 2, the binding of reduced enzyme to the monoclonal antibody is also accompanied by 50% quenching of p-pxy fluorescence.

The polarization of fluorescence results, taken together with the quenching, of p-pxy emission, are interpreted to mean that further immobilization of the p-pyridoxyl chromophode is the consequence of protein-protein interactions. These studies demonstrate that steady and nanosecond fluorescence techniques are suitable tools to study the interaction of vitamin B_6 dependent enzymes with specific monoclonal antibodies.

ACKNOWLEDGEMENTS

This work was supported by a grant from the National Science Foundation (BNS-8510237).

REFERENCES

Hearl, W. G., and Churchich, J. E., 1984, J. Biol. Chem., 259:11459.

Massey, J. B., and Churchich, J. E., 1979, Biophys. Chem., 9:157.

O'Connor, D. V., and Phillips, D., 1984, in: "Time Correlated Single Photon Counting", Academic Press, New York, London.

Viceps-Madore, D., Cidlowski, J. A., Kittler, J. M., and Thanassi, J. W., 1983, J. Biol. Chem., 258:2689.

INTRAMOLECULAR EXCIMER FLUORESCENCE OF PYRENE MALEIMIDE-LABELED

DITHIOTHREITOL (PYRENE-DTT)

S. S. Lehrer and Y. Ishii

Boston Biomedical Research Institute
Boston, MA 02114 USA

The fluorescence properties of pyrene-DTT were studied as a model
system for pyrene-tropomyosin for which excimer fluorescence has been
reported (Betcher-Lange and Lehrer, 1978). DTT in dimethyl formamide or
dimethyl sulfoxide was reacted at its 2 SH-groups with pyrene maleimide to
form $(S-[N-(1-pyrene)succinimido])_2$-DTT (pyrene$_I$-DTT). The product with
cleaved succinimido-rings, pyrene$_{II}$-DTT was also prepared by incubation of
pyrene$_I$-DTT in 1 mM NaOH. Pyrene-β-mercaptoethanol (βSH) adducts were
similarly prepared. For spectroscopic measurements, 1 μM solutions were
studied by dilution into appropriate solvents.

In water, the fluorescence spectra of the two pyrene-DTT types were
similar to those of the previously reported two types of pyrene-tropomyosin
with little or no excimer (E) observed for type I as compared to the high
excimer/monomer (E/M) ratio observed for type II (Figure 1). The dependence of E and M on the degree of labeling of DTT showed that intramolecular pyrene-pyrene interaction results in concentration quenching of M
fluorescence for both types; however, the quenching is accompanied by E
formation only in the case of type II. Absorption and excitation spectra
indicated ground state pyrene-pyrene interaction and that E originates from
a sub-set of pyrenes interacting in the ground-state. Lifetime studies of
pyrene$_{II}$-DTT confirmed the independent emission of E and M by showing that
there was no correlation between the E lifetime of ~40 nsec and the M decay
of the type II DTT (or βSH-adduct) which had short lifetimes (5-10 nsec).

In organic solvents, pyrene$_I$-DTT showed E fluorescence which only
changed slightly with polarity (E/M = 0.45, 0.59, 0.75 for dioxane, 2-
propanol and methanol, respectively. In these solvents excitation spectra
indicated that M and E originate from pyrenes in a similar environment. In
propanol, for example, the E peaked ~15 nsec after the M due to its formation by rotational diffusion. In glycerol the E/M ratio decreased almost
10X in agreement with the affect of viscosity on rotational diffusion.
These data indicate that the pyrenes in pyrene$_I$-DTT are capable of interacting in the excimeric configuration. It appears, however, that the lack
of E formation for the type I-DTT adduct in water is that the hydrophobic

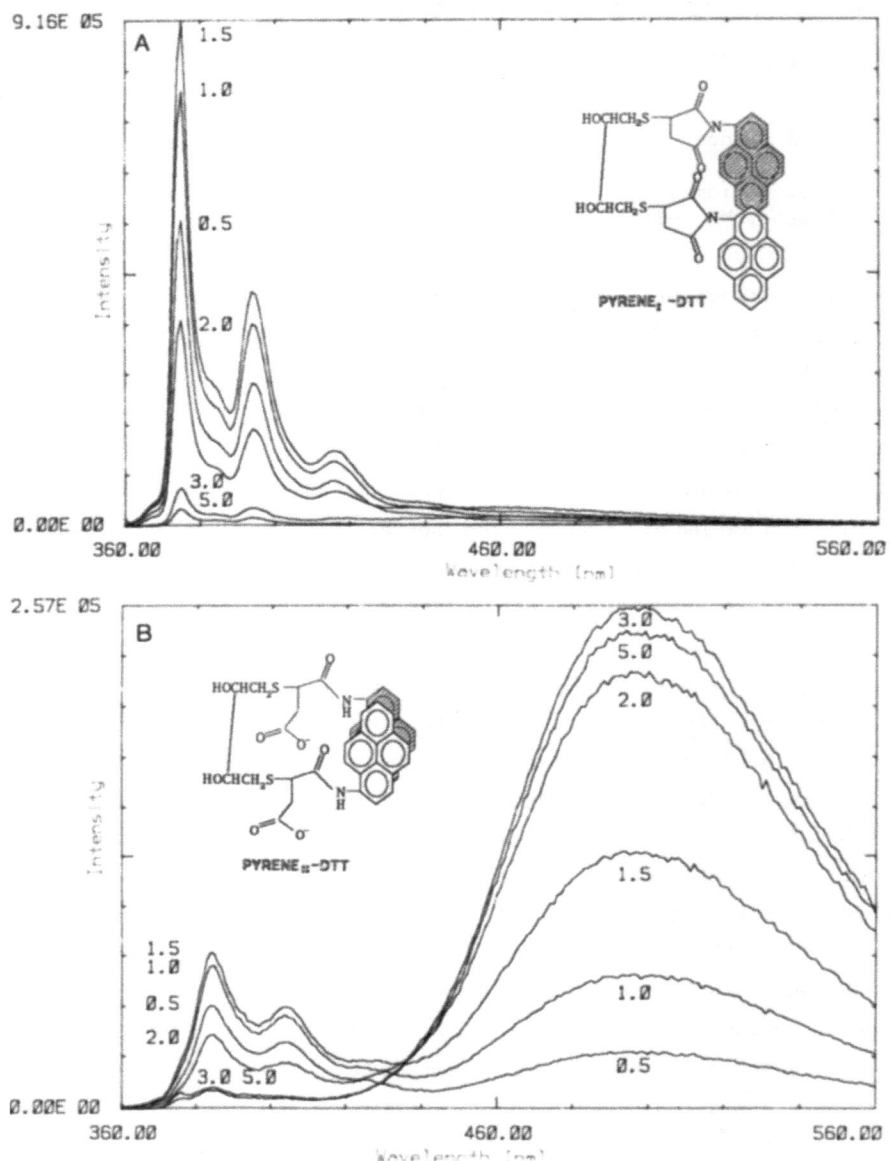

Figure 1. Fluorescence spectra of the 2 types of pyrene-DTT in H_2O at different ratios of pyrene maleimide to DTT. Upper, pyrene$_I$-DTT; Lower, pyrene$_{II}$-DTT; inserts, schematic structures of the 2 respective molecules. Pyrene maleimide was reacted with DTT at the ratios indicated to completion to form type I and a part was converted to type II as outlined in text. 25°, pH 7, 1 μM solutions. Unreacted pyrene maleimide does not contribute to the fluorescence since it is virtually non-fluorescent in H_2O.

pyrenes interact strongly in the ground state in a non-excimeric configuration which results in quenching of M before it can align into the proper E configuration. On the other hand, in relatively good solvents where there

424

is little ground state interaction, E can form by rotational diffusion during the long M lifetime (~90 nsec).

In contrast to the type I case, pyrene$_{II}$-DTT gave E/M ratios which increased with solvent polarity (E/M = 0.35, 0.45 and 1.65 in the same organic solvents). The much shorter M lifetime of the type II adducts (~10 nsec) decreases the probability that appreciable E could be formed by rotational diffusion in this system. It appears that the greater flexibility resulting from the succinimido-ring cleavage allows the pyrenes to stack via hydrophobic interactions closer to the excimeric configuration in polar solvents. The importance of polarity in this system was indicated by the effect of glycerol which gave a high E/M ratio appropriate to its polarity rather than its viscosity. Thus, in type II-DTT, the E/M ratio appears to be quite sensitive to the polarity of the environment in solvents of relatively high polarity whereas in type I-DTT it is quite sensitive to the viscosity of the solvent in relatively non-polar environments, reflecting the different origins of the E. This suggests that the two types of pyrene-DTT could act as probes of viscosity or polarity, respectively, in appropriate systems. (Supported by NSF, NIH and the MDA)

REFERENCES

Betcher-Lange and Lehrer, 1978, J. Biol. Chem., 253:3757.

TIME-RESOLVED FLUORESCENCE ENERGY TRANSFER IMMUNOASSAY

Richard B. Thompson and Anne W. Kusterbeck

Bio/Molecular Engineering Branch
Code 6190, Naval Research Laboratory
Washington, DC 20375-5000 USA

Time-resolved methods, including both time- and frequency-domain
techniques, are of growing importance in fluorescence immunoassay due to
their selectivity and ability to reject background fluorescence (Soini and
Hemmila, 1979). Phase methods are especially interesting because the
instrumentation is simple, and can potentially be made compact and solid
state. We decided to test the utility of variable-frequency phase fluorometry for measurement of an antigen-antibody binding reaction, a reaction
which could also be quantitated by a separate method. Thus antibodies and
antigens were labeled with fluorescence energy transfer acceptors and
donors, respectively. This allowed binding to be measured from changes in
the donor's decay kinetics (quenching), or from changes in the acceptor's
fluorescence excitation spectrum, which are caused by energy transfer.

Goat antibody (IgG specific for rabbit IgG, Cappel Worthington) was
labeled with either rhodamine-X isothiocyanate (XRITC) or tetramethyl-
rhodamine isothiocyanate (TRITC) as acceptors (Research Organics) according
to the manufacturer's directions. Fluorescein isothiocyanate (FITC)
labeled antigen (rabbit IgG, Cappel Worthington) was obtained commercially,
and mouse IgG was labelled with FITC to act as control. Steady-state
fluorescence and fluorescence anisotropy measurements were made on a Spex
Fluorolog 2, with carefully selected filtration to avoid Rayleigh and Raman
scattering. Multi-frequency phase fluorometric measurements were performed
on an ISS Greg-200 (Industria Strumentazioni Scientifiche, La Spezia,
Italy) essentially as previously described (Lakowicz et al., 1984).

We proposed to quantitate the proportion of antigen bound to antibody
using the well-known technique of Forster energy transfer (Forster, 1948).
The antigen labeled with donor will transfer its energy upon binding to the
antibody, which is labeled with the acceptor (Ullman et al., 1976). The
energy transfer process quenches the donor fluorescence, shortening its
lifetime. Thus we expect the phase data on the donor fluorescence to be
best fit by at least two decays: one for the free antigen, which can be
determined independently, and at least one for the bound antigen. The
antibody was labeled with acceptor instead of donor. If it were labeled

426

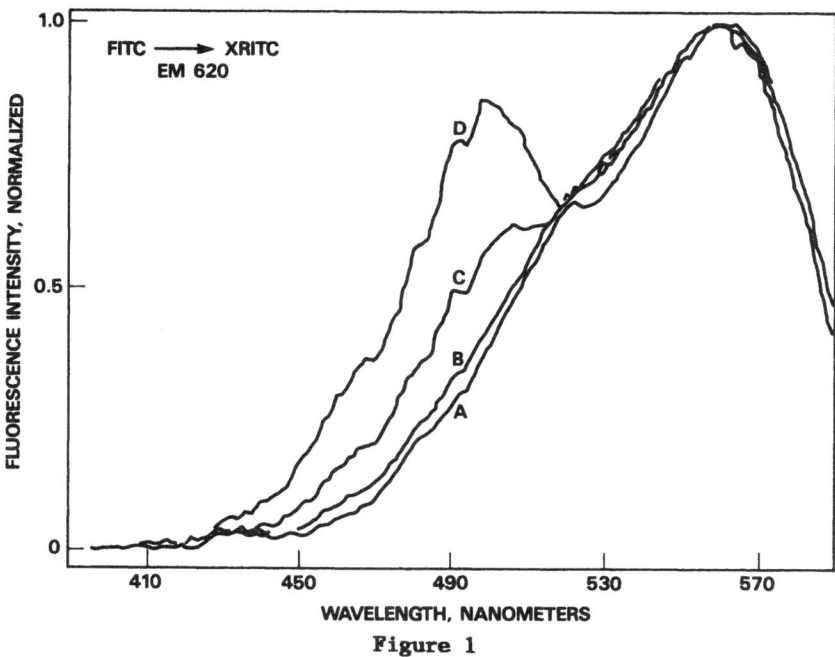

Figure 1

Normalized excitation spectra of XRITC-labelled goat anti-rabbit IgG in the presence of different concentrations of FITC-labelled IgG. Emission was monitored at 620 nm and the spectra are corrected for variation in lamp output with wavelength. Spectra A, C, and D correspond to 0.0, 62, and 75 nM FITC-rabbit IgG in the presence of 1.1, 0.94, and 0.68 uM XRITC-goat anti-rabbit IgG, respectively. In spectrum B, FITC-labelled mouse IgG, which is not bound by anti-rabbit IgG, shows negligible energy transfer, as expected. The binding constant obtained is 5.6×10^{-5} M.

with donor, the bivalent IgG would exhibit different decay kinetics depending on whether one or both sites were occupied, complicating the data analysis. We chose the acceptors XRITC and TRITC to themselves be fluorophores so that transfer (and thus binding) could be quantitated independently by the appearance in the acceptor excitation spectrum of a component corresponding to the donor absorbance at 495 nm (please see Figure). The apparent sensitization of the acceptor fluorescence in the Figure is due solely to energy transfer resulting from specific antigen-antibody binding. This is clear from spectrum B in the Figure, which was obtained from a sample containing an antigen (labelled with donor) which is not bound by the antibody; the lack of additional absorbance at 495 shows that nonspecific energy transfer is minimal.

We measured the phase and modulation of the donor fluorescence (isolated by an interference filter) at frequencies ranging from 1-150 MHz,

both free and in the presence of antibody labelled with one of the acceptors. In the absence of acceptor, the phase or modulation data are each
well fit by a monoexponential (fraction = 96+%, 3.50 $\pm$ .03 nsec, X^2=4).
Taken together, the data show a poorer fit, indicating a minor phase
artifact in this case. However, in the presence of TRITC-labelled and
XRITC-labelled antibody such that 50 and 57% of the donor would be bound,
respectively, the data were not fit by either two or three exponentials in
a meaningful fashion. In particular, the best fits gave a main component
with a longer lifetime than the free form, a high X^2, and calculated values
that differed systematically from the measured values; all criteria of an
inadequate model to fit the data.

It is well known (Bennett, 1964; Eisenthal and Siegel, 1964) that, in
the presence of energy transfer, donors can exhibit non-exponential
fluorescence decays. This occurs if either the donor or acceptor diffuses
rotationally or translationally during the donor lifetime. We examined
this possibility by measuring the fluorescence anisotropies of donor and
acceptors separately; the anisotropy is a sensitive measure of rotational
diffusion of the fluorescent moieties on the macromolecule (Weber, 1953).
In each case, the sample was excited at a wavelength where r_o is expected
to be near its maximum theoretical value of 0.4. For both acceptors and
donor, the anisotropy was low (<.10), indicating substantial motion of the
labels independent of the macromolecules. Although IgG's are known to
exhibit some segmental motion (Yguerabide et al., 1970), in fluid solvents
covalently-bound fluorophores usually exhibit anisotropies reflecting the
macromolecule's motion (Weber, 1953). While we cannot explicitly rule out
homotransfer among labels on the same molecule as the source of the low
anisotropies, the low labeling ratio (1 $\pm$ 015 label/macromolecule in all
cases) makes this unlikely.

Recent variable-frequency phase fluorometric measurements by Lakowicz,
et al. (1986), of a quenched system expected to display $\tau^{\frac{1}{2}}$ dependence
showed systematic deviations from an exponential fit similar to those we
observed. Current efforts are aimed at developing an algorithm to fit
nonexponential decays produced by energy transfer, and developing an
improved antibody labeling method for these studies.

REFERENCES

Bennett, R. G., 1964, J. Chem. Phys., 41:3037.

Eisenthal, K. B., and Siegel, S., 1964, J. Chem. Phys., 41:652.

Forster, Th., 1948, Ann. Phys., 2:55.

Gratton, E., Limkeman, M., Lakowicz, J. R., Maliwal, B. P., Cherek, H. and Laczko, G., 1984, Biophys. J., 46:479.

Lakowicz, J. R., Laczko, G., Cherek, H., Gratton, E., and Limkeman, M., 1984, Biophys. J., 46:463.

Lakowicz, J. R., Laczko, G., and Gryczynski, I., 1986, Rev. Sci. Instrum., 57:2499.

Soini, E., and Hemmila, I., 1979, Clin. Chem., 25:353.

Ullman, E. F., Schwarzberg, M., and Rubenstein, K. E., 1976, J. Biol. Chem., 251:4172.

Weber, G., 1953, Adv. Prot. Chem., 8:415.

Yguerabide, J., Epstein, H. F., and Stryer, L., 1970, J. Mol. Biol., 51:573.

PORPHYRIN MOTIONS IN Mb^{desFe} AND Hb^{desFe}

C. Royer and B. Alpert
Laboratoire de Biologie Physico-chimique
Universite Paris VII
2, Place Jussieu 75251 Paris, France

Porphyrin was bound to the heme pockets of apomyoglobin and apohemoglobin and the local rotations of this fluorescent probe were studied by fluorescence depolarization. Three temperature domains for the dynamics of porphyrin were observed. At very low temperatures the temperature dependence of the porphyrin rotations were found to follow the bulk solvent viscosity. At intermediate temperatures, porphyrin-heme pocket interactions were predominant. At temperatures above 0°C, the viscosity of the heme pocket appeared to be the determining factor. Large differences in the rotational parameters of the porphyrin were found between myoglobin and hemoglobin indicating different interactions between the porphyrin and these two heme pockets. For porphyrin in the hemoglobin heme cavity, the binding of the allosteric effector, inositol hexaphosphate resulted in immobilizing the probe in the low temperature regime, and in eliminating the intermediate temperature regime. This indicates that effector binding neutralizes certain porphyrin-pocket interactions, thus decoupling the fluorophore from the protein.

FLUORESCENCE ANISOTROPY EXPERIMENTS ON HOG RENAL BRUSH BORDER MEMBRANE VESICLES

M. Ameloot[*], H. De Smedt[o] and H. Hendrickx[*]

[*]Limburgs Universitair Centrum, Biophysics, Universitaire
Campus, B-3610 Diepenbeek, Belgium
[o]Laboratorium voor Fysiologie, Katholieke Universiteit Leuven,
B-3000 Leuven, Belgium

In a temperature study of the solute transport in hog renal brush border membrane (BBM) vesicles under tracer exchange conditions, discontinuities were observed in the Arrhenius plots for sodium-dependent D-glucose and phosphate uptake (De Smedt and Kinne, 1981). The most striking change in the slope of the Arrhenius plots occurred between 12 and 15°C. It has been suggested then that this behavior can be explained by an altered activity of intrinsic membrane proteins due to a change in membrane organization.

In the present study the temperature dependence of the membrane organization is investigated by measuring the fluorescence anisotropy of the extrinsic probes diphenylhexatriene (DPH) and its cationic analog with a trimethylamino group (TMA-DPH) embedded in BBM-vesicles. These molecules are expected to probe the membrane at different depth. In the investigated temperature range of 3-32°C, the steady state fluorescence anisotropy r_s of TMA-DPH is always higher than for DPH (see figure). The values of r_s range from 0.237 to 0.303 for DPH and from 0.274 to 0.314 for TMA-DPH. As for vesicles of pure lipid, the difference diminishes with decreasing temperature. By lowering the temperature, the temperature dependence of r_s also becomes different from 20°C for each of the two probes. Furthermore, in the temperature interval from 20 to 12°C, the r_s values for TMA-DPH are practically temperature independent where r_s for DPH steadily increases. This is the temperature region in which changes in the cotransport systems of the BBM-membrane have been observed.

Because of the small change in r_s the time resolved fluorescence anisotropy of DPH has been measured only at 6 and 30°C. The total fluorescence requires one to consider two relaxation times. The average lifetimes at the considered temperatures are almost the same. The time resolved anisotropy has been analyzed according to a recently described general model which allows the determination of the rotational diffusion constant of the probe and the order parameters of second and fourth rank, $\langle P_2 \rangle$ and

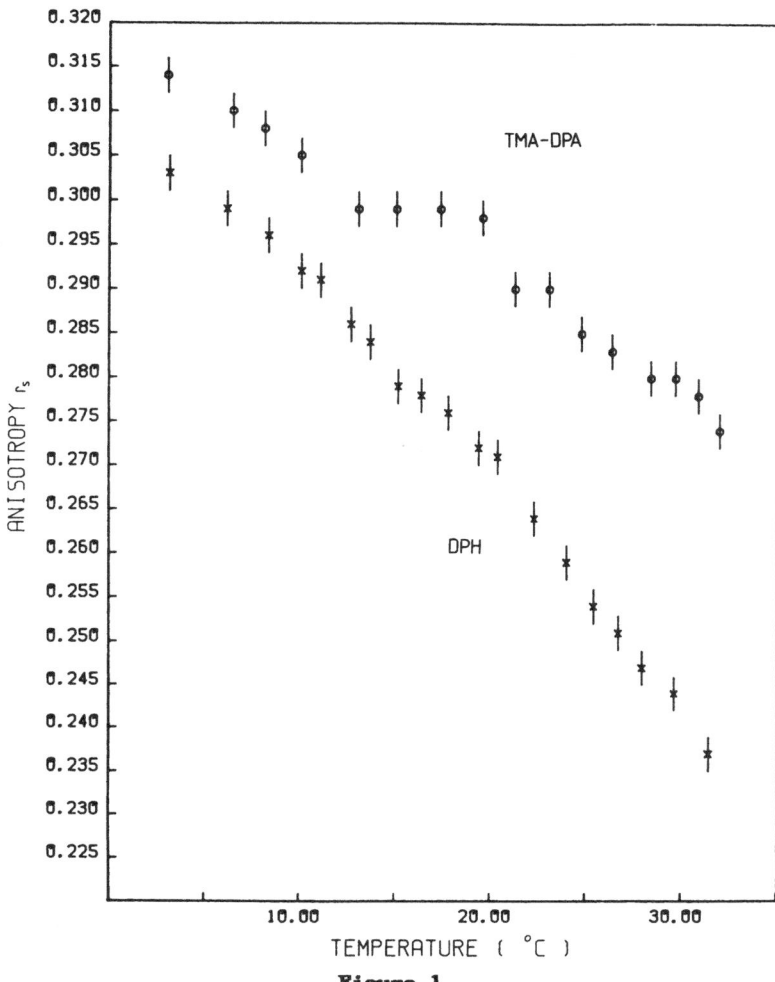

Figure 1

<P₄> [van der Meer, W. et al., Biophys. J. 46 (1984) 515; Ameloot, M. et al., Biophys. J. 46 (1984) 525]. This analysis reveals that the main contribution to r_s is due to a highly restricted motion of the probe. The high values for the order parameters at 6°C, <P₂> = 0.86 and <P₄> = 0.70, and at 30°C, <P₂> = 0.77 and <P₄> = 0.54, indicate that the long axis of the DPH molecule points along the normal to the membrane. From these results, it can be concluded that the order in the brush border membrane is high. Changes in order seem to be correlated with changes in activity of membrane proteins.

REFERENCES

Ameloot, M., Hendrickx, H., Herreman, W., Pottel, H., Cauwelaert, F. V., and van der Meer, W., 1984, Biophys. J., 46:525

De Smedt, H., and Kinne, R., 1981, Biochim. Biophys. Acta, 648:247

van der Meer, W., Pottel, H., Herreman, W., Ameloot, M., Hendrickx, H., and
     Schröder, H., 1984, Biophys. J., 46:515.

PHOTOPHYSICS OF THE TRYPTOPHAN RESIDUE IN RNaseT1 AND RNase T1-2'GMP COMPLEX

Lin X.-Q. Chen and Graham R. Fleming

Department of Chemistry and James Franck Institute
The University of Chicago
Chicago, IL 60637 USA

The photophysics of a single tryptophan residue in RNaseT1 was studied by time correlated single photon counting. A pH dependent transition in the fluorescence decay was observed, centered at pH 6.5-7.0. At pH values above this transition, tryptophan fluorescence decay can be described in terms of a double exponential, with τ_1 3.9 ns and τ_2 1.6 ns. Fluorescence was quenched with acrylamide, which quenches the short lifetime component five times faster than it does the long lifetime component. These results were interpreted with a two state model. At pH's below the transition, the fluorescence decay of tryptophan in RNaseT1 is a single exponential with τ_1 4.0 ns. Binding of 2'GMP, an enzyme inhibitor, to RNaseT1 caused a change in average fluorescence lifetime and in the conformation distribution of the tryptophan residue in the protein. The fluorescence anisotropy of RNaseT1 showed an apparent hydration volume at pH 7.4 which corresponds to a single layer of water molecules bound on the surface of the protein, whereas at pH 5.5, the fluorescence anisotropy did not follow the normal Stokes-Einstein hydrodynamic law.

432

THE EFFECT OF PHENYLHYDRAZINE ON THE MEMBRANE PHOSPHOLIPID ORGANIZATION

A. Arduini, J. Dejulia, V. Damonti, G. Mancinelli, A. Di Bartolomeo, M. Belfiglio, M. De Cesare and A. Stern*

Istituto di Scienze Biochimiche, Facolta di Medicina Universita degli Studi "G. D'Annunzio" Chieti, Italy
*Department of Pharmacology, New York University, NY USA

Phenylhydrazine (PH) represents an interesting probe in oxidative damage studies in human erythrocytes. We have recently demonstrated that spectrin, one of the most important cytoskeleton membrane proteins, was degraded after exposure of human red blood cells to PH. Because spectrin seems to be involved in the maintenance of phase-state asymmetry in the erythrocyte membrane, we decided to study the membrane phospholipid organization (MPO) in PH treated human red cells utilizing different methodological approaches.

Steady state fluorescence polarization measurements of diphenylexatriene (DPH) labeled red cells showed that PH induced only a slight increase of the lipid structural order parameter. In our experimental model hemoglobin did not interfere with the DPH fluorescence studies.

Fluorescence microscopic studies using Merocyanine 540 (M-540), a fluorophore sensitive to change in lipid packing, indicated that the outer leaflet of red cell membrane became more disordered upon exposure of intact erythrocytes to PH. These fluorescence studies indicate that major changes are observed in membrane lipids.

Analytical studies of membrane phospholipids (P) and cholesterol (C) revealed a significant decrease in phosphatidylethanolamine and cholesterol content. However, we did not observe a dose dependent effect in the degradation of cholesterol as the peak value was already reached at the lowest PH concentration.

A direct consequence of such changes is the decrease of the C/P which usually, in other experimental conditions, is often associated with an increase of the membrane fluidity. In fact, it has been reported that the incorporation of cholesterol into fluid-phase vesicles reduced the M-540 binding. Further studies on the phospholipid asymmetry utilizing the primary amino group reactive fluorescence probe, trinitrobenzene sulfonic acid, and the phospholipase A2 treatment suggest that a change in membrane phospholipid organization occurs in red cells treated with PH. The results

433

of the fluorescence and lipid studies, taken in association with our previously reported findings on spectrin and other cytoskeletal protein degradation in red cells exposed to phenylhydrazine, suggest that degradation of cytoskeleton membrane proteins is also responsible for changes in MPO.

THE USE OF DISULFONATONAPHTHALIMIDE FLUORESCENT DYES FOR THE FLUORESCENCE POLARIZATION IMMUNOASSAY OF STEROIDS

N. Cittanova, B. Desfosses, K. M. Rajkowski and N. Christeff

Unité Associée 586 du CNRS, Faculté de Médecine, 45 rue des Saints-Pères, Paris Cedex 06, France

A series of fluorescent disulfononaphthalimide (lucifer yellow) derivatives of testosterone and estriol have been synthesized: testosterone-3- and estriol-6-ureidooxime-lucifer yellow, testosterone-3- and estriol-6-(amidoureido)methoxyoxime lucifer yellow and testosterone-3-amidoheptylaminoethylsulfonylphenyl methoxyoxime lucifer yellow. The fluorescence lifetimes of these derivatives were higher than those of the free dye while the quantum yields were of the same order. The compounds were therefore compared in terms of their utility for steroid fluorescence polarization immunoassay. Because of the high Stokes' shift of these derivatives the blanks due to diffusion polarization were considerably reduced relative to those usually obtained with fluorescein derivatives. The sensitivity of the immunoassays were therefore improved and, with testosterone-3-(amidoureido)methoxyoxime lucifer yellow, the optimum of the testosterone derivatives, it was possible to assay 0.3 to 25 ng/ml testosterone. The use of monoclonal antibodies and simplification of the instrumentation should permit the development of clinically useful assays.

FOLLOWING INVERTED MICELLE BEHAVIOR BY INTRINSIC FLUORESCENCE PROBES

Tze-Chi Jao and Kenneth L. Kreuz

Texaco Research Center
P. O. Box 509
Beacon, NY   12508   U.S.A.

The aromatic moiety of calcium alkarylsulfonates was used as an intrinsic fluorescence probe to follow the behavior of inverted micelles formed by this general type of materials in hydrocarbon media. The rotational relaxation times of the intrinsic probe in different systems were determined from anisotropy decays obtained with a single photon counting system using a UV excitation source at 295 nm, generated by a Spectra-Physics Model 375 synchronously-pumped, cavity-dumped dye laser system coupled with an INRAD frequency doubler. They provide information on the rigidity of sulfonate molecules adsorbed on the surface of basic inorganic materials that are components of the polar core of the inverted micelles.

The rotational relaxation times of probe/micelle of petroleum sulfonate and mixed petroleum/synthetic sulfonates in heptane in the range

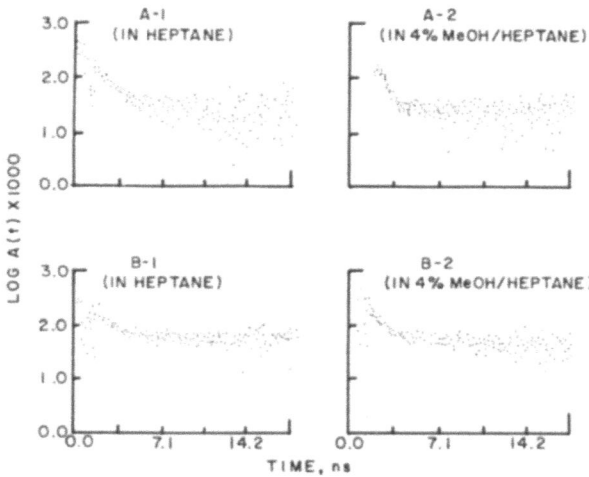

Figure 1. Anisotropy decays of mixed petroleum/synthetic sulfonate in methanol/heptane solutions. All sulfonate solutions are 10^{-3}M. A-1 and A-2 contain 0.5 mole inorganic Ca Base per mole of sulfonate, and B-1 and B-2 contain 20 moles of inorganic Ca base per mole of sulfonate.

of 8-11 nanoseconds were obtained depending on the amount of inorganic calcium base. Methanol has a large effect on the rigidity of micelles (see Figure 1). It reduces the rotational relaxation time of the intrinsic probe by 2-5 nanoseconds. The magnitude of reduction seems to be inversely proportional to the amount of inorganic content in the polar core.

TRYPTOPHAN DYNAMICS AND COOPERATIVITY IN ATCase FROM E. coli

C. Royer, J.-C. Brochon*, P. Tauc, and G. Hervé

Laboratoire d'Enzymologie CNRS Gif-sur-Yvette, 91190 France
*Laboratoire d'Utilization du Rayonnement
Electromagnétique 91405 Orsay France

The temperature and viscosity dependence of the rotations of the intrinsic tryptophan moieties in ATCase from E. coli was studied as a function of temperature. The two tryptophan residues were differentiated in the decay curve analysis, as a short lifetime (1-3 ns) and a long lifetime (4-6 ns) component were found. The two protein slopes observed in the Y-plots were interpreted as corresponding to the two tryptophan residues which were differentiated by their surrounding free space. Using the fractional intensity data obtained from fluorescence intensity decay measurements the observed slopes were decomposed into the protein slopes for each tryptophan residue. In ATCase alone, the two were similarly coupled to the protein matrix. However, upon binding of the substrate analogue, PALA, one tryptophan was decoupled from the protein, whereas the other appeared to be more restrained. The fixation of the nucleotide effector molecules elicited contrary results, despite the fact that they bind to the same site. ATP increases the free space around one tryptophan and slightly decoupled both their motions, whereas CTP binding gave similar results as those obtained in presence of PALA. These studies have shown that the fixation of small molecules at a large distance from a given amino-acid residue can have important and specific effects upon its dynamic properties.

436

MULTIFREQUENCY PHASE/MODULATION FLUORESCENCE LIFETIME MEASUREMENTS WITH RIBONUCLEASE T$_1$, ALCOHOL DEHYDROGENASE, AND SELECTED TWO-TRYPTOPHAN PROTEINS

Maurice R. Eftink, Zygmunt Wasylewski, and Camillo A. Ghiron

Department of Chemistry, University of Mississippi, University, MS 38677 USA
Department of Biochemistry, University of Missouri, Columbia, MO 65211 USA

Here we report frequency domain (10-200 MHz) fluorescence studies of the single-trp containing protein, RNAse T$_1$, and several two-trp proteins, horse liver alcohol dehydrogenase (LADH), yeast 3-phosphoglycerate kinase (PGK), a metalloprotease from Staph. aureus, and ribulose-5-phosphate kinase (RPK) from spinach. We will demonstrate the ability of multi-frequency phase/modulation fluorometry to separate the fluorescence components of these proteins. Table 1 summarizes our findings; below we will comment on the various proteins.

Chen and Fleming (in this volume) have reported that the decay of trp-59 of RNAse T$_1$ is a single exponential at pH 5.5 and a double exponential at pH 7.4. We confirm their results and we have studied the temperature dependence of the lifetime at both pHs. At pH 7.4 we find an activation energy, E$_a$, equal to 4 kcal/mole for the 1 ns component and E$_a$ = 1 kcal/mole for the 3.5 ns component. Alternatively we have described the data via a Lorentzian distribution and find the width of this distribution to increase with temperature (5-45°C) at pH 7.4.

For LADH we have fitted frequency domain data to a two component decay with τ_1 = 3.0 ns (Trp-314) and τ_2 = 6.4 ns (Trp-15), in agreement with previous pulse studies (Ross et al. 1981). We have studied the temperature dependence of these lifetimes and the effect of coenzyme binding. Phase resolved spectra (Gratton and Jameson, 1985) for LADH were obtained, as shown in Figure 1. The spectra corresponding to the 3 ns component (Trp-314) is clearly blue shifted, as compared to that for the 6.4 ns component (Trp-15). Such phase resolved spectra are analogous to decay associated spectra (Knutson et al., 1982).

For PGK our frequency domain data do not allow a three component fit, as obtained by Privat et al. (1980), but the two component fit is in general agreement. The binding of ATP and 2-phosphoglycerate slightly lengthens both lifetimes and the phase resolved spectrum reveals a red shift of both components upon ternary complex formation.

Table 1. Fits to Multifrequency Phase/Modulation Data (Lifetimes in ns).

Protein	τ_1	f_1	τ_2
RNAse T_1, pH 5.5	3.82	1.0	--
RNAse T_1, pH 7.4	3.74	0.94	1.05
LADH, pH 7.2	3.05	0.35	6.40
PGK, pH 7.4	0.31	0.70	3.09
RPK, pH 7.0	1.14	0.70	5.40
metalloprotease, pH9.0	1.30	0.50	5.67

In a similar manner, frequency domain data for RPK and metalloprotease have been analyzed in terms of a two component model and phase resolved spectra for the components have been obtained. Each of these proteins has a very blue (~320 nm) component and a red component. Removal of the calcium from the metalloprotease results in a selective red shift of this blue component from <320 to 345 nm.

These studies illustrate that multi-frequency phase/modulation and phase resolved spectra measurements can separate the fluorescence of two-trp proteins into two components. In the case of LADH auxillary evidence (i.e., solute quenching) strongly supports the assignment of the two lifetimes to the two individual trp residues in this protein. For the

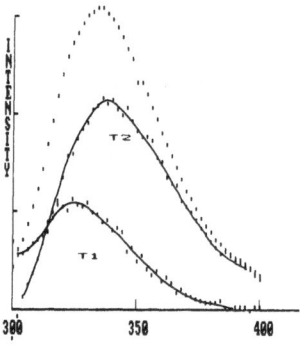

Figure 1. Phase resolved spectrum of LADH at 20°C pH 7.2. T1 indicates 3.05 ns component. T2 indicates the 6.4 ns component. Excitation at 290 nm.

other two-trp proteins studied here the assignment of lifetimes to specific trp residues is not well established as yet. Also our studies with the single-trp protein, RNAse T$_1$, reemphasize the well known fact that the decay of individual trp residues in proteins is not always a single exponential (i.e., distribution of lifetimes fits to such decays must also be considered).

REFERENCES

Gratton, E., and Jameson, D., 1985, Anal. Chem., 57:1694.

Knutson, J. R., Walbridge, D. G., and Brand, L., 1982, Biochemistry, 21:4671.

Privat, J. P., Wahl, P., Auchet, J. C., and Pain, R., 1980, Biophys. Chem., 11:239.

Ross, J. B. A., Schmidt, C. J., and Brand, L., 1981, Biochemistry, 20:4369.

APPLICATIONS OF THE URANYL ION AS A FLUORESCENT PROBE IN MODEL BIOMEMBRANES

M. G. Miguel, S. J. Formosinho and H. D. Burrows

Departamento de Quimica
Universidade de Coimbra
3049-Coimbra, Portugal

The uranyl ion (UO_2^{2+}) presents a well characterized luminescence, whose spectrum and lifetime are strongly dependent on temperature, pH and the presence of coordinated ligands (Jorgensen and Reisfeld, 1982; Miguel et al., 1984; Formosinho and Miguel, 1984; M. Moriyasu et al., 1977). In addition, this luminescence is strongly quenched by a variety of organic (Matsushima, 1972; Ambroz et al., 1984 and references therein) and inorganic (Matsushima et al., 1974; Marcantonatos, 1979) species. Previous studies from this laboratory have shown that this luminescence can be used to probe micellization in both ionic and non-ionic surfactants (Formosinho and Martinho do Rosario, 1978; Formosinho and Miguel, 1985).

Such behavior also makes uranyl ion a potentially useful probe of other amphiphile aggregates, including model biomembranes. We report the study of the luminescent behavior of uranyl ion in W/O microemulsions, and in synthetic vesicles. The possibility of using changes in uranyl luminescence as a pH indicator in the region 1-4 is studied. In addition, the quenching of uranyl fluorescence by a variety of cations and anions is examined as a potential ion probe in these systems.

Two distinct types of behavior are observed. In some systems the uranyl ion does not complex significantly with the amphiphile. In these cases its luminescence may be used as a fairly reliable probe of the behavior of aqueous regions. In other cases, including many phospholipids, there is strong complexing between amphiphile and uranyl. In these systems, the application of UO_2^{2+} as a luminescent probe is limited due to strong quenching by the phospholipid. The potential of using different uranyl complexes in these cases is examined.

REFERENCES

Ambroz, H. B., Butter, K. R., and Kemp, T. J., 1984, Faraday Disc. Chem. Soc., 78:107.

Formosinho, S. J., and Martinho do Rosario, L. M., 1978, Rev. Port. Quim., 20:88.

Formosinho, S. J., and Miguel, M. G., 1984, <u>J. Chem. Soc., Faraday Trans 1</u>, 80:1745.

Formosinho, S. J., and Miguel, M. G., 1985, <u>J. Chem. Soc., Faraday Trans. 1</u>, 81:1891.

Jorgensen, C. J., and Reisfeld, R., 1982, <u>Struct. Bonding (Berlin)</u>, 50:121.

Marcantonatos, M. D., 1979, <u>J. Chem. Soc., Faraday Trans. 1</u>, 75:2252.

Matsushima, R., 1972, <u>J. Am. Chem. Soc.</u>, 94:6010.

Matsushima, R., Fujimori, H., and Sakuraba, S., 1974, <u>J. Chem. Soc., Faraday Trans. 1</u>, 70:1702.

Miguel, M. G., Formosinho, S. J., Cardoso, A. C., and Burrows, H. D., 1984, <u>J. Chem. Soc., Faraday Trans. 1</u>, 80:1735.

Moriyasu, M., Yokoyama, Y., and Ikeda, S., 1977, <u>J. Inorg. Nucl. Chem.</u>, 39:2199.

VARIATION OF 1,6-DIPHENYL-1,3,5-HEXATRIENE LIFETIME DISTRIBUTIONS IN DIFFERENTIATING PROERYTHROBLASTS. A MULTIFREQUENCY PHASE AND MODULATION FLUOROMETRY STUDY

Tiziana Parasassi[x], Filippo Conti[+], Enrico Gratton[o] and Orazio Sapora[y]

[x]Istituto di Medicina Sperimentale, CNR, Via Monti Tiburtini 509, 00157 Roma
[+]Dipartimento di Chimica, Universita La Sapienza, Piazzale A. Moro 5, 00185 Roma
[o]Department of Physics, University of Illinois, 1110 W. Green St., Urbana, IL 61801 USA
[y]Istituto Superiore di Sanita, Viale Regina Elena 160, 00185 Roma

During the cell differentiation process structural and functional properties peculiar to the mature, specialized cell are acquired. Although the molecular basis of the process are still largely unknown, the appearance of specific structural membrane components (Seiser, 1973; Fukuda et al., 1981) leaded to investigations on the cell membranes involvement in differentiation. Moreover, the modulation of the activity of some membrane enzyme during physiological processes could be achieved by structural and dynamical modification of the membrane matrix (Kimelberg, 1977). Since the heterogeneity of cell membranes composition many efforts have been devoted to the detection and quantitation of different coexisting lipid phases (Klausner et al., 1980; Karnowsky, 1979; Parasassi et al., 1984). The fluorescence lifetime of 1,6-diphenyl-1,3,5-hexatriene (DPH) shows marked values differences when inserted in phospholipid vesicles of different phases. Characteristic lifetime values have been determined for the pure gel and liquid-crystalline phase of the vesicles, and the quantitation of the phospholipid physical state has been possible also in the case of coexisting different phases (Parasassi et al., 1984). Nevertheless, this approach gave ambiguous results when applied to cell membranes. In the case of natural samples a separation of the membrane matrix between the gel and the liquid-crystalline phase could represent an oversimplification because of the heterogeneity of the sample and the multiplicity of environments and conformations of the fluorophore.

A recently developed model to describe the fluorescence decay assumes a continuous distribution of lifetimes instead of a sum of discrete exponentials. The distribution is characterized by a lorentzian shape centered at a decay time, C, and having a width, W.

We report results obtained by labeling the K562 cell membranes with

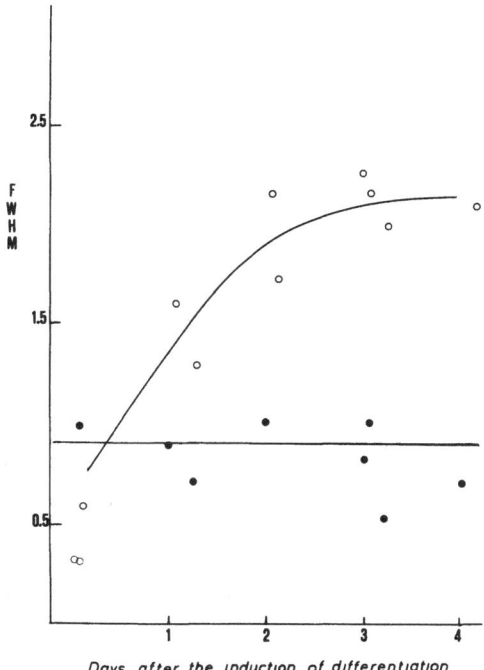

Days after the induction of differentiation

Figure 1. DPH lifetime distribution in A. non-differentiated and B. after four days from the induction of differentiation.

DPH. K562 cells are a human proerythroblastic line which can be induced to differentiate along the erythroid pathway by various substances, butyric acid and hemin for instance (Cioe et al., 1981; Conti et al., 1984). For the determination of DPH lifetimes we used sinusoidally modulated exciting light at frequencies variable from 2 to 160 MHz (Gratton and Limkeman, 1983). The phase and modulation data were analyzed using both a model of discrete exponential components and of continuous lifetime distribution.

In figures 1a and 1b the DPH lifetime distribution in non-differenti-ated and in differentiated K562 cells is reported. Superimposed are the lifetime values obtained by a three component analysis. Fits obtained by the discrete exponentials analysis always showed chi-square values larger of about 40% with respect to the chi-square values obtained by the continuous lifetime distribution analysis. Moreover, following the cells differentiation process the values obtained by the discrete exponentials analysis did not show any significant variation.

The DPH lifetime distribution in non-differentiated cells is relatively

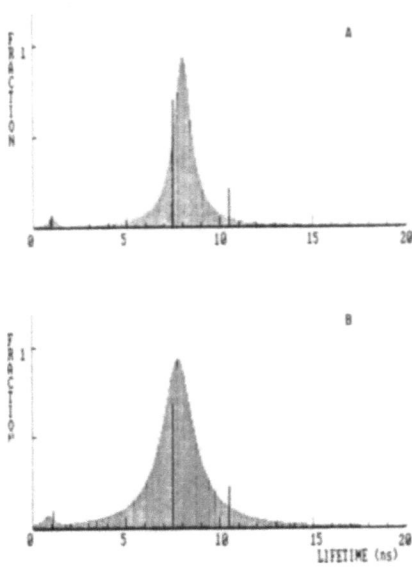

Figure 2. Full width at half maximum (FWHM) value variation of DPH distribution during erythroid K562 differentiation.

broad (full width at half maximum -FWHM- of around 0.8 nsec - figure 2) and centered around 8 nsec. The observed width of the distribution can qualitatively explain the necessity of using more than one exponential term in the case of the discrete components approach can be deduced from the observed width of the distribution. Following the differentiation process the FWHM values of the DPH lifetime distribution increased at about a factor of two. In figure 1b the DPH lifetime distribution in K562 cells during the fourth day after the induction of the differentiation is reported. The FWHM value increased to 2.1 nsec. In figure 2 the behavior of the FWHM value during the four days after the butyric acid induction of differentiation is reported.

The increase of the FWHM value of the DPH lifetime distribution during K562 cells differentiation can indicate a heterogeneity increase in the cell membranes with the consequent increase of the different environments the probe molecules can experience.

REFERENCES

Cioe, L., McNab, A., Hubbel, H. R., Meo, P., Curtis, P., and Rovera, G.,

1981, *Cancer Res.*, 41:237.

Conti, F., Parasassi, T., Rosato, N., Sapora, O., and Gratton, E., *Biochem. Biophys. Acta*, 805:117.

Fukuda, M., Koeffler, H. P., and Minowada, J., 1981, *Proc. Nat. Acad. Sci. USA*, 78:6299.

Gratton, E., and Limkeman, M., 1983, *Biophys. J.*, 44:315.

Karnowsky, M. J., 1979, *Am. J. Pathol.*, 97:212.

Kimelberg, H. K., 1977, *Cell Surf. Rev.*, 3:205.

Klausner, R. D., Bhalla, D. K., Dragsten, P., and Hoover, R. L., 1980, *Proc. Nat. Acad. Sci. USA*, 77:437.

Parasassi, T., Conti, F., Glaser, M., and Gratton, E., 1984, *J. Biol. Chem.*, 259:14011.

Weiser, M. M., 1973, *J. Biol. Chem.*, 248:2536.

PARTICIPANTS

Nando Acerbi, Centro di Studi Sociali, La Spezia, Italy

A. Ulises Acuna, Instituto de Quimica Fisica, Rocasolano, Consejo Superior de Investigationes Cientificas, Serrano 119, Madrid 6, Spain

Giancarlo Agostini, Istituto di Chimica Fisica, Università di Padova, Padova, Italy

J. Ricardo Alcala, Department of Physics, University of Illinois, 1110 W. Green, Urbana, Illinois 61801, USA

Marcel Ameloot, Department of Biophysics, Limburgs Universitair Centrum, Diepenbeek B-3610, Belgium

Sonia Anderson, Department of Biochemistry and Biophysics, Oregon State University, Corvallis, Oregon 97331-6503, USA

Alberto Arcioni, Dipartimento di Chimica Fisica, Università di Bologna, Viale Risorgimento, 4, 40136 Bologna, Italy

Arduino Arduini, Instituto di Scienze Biochimiche, Università di Chieti, Chieti, Italy

Maria Luisa Barcellona, Istituo di Biochimica, Università di Catania, Viale Andrea Doria, 6, 95125 Catania, Italy

Francis Baros, Grapp of U.A. 328 of CNRS, ENSIS-INPL 1, Rue Grandville F-54042 Nancy, France

Franco Bassani, Scuola Normale Superiore di Pisa, Piazza dei Cavalieri, Pisa, Italy

Peter M. Bayley, Department of Physical Biochemistry, National Institute for Medical Research, Mill Hill, London NW71AA, United Kingdom

Sara Benci, Istituto di Scienze Fisiche, Università degli Studi di Parma, Via M. d'Azeglio, 85 Parma, Italy

Mario N.M.S. Berberan Santos, Complexo Interisciplinar, Centro de Quimica Estrutural, Av. Rovisco Pais, 1096 LISBOA, Portugal

Ettore Bismuto, Istituto di Chimica Biologica, Università degli Studi di Napoli, I Facoltà di Medicina, Via Costantinopoli, 16, 80138 Napoli, Italy

Ludwig Brand, Department of Biology, Mudd Hall, 30th and Charles St., Johns Hopkins University, Baltimore, Maryland 21218, USA

Jean-Claude Brochon, L.U.R.E., Universite Paris Sud, Bat. 209 D, 91405 Orsay, France

Augusta Brovelli, Dipartimento di Biochimica, Università di Pavia, Piazza Botta, 27100 Pavia, Italy

Enrico Bucci, Department of Biological Chemistry, University of Maryland Medical School, 660 W. Redwood St., Baltimore, Maryland 21201, USA

Jan Bus, Department of Structure Analysis, Unilever Research Laboratory, P. O. Box 114, 3130 AC Vlaardingen, The Netherlands

Marcello Cacace, Istituto di Biochimica delle Proteine ed Enzimologia, Via Toiano 6, 80072 Arcofelice, Napoli, Italy

Salvatore Califano, Istituto di Chimica Fisica, Università di Firenze, Firenze, Italy

Paola Cavatorta, Istituto di Fisica, Università degli Studi di Parma, Via M. D'Azeglio 85, 43100 Parma, Italy

Lin Y-Q. Chen, Department of Chemistry, University of Chicago, 5735 S. Ellis Avenue, Chicago, Illinois 60637, USA

Parkson Chong, University of Virginia, Department of Biochemistry, Jordan Hall, Room 6-17, Charlottesville, Virgina 22908, USA

Jorge Churchich, Department of Biochemistry, University of Tennessee, Knoxville, Tennessee 37996, USA

Nicole Cittanova, Unite Associe 586 du CNRS, UER Biomedicale des Saints Peres, 45 Rue des Saints Peres, 75270 Paris, France

Koen J. M. Clays, Laboratory of Chemical and Biological Dynamics, Katholieke Universiteit of Leuven, Celestijnenlaan 200 D, B-3030 Heverlee, Belgium

Raffaele Colonna, Istituto di Patologia Generale, Università di Padova, Via Loredan, 16, 35100 Padova, Italy

Robert Cundall, MRC Radiobiology Unit, Chilton, Didcot, United Kingdom

Giovanna Curatola, Istituto di Biochimica, Facoltà di Medicina, Università di Ancona, Via Ranieri, 60131 Ancona, Italy

Robert Dale, Department of Biophysics and Instrumentation, Paterson Laboratories, Christie Hospital and Holt Radium Institute, M20 9BX Manchester, United Kingdom

Maurice R. Eftink, Department of Chemistry, University of Mississippi, University, Missippi 38677, USA

Josef Eisinger, Molecular Biophysics Research Department, Room IC 423, AT&T Bell Laboratories, Murray Hill, New Jersey 07974, USA

Roger P. Ekins, Middlesex Hospital, London, United Kingdom

Yves Engelborghs, Laboratory of Chemical and Biological Dynamics, Katholieke Universiteit of Leuven, Celestijnenlaan 200 D, B-3030 Heverlee, Belgium

Romana Fato, Dipartimento di Biochimica, Istituto di Chimica Biologica, Via Irnerio, 48, 40126 Bologna, Italy

M. Bernadetta Ferrari, Istituto di Chimica Biologica, Università di Parma, Ospedale Maggiore, Viale Gramsci, 14, 43100 Parma, Italy

Alessandro Finazzi Agrò, Dipartmento di Medicina Sperimentale e Scienze Biochimiche, II Università di Roma, Tor Vergata, Via Orazio Raimondo, Roma, Italy

Rosamaria Fiorini, Istituto di Biochimica, Facoltà di Medicina, Università degli Studi di Ancona, Via Ranieri, 60131 Ancona, Italy

Nora Folena, Istituto di Patologia Generale, Università degli Studi di Padova, Via Loredan, 16, 35100 Padova, Italy

P. A. George Fortes, Department of Biology, University of California at San Diego, La Jolla, California 92093, USA

Robert Gennis, Department of Biochemistry, University of Illinois, 1209 W. California, Urbana, Illinois 61801, USA

Giorgio Giacometti, Cattedra di Chimica Biologica, Facoltà di Scienze, Padova, Italy

Enrico Gratton, Department of Physics, University of Illinois, 1110 W. Green, Urbana, Illinois 61801, USA

Settimio Grimaldi, Istituto di Medicina, Sperimentale del CNR, II Clinica Medica Policlinico Umberto I, Roma, Italy

Carlo Giunta, Dipartimento di Biologia Animale, Università degli Studi di Torino, Via Accademia Albertina, 17, 10123 Torino, Italy

Robert Hall, Laboratory of Molecular Biophysics, National Institute of Environmental Health Sciences, Research Triangle Park, North Carolina 22709, USA

Myun K. Han, The Johns Hopkins University, Department of Biology, Mudd Hall, Baltimore, Maryland 21205, USA

Albin H. Hermetter, Institut fur Biochemie, Technical University Graz, Schlogelstrasse 9, A-8010 Graz, Austria

Jim Herron, Department of Bioengineering, University of Utah, Salt Lake City, Utah 84112, USA

Guy Herve, Institute of Enzymology, CNRS, 91190 Gif-Sur-Yvette, France

Gaetano Irace, Istituto di Chimica e Chimica Biologica, I Facoltà di Medicina Università di Napoli, Via Constantinopoli, 16, 80138 Napoli, Italy

Tze C. Jao, Texaco Research Center, Texaco, Inc., P. O. Box 509, Beacon, New York 12508, USA

Ken Jacobson, Laboratories for Cell Biology, Department of Anatomy, Chapel Hill, North Carolina 27514, USA

David Jameson, Department of Pharmacology, University of Texas Southwestern Medical Center at Dallas, 5323 Harry Hines Blvd., Dallas, Texas 75235, USA

Ruud Kraayenhof, Department of Molecular and Cellular Biology, Biological Laboratory, Vrije Universiteit de Boelelaan 1087, 1081 HV Amsterdam, The Netherlands

Ann Kusterdeck, Code 6190, Naval Research Laboratory, Washington D.C. 20375, USA

Sherwin S. Lehrer, Muscle Research Department, Biomedical Research Institute, 20 Staniford St., Boston, Massachusetts 02114, USA

Giampaolo Littarru, Istituto di Biochemica, Facoltà di Medicina, Università degli Studi di Ancona, 60131 Ancona, Italy

James Longworth, Department of Physics, Illinois Institute of Technology, Chicago, Illinois 60616, USA

Giulio Lupidi, Dipartimento di Biologia Cellulare, Università di Camerino, Camerino, Italy

Henri Magdelenat, Laboratoire de Radiopathologie, Institut Curie, 26 Rue d'Ulm, 75231 Paris, France

William W. Mantulin, Laboratory for Fluorescence Dynamics, Physics Department, University of Illinois, Urbana, Illinois 61801, USA

Gerard Marriott, Department of Biochemistry, University of Illinois, Urbana, Illinois 61801, USA

Giuseppe Ettore Martorana, Istituto di Chimica Biologica, Università Cattolica, Largo F. Vito 1, 00168 Roma, Italy

Lanfranco Masotti, Istituto di Chimica Biologica, Facoltà di Medicina, Università di Parma, Parma, Italy

Giampiero Mei, Dipartimento di Scienze Biochimiche, Università di Roma, La Sapienza, Roma, Italy

Maria Da Graca Miguel, Department of Chemistry, University of Coimbra, 3049 Coimbra, Portugal

Roberto Morelli, Istituto di Chimica Fisica, Università di Milano, Via C. Golgi, 19, 20133 Milano, Italy

Christopher C. Morgan, Department of Biological Sciences, Salford University, Salford MS 4WT, United Kingdom

Andrea Mosca, Dipartimento di Scienze e Tecnologie Biomediche, Università di Milano, Via Olgettina, 60, 20132 Milano, Italy

Guido Motolese, I.S.S. s.r.l., Ceparana La Spezia, Italy

Alejandro Paladini, INGEBI, Obligado 2490, 1428 Buenos Aires, Argentina

Tiziana Parasassi, Dipartimento di Chimica, Istituto di Medicina Sperimentale del CNR, Piazzale Aldo Moro, 5, 00185 Roma, Italy

Francesco Pennisi, Cremascoli S.p.a., Via C. Prudenzio, Milano, Italy

Licinio C. Pereira, Universidade do Minho, Largo do Paco, 4719 Braga, Portugal

Mario Piacentini, Istituto Struttura della Materia del CNR, Frascati (Roma), Italy

David Piston, Department of Physics, University of Illinois, Urbana, Illinois 61801, USA

Deleana Pozzi, Dipartimento di Medicina Sperimentale, Policlinico Umberto I, Roma, Italy

Gregory Reinhart, Chemistry Department, University of Oklahoma, Norman, Oklahoma 73019, USA

Renato Angelo Ricci, Dipartimento di Fisica, Università di Padova, Padova, Italy

Nicola Rosato, Centro di Biologia Molecolare del CNR, c/o Dipartimento di Scienze Biochimiche, Università di Roma, La Sapienza, Roma, Italy

Jane Rosen, 180 Thompson St., Apt. 6D, New York, New York 10012, USA

Catherine Royer, Laboratoire de Biologie Physico-Chimique, Paris VII - CNRS, 10 rue des Plantes, 91230 Montgeron, France

Gaetano Saitta, Istituto di Struttura della Materia, Università di Messina, Via dei Verde, Messina, Italy

Gautam Sanyal, Department of Chemistry, Hamilton College, Clinton, New York 13323, USA

Meir Shinitzky, Department of Membrane Research, Heizmann Institute of Science, Rehovot, Israel

Jerson Silvia, Department of Biochemistry, University of Illinois, Urbana, Illinois 61801, USA

Peter Sims, Thrombosis/Hematology Program, OMRF, 825 N.E. 13th St., Oklahoma City, Oklahoma 73104, USA

Angelo Spinedi, Dipartimento di Biologia, II Università di Roma, Tor Vergata, Via O. Raimondo, 173, 00173 Roma, Italy

Arthur G. Szabo, Department of Biological Sciences, National Research Council, 100 Sussex Dr., Ottawa, K1AOR6 Canada

Patrick Tauc, Institute d'Enzymologie du CNRS, Gif-Sur-Yvette, France

Richard Thompson, Code 6190, Naval Research Laboratory, Washington D.C. 20375-5000, USA

Bernard Valeur, Conservatoire National des Arts et Metiers, Chimie General, 292 Rue Saint-Martin, 75141 Paris, France

Arnaldo Vecli, Dipartimento di Fisica, Università degli Studi di Parma, Via M. D'Azeglio, 85, 43100, Parma, Italy

Paola Viani, Dipartimento di Chimica e Biochimica Medica, Facoltà di Medicina Università di Milano, Via Saldini, 50, 20133 Milano, Italy

Michel Vincent, Department of Biochemistry and Biophysics, ER 64-01 CNRS, 45 rue des Saints-Pères, Paris, France

A.J.W.G. Visser, Agricultural University, 6703 BC Wageningen, The Netherlands

Edward Voss, Jr., Department of Microbiology, 131 Burril Hall, 407 S. Goodwin St., University of Illinois, Urbana, Illinois 61801, USA

Therese Wiedmer, Thrombosis/Hematology Program, OMRF, 825 N.E. 13th St., Oklahoma City, Oklahoma 73104, USA

Gregorio Weber, Department of Biochemistry, University of Illinois, 1209 W. California St., Urbana, Illinois 61801, USA

Claudio Zannoni, Istituto di Chimica Fisica, Università di Bologna, Viale Risorgimento, 4, Bologna, Italy

Lello Zolla, Dipartimento di Bologia Cellulare, Università di Camerino, 62032 Camerino, Macerata, Italy

AUTHOR INDEX

CBP2, 401
CCD, 141
Cell differentiation, 442
Cell locomotion, 145
Chirality, 133
Chiron, 138
Chi-square (χ^2), 39, 70, 89, 181, 311
Cholesterol, 433
Circular Dichroism, 136, 176, 229, 254, 383, 399, 401
Colchicine, 381
Color centers, 12, 91
Conformational States, 261
Continuous lifetime distribution models, 17, 70, 389, 403, 411
Coupling enthalpy, 211
Coupling entropy, 211
Coupling free energy, 197
Correlation matrix, 181
Critical transfer distance, 272, 346
Cytochrome b-c1, 162
Cytoskeleton, 433

Debye rotational relaxation time, 68
Debye-Sears tank, 69
Decay-associated spectra (DAS), 33
Dehydroergosterol, 374
Deibler-MBP, 383
Dexter's theory, 274
Diffusion
  Fick's law, 155
  in membranes, 144, 151
  lateral, 118
  mean-time-to-capture, 167
  obstructed, 162
  rotational, 154, 345
  translational, 154, 274, 345
Diffusive transport, 166
Digitized Fluorescence Microscopy (DFM), 139
Dipole-dipole interaction, 272
Distribution of distances, 296
Distribution of rate constants, 296
Dityrosine, 233
Durbin-Watson parameter, 330, 390
Dynamic averaging, 273
Dynamic polarization equations, 72

Einstein coefficients, 3
Electrophoresis (slab gel), 101
Elongation factor Ts (EF-Ts), 65
Elongation factor Tu (EF-Tu), 61
  intrinsic fluorescence, 64
  polarization, 68
  ternary complex, 62
  time-resolved fluorescence, 71
Enantiomers, 135
β-Endorphin, 218
Energy transfer, 88, 253, 269, 314, 330, 345, 426
Equilibrium sink, 264
Erythrocytes, 153, 433
Estriol, 434
Eximer, 122, 269, 423

Linkage,
  between multiple ligands, 203
  and dynamics of rat liver PFK, 210
  and PFK aggregation, 198
  thermodynamic, 195
Lipid structural order parameter, 433
Lipid transfer, 116
Liquid crystal (thermotropic), 393
Lysoamide dehydrogenase, 330

Marquard algorithm, 175, 379, 389
Membranes
  biological, 357, 433, 440, 442
  brush border, 430
  fluid mosaic model, 151
  fluidity, 127, 151, 357, 433
  phase-state asymmetry, 433
  water interface, 358
Metalloprotease, 437
Micellization, 440
Microchannel plates, 52, 141, 175, 308
Microemulsions (w/o), 440
Microscopy
  DFM instrument, 140
  digitized fluorescence (DFM), 139
  fluidity, 152
  fluidity imaging, 168
  FRAP, 144, 151
  video FRAP, 144
Milling crowd model, 157
Mode-locked lasers, 23, 35, 69, 307, 378
Monoclonal antibody, 247
Multifrequency phase and modulation fluorometry, 12, 69, 94, 279, 376,
  378, 381, 403, 426, 437, 442
Multistate models, 263
Myelin basic protein, 383
Myoglobin, 18, 105

Nd:YAG laser, 36
Non-linear least squares analysis, 353, 360
Non-radiative transfer, 271

Optical rotation, 135
Order parameter, 430
Orientation factor, 273
Overlap integral, 273, 281

Parity, 133
Peptides
  amphiphilic, 401
  bombesin, 173
  crabrolin, 227
  dermorphin, 173
  dynorphin, 229
  helodermin, 227
  hormones, 397
  mastoparan polistes, 218, 401
  mastoparan x, 222, 401
  melittin, 226, 401
  sauvagine, 173